# QUANTITATIVE X-RAY SPECTROMETRY

# QUANTITATIVE X-RAY SPECTROMETRY

**Ron Jenkins**

Philips Electronic Instruments, Inc.
Mahwah, New Jersey

**R. W. Gould**

University of Florida
Gainesville, Florida

**Dale Gedcke**

EG&G ORTEC
Oak Ridge, Tennessee

*MARCEL DEKKER, INC.*        *NEW YORK AND BASEL*

Library of Congress Cataloging in Publication Data

Jenkins, Ron, [date]
  Quantitative x-ray spectrometry.

  Includes bibliographical references and index.
  1. X-ray spectroscopy.   I. Gould, Robert William.
II. Gedcke, Dale, [date]     III. Title.   [DNLM:
1. Spectrometry, X-Ray emission.   QD 96.X2 J52q]
QD96.X2J46      545'.836       81-3289
ISBN 0-8247-1266-8            AACR2

MARCEL DEKKER, INC.

270 Madison Avenue, New York, New York 10016

Current printing (last digit):
10  9  8  7  6  5  4  3  2  1

PRINTED IN THE UNITED STATES OF AMERICA

## Preface

X-ray fluorescence spectrometry is now well into its third decade
of use, initially utilizing the wavelength-dispersive spectrometer
and more recently being supplemented with the energy-dispersive
systems.  Like most other analytical methods, it has seen the birth
pains of innovation and has struggled through a period of misunder-
standing and overexploitation before finally reaching the stable
plateau of a reliable and well-accepted analytical technique.  There
are probably in excess of 10,000 users of the x-ray fluorescence
method in the world today, and these are found in a wide range of
disciplines including process control, materials analysis, explora-
tion, mining, metallurgy, and almost every other major branch of
science.

Da Vinci is reported to have said "Experiment is the interpreter of science, experiment never deceives, it is only our judge-
ment which sometimes deceives itself when it expects results which
experiment refuses."  The irony of this statement will certainly
not be lost on any spectroscopist who in the early days attempted
to solve a complex matrix problem and finished up with an unreasonable set of "alpha" coefficients.  The impact of the minicomputer
in the mid-1960s did much to replace art by science, and we are fast

approaching the chemist's dream of elemental analysis without standards.  Certainly in the case of the energy-dispersive spectrometer, it has become a necessary and integral part of the system,
and most modern wavelength-dispersive spectrometers boast some degree of automation for hardware control and data manipulation.

All three of us devote a large fraction of our working lives
teaching mature scientists and technicians details of the x-ray
method from the basic principles to a detailed discussion of the
subtleties and idiosyncrasies of the two types of instruments.  We
are thus well aware of the needs of the practicing spectroscopists
and the requirement for a detailed discussion of the practical aspects of the application of quantitative methods.  We hope that
this book will fill a gap in the existing literature as well as
supplement certain areas already covered.  As an example, texts
are available covering either energy-dispersive or wavelength-
dispersive instruments, but none of these give detailed coverage of
the similarities and contrasts of the two methods.  We have chosen
to discuss instrumentation in great detail specifically in the area
of photon counting and its associated errors, for it is in this
basic process where many systematic errors in quantitative work
arise.

We would each like to thank the many who have helped in the
preparation and proofreading of the manuscript and especially Dr.
Bruce Artz, Professor Jim Brown, Dr. Michael Short, and Dr. Russ
Westberg for their comments and suggestions.  R. J. would like to
thank Philips Electronic Instruments, Inc., and D. G. would like
to thank EG&G ORTEC for allowing us to draw on our experience gained
with these companies and for their liberal allowance of time and
facilities.  We would like to thank our secretaries Rose Gaw,
Patricia Kretzmer, Cher McDonald, Mary McKeethan, Cathy Walton,
and Janice Winslett without whose painstaking work the manuscript
could not have been completed.

                                            Ron Jenkins
                                            Bob Gould
                                            Dale Gedcke

# Contents

# QUANTITATIVE X-RAY SPECTROMETRY

# 1

## Introduction

### 1.1 EARLY DEVELOPMENT OF X-RAY EMISSION ANALYSIS

The use of x-ray emission spectrometers for elemental analysis became widespread in the late 1950s and early 1960s.  X-ray spectroscopy itself dates back to around 1910 when Barkla [1] obtained the first positive evidence of characteristic x-ray emission spectra. Three years later, Moseley [2-4] established the relationship between frequency and atomic number which laid the foundation of the technique for the identification of elements by x-ray emission analysis.  Although the potential of the new technique was quickly appreciated, the practical difficulties in the use of the equipment available at that time severely limited the application of the method.  A major difficulty stemmed from the use of electrons as the excitation source.  In addition to the requirement for a high vacuum and a conducting specimen, most of the electron energy is converted to heat, causing problems of volatility following excessive heating of the specimen.  In the mid-1920s, several workers [5-10] pointed out that the use of an x-ray source would obviate many of the practical problems encountered with the electron source, but unfortunately, the lower efficiency of photon excitation coupled

1

with the rather primitive detection instrumentation made this approach of x-ray *fluorescence* analysis impracticable. Nevertheless, it was the fluorescence approach that was to be employed in the commercial instrumentation that became available in the early 1950s.

Until the late 1960s, nearly all spectrometers were the so-called wavelength-dispersive spectrometers in which wavelengths are separated by Bragg diffraction from a single crystal. More recently, *energy-dispersive spectrometers* have become available in which a lithium-drifted silicon or germanium detector is used to give a distribution to voltage pulse amplitudes proportional to the distribution of photon energies. Electronic separation of the pulse height distribution then gives a photon energy spectrum. Each of these x-ray photon separation techniques has its own advantages and disadvantages, and a detailed description of each method will be found in the succeeding chapters of this book.


## 1.2  BASIS OF QUALITATIVE X-RAY EMISSION ANALYSIS

There are always four stages in any scheme of x-ray emission analysis. The excitation of characteristic radiation from the specimen by bombardment with high-energy photons, electrons, protons, etc.; the selection of a characteristic emission line from the element in question by means of a wavelength or energy-dispersive spectrometer; the detection and integration of the characteristic photons to give a measure of characteristic emission line intensity; and finally, the conversion of the characteristic emission line intensity to elemental concentration by use of a suitable calibration procedure.

All elements will give one or more sets (series) of characteristic emission lines and the number of these lines measurable by a spectrometer will depend upon the range and separating ability (resolution) of the spectrometer. Most commercial spectrometers operate over the range 0.2 to 20 $\overset{o}{A}$ (60-0.6 keV) and as such are able to detect the majority of the K and L series lines, plus a few lines from the M series of the higher atomic number elements. In practical terms, this means the detection of characteristic lines from all elements above atomic number 8 (oxygen) with the number of

lines per element varying from two or three for the low atomic numbers, to 20-30 for the high atomic numbers.  One or two lines are
normally sufficient for the positive identification of an element
in qualitative terms.  The Moseley law relates the frequency $\nu$ with
the atomic number Z.

$$\nu = K(Z - \sigma)^2 \tag{1.1}$$

in which K and $\sigma$ are both constants which vary with the spectral
series.  A simple relationship also exists between the wavelength $\lambda$
and energy E of an x-ray photon

$$E = \frac{hc}{\lambda} \tag{1.2}$$

in this expression h is Planck's constant ($6.626 \times 10^{-27}$ erg·s) and
c the velocity of light ($3 \times 10^{10}$ cm/s).  When E is expressed in
thousands of electron volts (keV) and $\lambda$ in angstrom units
($1 \text{ Å} = 10^{-8}$ cm) the following approximate relationship is applicable:

$$E = \frac{12.398}{\lambda} \tag{1.3}$$

It is a common practice to use tables relating either wavelength or energy with atomic number for the qualitative identification of elements from their emission spectra.  Most spectrometers
lose sensitivity at longer wavelengths (lower energies) and their
ability to detect the low-atomic-number elements generally falls off
sharply around atomic number 12.  This combined with the low penetration of long-wavelength x-rays and the fact that they arise
from molecular rather than atomic orbital transitions (and are thus
dependent upon chemical bonding) makes the x-ray emission method
largely impracticable for elements below atomic number 9 (fluorine).

## 1.3  QUANTITATIVE ANALYSIS

For the quantitative analysis of a given element, it is sufficient
to measure one selected emission line and to relate the intensity
of this line to concentration.  Such a relationship is linear only

over a very limited concentration range and in most cases the intensity of the emitted line is dependent not only on the concentration
of the excited element but also upon the influence of other elements
making up the specimen.  These so-called matrix effects are both
predictable and correctable by careful calibration and computation,
and the advent of the small digital computer has done much to simplify the otherwise unwieldy and time-consuming calculations.  Since
the penetration of characteristic x-ray lines is small (typically
1-1000 μm), it is vital that the specimen be homogeneous over the
depth of specimen contributing to the measured signal.  This in turn
may require a specimen preparation procedure that varies from being
minimal to relatively sophisticated.  Finally, accurate results can
only be obtained when the source, spectrometer, and counting equipment are themselves stable and free from systematic errors.  Modern
spectrometers have potential precisions of around one-tenth of a
percent and analytical accuracies approaching this value are obtainable where correct methodology is employed.


## 1.4   TYPES OF X-RAY SPECTROMETERS

All spectrometers comprise an excitation source, a means of separating and isolating characteristic lines, plus a device for measuring characteristic line intensities.  More sophisticated systems
may include a dedicated computer or hardwired data processor for
data manipulation and even the calculation of elemental composition.
     There are many types of x-ray spectrometers but several broad
categories exist.  Single-channel instruments are spectrometers
which are designed to measure one element at a time, or many elements sequentially.  The majority of flat-crystal wavelength-
dispersive scanning spectrometers fit into this category, which is
probably the most commonly employed.  They have the advantage of
great flexibility and a multiplicity of selectable single crystals
is generally provided to offer good dispersion over a wide range of

wavelengths.  Such spectrometers are usually provided with a high-
power x-ray tube as the source, which may offer broad-range excitation up to 100 kV.  Inexpensive single-channel instruments may also
be constructed using a proportional detector with suitable bandpass
filters, using a radioisotopic source emitting $\gamma$-rays or x-rays.
Such systems are particularly useful for the determination of single
elements.  All single-channel sequential spectrometers suffer the
disadvantage of being slow relative to their multichannel counter-
parts.

   The multichannel x-ray spectrometer provides the means of measuring many elements simultaneously.  A multichannel spectrometer
might consist of a series (7-30) of single-channel wavelength-
dispersive spectrometers grouped in a semicircle around the specimen.  Since fixed optics are being employed, the crystals employed
are often curved to constant radius or in a logarithmic spiral to
give the optimum between resolution and line intensity.  Such a
spectrometer is very fast and able to provide a high specimen
throughout where the analytical problem is well defined.  It is,
however, rather inflexible and expensive.  All energy-dispersive
spectrometers are by definition multichannel but retain more flexibility than fixed-channel wavelength-dispersive systems.  This is
because the energy-dispersive system always measures the whole
spectrum and the required portions can be selected from the acquired
spectrum.  Energy-dispersive systems thus combine many of the advantages of the single-channel and multichannel systems since they
are fast but still retain a high degree of flexibility.  They are,
however, count-rate limited and many commercial systems employ a
low-power x-ray tube as the source to deliberately limit the radiation flux from the specimen.  Pulsed x-ray tubes may also be employed to reduce the counting loss of the scaling system.  More
sophisticated energy-dispersive spectrometers can employ one or more
secondary emitters and/or filters, which limits the number of elements being excited thus making the excitation process more specific.

## 1.5    USE OF THE X-RAY SPECTROMETER IN THE ROUTINE ANALYTICAL LABORATORY

There are currently many thousands of x-ray spectrometers being utilized for routine qualitative and quantitative analysis. Sensitivities available for most elements reach the low part-per-million range, and the method is equally as applicable at high or low concentration levels. Accuracies of the order of a few tenths of a percent are achievable with analysis times on the order of minutes for scanning spectrometers and even less for multichannel instruments.

The x-ray spectroscopic technique is essentially nondestructive, and measurable signals can be obtained from as little as 1 mg of specimen. Optimum specimen sizes range from 0.1 to 5 g of solid material, and the technique is equally applicable to solids, liquids, and even gases. A wide range of well-analyzed calibration standards has become available over the past few years and this, combined with the development of "fundamental"-type algorithms relating x-ray emission intensity and elemental composition, is allowing the use of the x-ray method in a wide variety of analytical disciplines.

X-rays themselves pose a particular health hazard, and although a significant number of accidents did occur with earlier spectrometers, there is now strict governmental control over equipment design and acceptable dose levels. When correctly employed, the modern spectrometer is probably less of a radiation health hazard than the average color television receiver. Nevertheless, equipment must be used with proper precautions and a separate chapter of this text is devoted to this area.

REFERENCES

1.  C. G. Barkla, *Phil. Mag., 22*:396 (1911).
2.  H. G. J. Moseley, *Phil. Mag., 26*:1024 (1912).
3.  H. G. J. Moseley, *Phil. Mag., 27*:703 (1913).
4.  B. Jaffe, in *Moseley and the Numbering of the Elements,* Doubleday, New York, 1971.

5.  R. Whiddington, *Proc. Roy. Soc., Ser. A., 85*:323 (1911).
6.  R. T. Beatty, *Proc. Roy. Soc., Ser. A., 87*:511 (1912).
7.  S. K. Allison and W. Duane, *Proc. Nat. Acad. Sci., 11*:485 (1925).
8.  G. Von Hevesey, *Chemical Analysis by X-rays and Its Applications*, McGraw-Hill, New York, 1932.
9.  D. Coster and J. Nishina, *Chem. News, 130*:149 (1925).
10. R. Glocker and H. Schrieber, *Ann. Physik, 85*:1085 (1928).

# 2

# The Interaction of X-rays with Matter

## 2.1 GENERAL

The interaction of electromagnetic radiation with matter is a complex subject encompassing many aspects of modern physics. This
chapter, however, will only present those concepts that are directly
relevant to the production of analytically measurable x-radiation.
For a more complete survey of this area, the interested reader is
referred to Refs. 1 to 4 at the end of this chapter. Figure 2.1
illustrates the various processes which occur when x-ray photons
impinge upon matter. The relative importance of these processes
depends upon the energies (wavelengths) present in the primary
radiation and the composition of the scattering substance.

As indicated in Eq. (1.3) there is a simple relationship between wavelength and energy; Fig. 2.2 gives a useful scale for
conversion.

If a beam of x-rays interacts with a substance, the x-ray beam
intensity is attenuated. When this process of attenuation is
examined carefully, it is found that several distinct types of
interaction can occur, all of which result in a decrease in the intensity of the incident beam of x-rays. The magnitudes of these

9

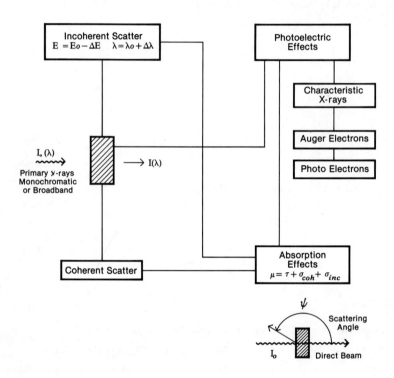

Figure 2.1  Interaction of electromagnetic radiation with matter.

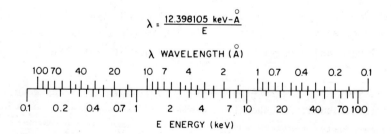

Figure 2.2  Conversion from energy to wavelength.  Reprinted by courtesy of EG&G ORTEC.

subinteractions are strongly influenced by (1) the energy of the
incident x-ray beam, (2) its degree of monochromatization, and
(3) the average atomic number and crystalline structure of the
scattering substance.  Unfortunately, there is no simple, general
"rule of thumb" which will adequately describe these interactions
if these three variables are allowed to assume their total range of
typical values.

   Analytically useful x-rays interact almost exclusively with the
electrons in matter.  For this reason, any discussion of the interaction of x-rays with matter should begin with a description of the
interaction of an x-ray photon with a single free electron and proceed to multielectron atoms and thence to multiatom solids.  It
will be shown that two basic interaction processes dominate:
(1) the photoelectric effect and (2) x-ray scattering.

## 2.2   INTERACTION OF X-RAY PHOTONS WITH A SINGLE FREE ELECTRON

Electrons are responsible for the scattering of x-rays by matter.
All scattering events involving x-rays and matter can for convenience be referred or compared to the scattered intensity ($I_e$) from a
single electron.  $I_e$ is the energy scattered per unit solid angle
per second by a classical free electron struck by an unpolarized
x-ray beam having an energy flux $I_o$ per square centimeter of area.
Equation (2.1), or *Thompson's law* as it is called, expresses $I_e$ in
terms of $I_o$ and the scattering angle $\psi$ (see Fig. 2.1).

$$I_e = 7.90 \times 10^{-26} \frac{1 + \cos^2\psi}{2} I_o \qquad (2.1)$$

   It is necessary at this point to distinguish two types of x-ray
scattering.  When the scattered photon has the same energy or
wavelength as the incident photon, the scattering is called *coherent*.  If the energy or wavelength of the scattered photon has been
modified, the scattered x-ray is said to be *incoherently* or *Compton
scattered*.  Compton [1] was the first to show that the x-radiation
scattered by a single free electron is changed in wavelength

according to the relationships shown in Eqs. (2.2a) and (2.2b).

$$\Delta\lambda = \lambda_{inc} - \lambda_o = 0.024(1 - \cos \psi) \qquad (2.2a)$$

(where $\Delta\lambda$ is given in angstrom units), or in terms of energy,

$$E_{inc} = \frac{E_o}{1 + (E_o/m_o c^2)(1 - \cos \psi)} \qquad (2.2b)$$

where $m_o c^2 \simeq 511$ keV.

## 2.3  INTERACTION OF X-RAY PHOTONS WITH ATOMIC ELECTRONS

Quantum mechanical calculations as well as experimental evidence have shown that scattering of x-ray energies (E $\simeq$ 10 keV) from atomic electrons obeys the following general principles:

1.  Both incoherent and coherent scattering occur.
2.  The total scattering per electron (incoherent plus coherent) is given by the classical Thompson equation.

The coherent portion of the scattering by an atom is found by summing the coherent scattering amplitude from each of the electrons in the atom.  If $f_e$ is the coherent scattering amplitude of one bound electron expressed in electron units (eu), then by definition

$$f_e = \frac{\text{coherent scattering amplitude by one atomic electron}}{\text{amplitude scattered by classical free electron}}$$

If the individual electron cloud is assumed to be spherically symmetrical, we have

$$f_e = \int_0^\infty 4\pi r^2 \rho(r) \frac{\sin kr}{kr} \, dr \qquad (2.3)$$

where r is the distance from the center of the electron cloud, $\rho(r)$ is the electron density distribution, and $k = (4\pi \sin \theta)/\lambda$, where $\theta = \psi/2$.  The scattering amplitude for the entire atom containing Z electrons is given by

$$f = \sum_{n=1}^{Z} (f_e)_n = \sum_{n=1}^{Z} \int_0^{\infty} 4\pi r^2 \rho_n(r) \frac{\sin kr}{kr} dr \qquad (2.4)$$

Thus for any atom, f is a function of $(\sin \theta)/\lambda$, and when $(\sin \theta)/\lambda$ approaches zero, f approaches

$$\sum_n \int_0^{\infty} 4\pi r^2 \rho_n(r) \, dr \longrightarrow Z$$

Note that $(\sin \theta)/\lambda$ can approach zero if $\theta$ approaches zero (low scattering angles) or if $\lambda$ becomes large (low energies).

## 2.4  INCOHERENT SCATTERING FROM ATOMS

For incoherently scattered radiation there is no interference bet-ween scattered waves, and the total intensity scattered by the elec-trons in the electron cloud surrounding the nucleus is given by the sum of the scattering intensities of the individual electrons:

$$i_{inc} = \sum_{n=1}^{Z} [1 - (f_e)^2] I_e = [Z - \sum_{n=1}^{Z} (f_e)^2] I_e = i_{inc} I_e \qquad (2.5)$$

$i_{inc}$ has been calculated [4] for lithium ($Z = 3$) using an as-sumed electron density for the two K electrons and the one L elec-tron.  Figure 2.3 shows the dependence of $f_{Li}$ and $i_{inc(Li)}$ on $(\sin \theta)/\lambda$.  The coherent scattering intensity equals $f^2 I_e$, and for the lithium atom shown in Fig. 2.3, $f_{Li}^2$ (coherent) approaches $Z^2$ as $(\sin \theta)/\lambda$ approaches zero.  The total scattered intensity for the lithium atom thus consists of $f_e^2 I_e$, (or $I_{coh} I_e$) plus $i_{inc} I_e$.  For lithium at small values of $(\sin \theta)/\lambda$, the coherent scattering in-tensity dominates, while at large values of $(\sin \theta)/\lambda$, the incoher-ent scatter intensity will be larger than the coherent fraction. That this is not a completely general statement for all atom types can be seen from Fig. 2.4 where $I_{coh}$ and $i_{inc}$ are plotted for lithium ($Z = 3$), aluminum ($Z = 13$), and copper ($Z = 29$).

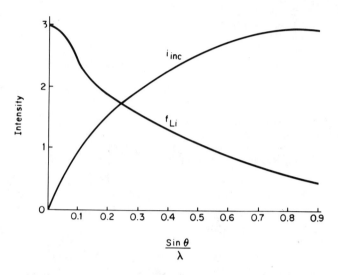

Figure 2.3  The incoherent scattering intensity $i_{inc}$ and the coherent scattering amplitude $f_{Li}$ for Lithium, atomic number 3 (after
Warren [4]).


## 2.5  SCATTERING FROM ATOMIC AGGREGATES

When scattering occurs from a solid or liquid consisting of an ag-
gregation of atoms, other scattering effects may be noted.  Figure
2.5 shows the corrected coherent and incoherent intensities for
vitreous quartz ($SiO_2$).  Even for this noncrystalline solid, there
are pronounced maxima in the coherent scattering intensity.  These
maxima occur because of constructive interference (diffraction)
between scattered coherent waves.  This effect is most pronounced
in a crystalline substance in which nearly all of the coherent scattering is "gathered up" into sharp peaks.  As a result of diffraction
effects, the coherent/incoherent intensity ratio will vary widely
with scattering angle, as shown in Fig. 2.5, and will also depend
upon the crystallinity and average atomic number of the specimen,
as well as the energy of the scattering radiation.  In specimens of
finite thickness, absorption of the incident and scattered radiation
in the specimen must also be considered.  If the specimen is

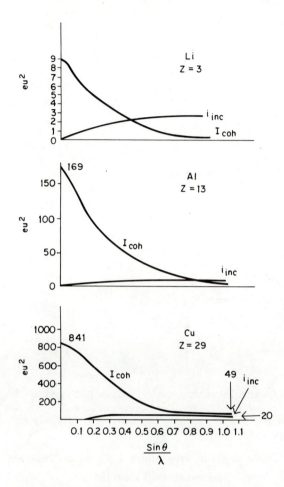

Figure 2.4   Coherent and incoherent scatter for lithium, aluminum, and copper.

amorphous, the absorption effect is similar to that derived for the fluoresced radiation in Sec. 2.9.

The x-ray diffraction effect is used in the wavelength-disper-sive spectrometer as a basis for spectral separation.  A single crystal is cleaved such that a selected set of atomic planes (hkl) of interplanar spacing d are parallel with the surface.  This crystal is used to diffract the polychromatic beam of fluorescence emission from the specimen, and rotating the crystal to an

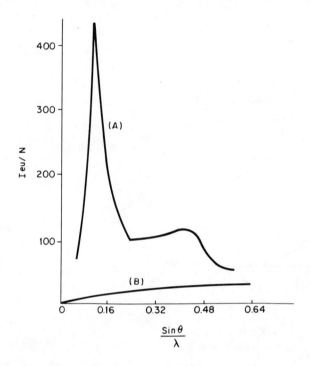

Figure 2.5  Measured scattering intensity for vitreous $SiO_2$ using
Cu K$\alpha$ radiation (corrected for absorption, polarization, and
normalized to electron units; see ref. 4):  (a) Coherent intensity
I eu/N; (b) incoherently scattered intensity.

appropriate angle $\theta$ will cause a given wavelength $\lambda$ to be diffracted
at angle $2\theta$, provided that the Bragg relation is satisfied:

$$n\lambda = 2d \sin \theta \qquad\qquad\qquad (2.6)$$

in which n is the order of diffraction.

2.6  THE ENERGY AND WAVELENGTH OF INCOHERENTLY SCATTERED RADIATION

It was shown in Eq. (2.2a) that there is a relationship between the
coherently scattered ($\lambda_o$) and incoherently scattered ($\lambda_{inc}$) wavelengths and the scattering angle $\psi$.  Figure 2.6 shows the scattering
angle $\psi$ and its relationship to the average angle of incidence $\psi_1$

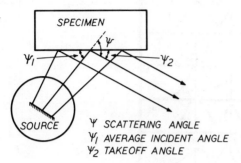

Figure 2.6   Incident and takeoff angles of the spectrometer.

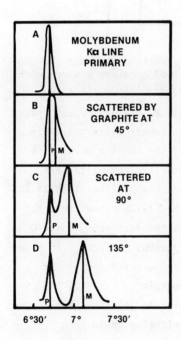

Figure 2.7   Spectra of x-rays scattered by graphite at different
angles showing modified lines wider than primary P and displaced to
the theoretical position M (after Compton and Allison [1]).

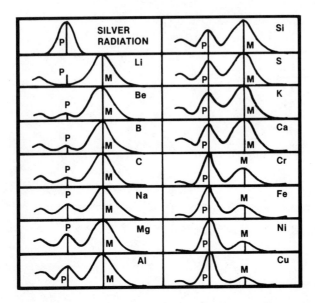

Figure 2.8   Spectra of Ag Kα scattered by various elements showing the increase in the prominence of unmodified coherent scatter with increase of atomic number (after Compton and Allison [1]).

of the primary beam, and the takeoff angle $\psi_2$ defined by the optics of a typical x-ray spectrometer.   In most commercially available x-ray spectrometers, $\psi$ approaches $90^o$ and the Compton wavelength shift is fixed since it depends only on the scattering angle $\psi$, according to Eq. (2.2a), and thus will equal approximately 0.24 $\overset{o}{A}$. For the energy dispersive spectrometer, Eq. (2.2b) shows that the Compton shift in energy depends upon both the scattering angle $\psi$ and the energy $E_o$ of the incident x-ray photons.   Figure 8.15 shows the magnitude of the Compton energy shift for three scattering angles over a range of incident photon energies.

Figure 2.7 taken from Compton and Allison [1] shows the Compton wavelength shift as a function of the scattering angle when molybdenum radiation is scattered from graphite.   Figure 2.8 shows the effect on the incoherent scattered intensity caused by the atomic number of the scatterer [5].   In Fig. 2.8, silver Kα radiation has been scattered at $\psi = 120^o$ from a variety of scattering

substances ranging from lithium (Z = 3) to copper (Z = 29). Note
that the ratio of the incoherent scattered intensity to the coherent scattered intensity of the silver Kα line decreases as the
atomic number of the scatterer increases.


## 2.7  PHOTOELECTRIC INTERACTION

If a photon strikes a bound electron and the energy of the photon
is greater than the binding energy of the electron in its shell,
then it is possible for the electron to absorb the total energy of
the photon.  The photon disappears in this process and its energy
is transferred to the electron which is ejected from its shell.
The ejected electron is called a *photoelectron*, and the interaction
is called the *photoelectric effect*.  Figure 2.9 schematically represents the process.  The photoelectron is emitted with an energy
E - φ, where E is the original photon energy and φ is the binding
energy of the electron in its shell.  In Fig. 2.9, one of the K
electrons has been removed.  Thus φ is the binding energy of a K-
shell electron, i.e., $\phi_K$.  The vacancy left in the K shell represents an unstable situation.  Consequently, an electron from a shell
with a lower binding energy will transfer to the K shell to fill
the vacancy.  The difference in binding energies between the two
shells can be given off in the form of a characteristic x-ray photon.
For example, if an electron from the L shell fills the vacancy, the
difference in binding energies, $\phi_K - \phi_L$, can be emitted as a characteristic K x-ray photon.

   A competing process for consuming the difference in binding
energies following interaction is the emission of Auger electrons
from the outer shells.  In this case no characteristic x-ray is
emitted.  The probability that a characteristic x-ray will be
emitted once a vacancy has been created is described by the fluorescence yield ω.  The fluorescence yield lies between 0 and 1.  For
low-atomic-number elements, Auger electron emission is more probable.
For high-atomic-number elements, characteristic x-ray emission becomes more probable.

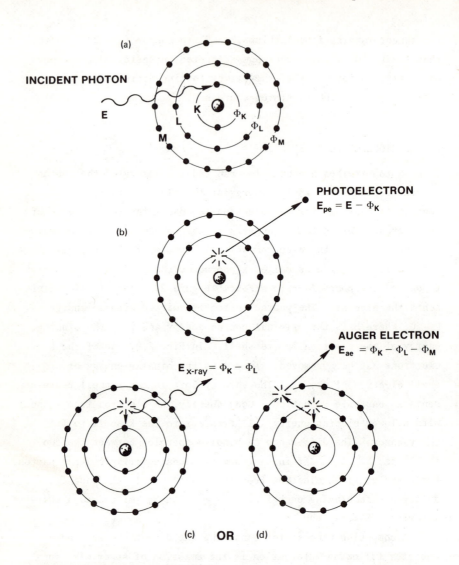

Figure 2.9  The photoelectric interaction.  (a)  Before photoelectric interaction a photon of energy E encounters the atom.
(b)  In the photoelectric interaction the photon is absorbed by a
K-shell electron, and the electron is ejected with an energy equal
to the photon energy less the K shell electron-binding energy.
(c)  The K-shell vacancy is filled by an L shell electron, and the
difference in binding energies is given off as either (c) a characteristic x-ray photon or (d) as an Auger electron.  Reprinted by
courtesy of EG&G ORTEC.

Whereas the majority of the scatter arises from the outer, loosely bound, electrons, the greatest energy losses occur through photoelectric absorption on the tightly bound inner-shell electrons. This is an important point to realize in quantitative x-ray spectrometry since one immediate effect will be a high scattering intensity from low-average-atomic-number specimens resulting in high background, because the ratio of loosely bound to tightly bound electrons is relatively large.  From the point of view of total absorption, however, the greatest absorption loss will occur where the energy of the initial x-ray photon is just greater than the binding energy of an electron in a given level of an atom.

## 2.8   THE MASS ATTENUATION COEFFICIENT

As has been discussed in the previous sections, when a beam of x-ray photons passes through a material, some of the photons will suffer interactions with the atoms composing the material.  The interactions which take place are the photoelectric effect, incoherent scattering, coherent scattering, and diffraction, which is a special case of coherent scattering.  The fraction of the photons which pass through the material without interacting is conveniently described using the concept of a *mass attenuation coefficient*.

Figure 2.10(a) illustrates the definition of the mass attenuation coefficient for a material of density $\rho$ and infinitesimal thickness dx.  An x-ray beam of intensity $I_o(E)$ photons per second strikes the differential slab of material perpendicular to the surface.  The energy of each incident photon is E.  The number of photons per second which interact with the atoms in the slab is $-dI(E)$.  Consequently, the rate of photons which is transmitted through the slab without interacting with the material is $I_o(E) + dI(E)$, where $dI(E)$ is expected to be a negative number. That is, the transmitted rate is less than the incident rate.

The number of photons per second which interact in the slab is expected to be proportional to both the incident photon rate $I_o(E)$

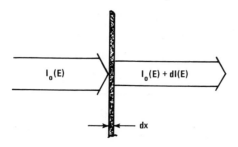

(a)  Infinitesimal Thickness

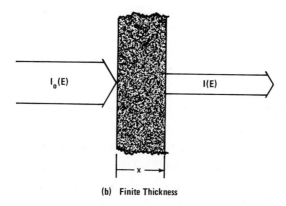

(b)  Finite Thickness

Figure 2.10  (a)  Definition of the mass attenuation coefficient and (b) illustration of Beer's law.  Reprinted by courtesy of EG&G ORTEC.

and the mass per unit area of the slab which is given by $\rho \, dx$. This can be stated as

$$-dI(E) = \mu(E)I_o(E)\rho \, dx \qquad\qquad (2.7)$$

where the proportionality constant is $\mu(E)$.  The constant $\mu(E)$ is called the *mass attenuation coefficient*.  It is characteristic of the material in the slab and the energy of the x-ray photons.  The units of $\mu(E)$ are square centimeters per gram $(cm^2/g)$ when the density $\rho$ is expressed in grams per cubic centimeter $(g/cm^3)$ and thickness is specified in centimeters (cm).

For a piece of material of finite thickness x, such as shown in Fig. 2.10(b), integration of Eq. (2.7) shows that the transmitted intensity of photons which have not suffered interactions in the material is given by

$$I(E) = I_o(E)e^{-\mu(E)\rho x} \tag{2.8}$$

This is the Beer-Lambert law. The mass attenuation coefficient $\mu(E)$ accounts for the various interactions which can occur in the specimen. Thus $\mu(E)$ is composed of three major components:

$$\mu(E) = \tau(E) + \sigma_{coh}(E) + \sigma_{inc}(E) \tag{2.9}$$

$\tau(E)$ is the photoelectric mass absorption coefficient, which describes the photoelectric effect. The coherent scattering process is defined by the total coherent mass scattering coefficient $\sigma_{coh}(E)$, and $\sigma_{inc}(E)$ is the total incoherent mass scattering coefficient. Both $\sigma_{coh}(E)$ and $\sigma_{inc}(E)$ include all possible scattering angles.

Once the functional dependence on energy or wavelength is clearly understood, it is no longer necessary to include the functional form in the notation and $\mu(E)$, $\tau(E)$, $\sigma_{coh}(E)$, and $\sigma_{inc}(E)$ can be written simply as $\mu$, $\tau$, $\sigma_{coh}$, and $\sigma_{inc}$.

Since the photoelectric mass absorption coefficient includes the probability for ionizing all the shells in an atom, it can be broken down into a sum of the probabilities of ionizing each shell. In other words,

$$\begin{aligned}
\tau = \tau_K &+ \left( \tau_{L_I} + \tau_{L_{II}} + \tau_{L_{III}} \right) \\
&+ \left( \tau_{M_I} + \tau_{M_{II}} + \tau_{M_{III}} + \tau_{M_{IV}} + \tau_{M_V} \right) + \cdots
\end{aligned} \tag{2.10}$$

where each term expresses the photoelectric mass absorption coefficient for a particular subshell of the atom. If the energy of the incoming photon is less than that required to ionize a particular shell, then the term for that shell will be zero. Thus there are abrupt discontinuities in $\tau$ as a function of energy where the photon energy becomes less than the binding energy $\phi$ of a particular shell.

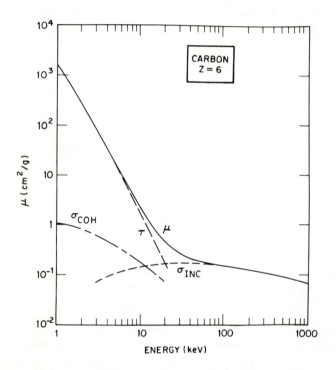

Figure 2.11  Mass attenuation coefficient and its components for
carbon as a function of incident photon energy.  Reprinted by
courtesy of EG&G ORTEC.

Figures 2.11 to 2.14 taken from McMaster et al. [6] show the
various components of the mass attenuation coefficient for the pure
elements carbon (Z = 6), aluminum (Z = 13), iron (Z = 26), and lead
(Z = 82) as a function of the incident photon energy.  The absorption discontinuities (absorption edges) are clearly shown.  Notice
that there is only one K absorption edge (since there is only one K
energy level), while close inspection will reveal three closely
spaced L edges and five M edges corresponding to the multiple L and
M energy levels.  Since Figs. 2.11 to 2.14 are plotted as a function
of the energy E, they may appear peculiar to wavelength dispersive
x-ray spectroscopists who are accustomed to plotting μ versus λ.
Using Fig. 2.11 (carbon) and assuming an incident photon energy such
as copper Kα, (8046 eV), we obtain the coefficients shown in
Table 2.1.

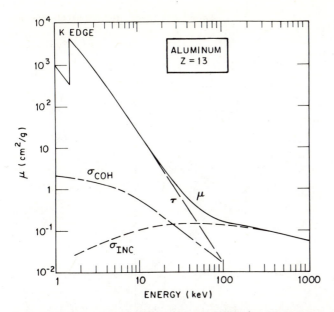

Figure 2.12  Mass attenuation coefficient and its components for aluminum as a function of incident photon energy.  Reprinted by courtesy of EG&G ORTEC.

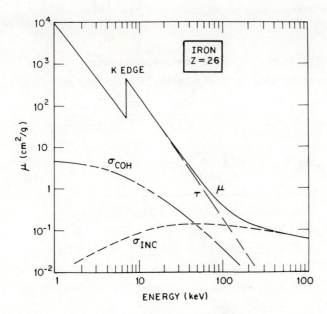

Figure 2.13  Mass attenuation coefficient and its components for iron as a function of incident photon energy.  Reprinted by courtesy of EG&G ORTEC.

25

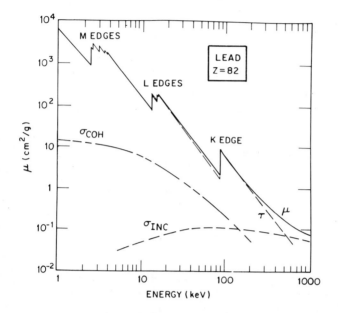

Figure 2.14  Mass attenuation coefficient and its components for lead as a function of incident photon energy.  Reprinted by courtesy of EG&G ORTEC.

When a material is made up of a homogeneous mixture of pure elements, the resultant mass attenuation coefficient may be calculated from

$$\mu = \sum_{j} W_{j} \mu_{j} \tag{2.11}$$

Table 2.1  Scattering and Photoelectric Coefficients for Carbon When $E_{o}$ = 8046 eV (Cu K$\alpha_1$)

|                      | $cm^2/g$ | Fraction of Total |
|----------------------|----------|-------------------|
| Incoherent scatter   | 0.133    | 0.029             |
| Coherent scatter     | 0.231    | 0.051             |
| Total scatter        | 0.364    | 0.081             |
| Photoelectric        | 4.15     | 0.919             |
| Total                | 4.51     | 1.000             |

where $\mu_j$ is the mass attenuation coefficient of element j present in the material with a weight fraction $W_j$. The sum is carried over all elements in the material such that

$$\sum_j W_j = 1 \tag{2.12}$$

## 2.9 THE PRIMARY FLUORESCED INTENSITY

In x-ray fluorescence spectrometry, it is the intensity of the fluoresced characteristic x-ray from the specimen which provides the analytical signal for qualitative and quantitative analysis. The characteristic x-rays are generated as a result of a photoelectric interaction. On the other hand, coherent scattering and incoherent scattering of the excitation spectrum generally provides background contributions which tend to interfere with the analysis of the characteristic x-rays. In this section, a simple derivation of the primary fluoresced intensity for the characteristic x-rays is presented. This derivation is useful in that it illustrates the fundamental interactions involved in generating the signal of interest and it forms a foundation for understanding a variety of effects treated in later chapters of this book. Although the derivation is done in the energy coordinate system, the equivalent equations in the wavelength representation may be obtained by simply making the substitutions listed in Table 2.2. The equations will be written in terms of differential solid-angle elements to simplify the illustration.

Figure 2.15 defines the geometry of the excitation source, the specimen, and the detector. The excitation source may be an x-ray tube anode or a radioisotope. To simplify the discussion the source is considered to be a point source. The number of x-ray photons per second emitted by the excitation source in the energy interval $E_o$ to $E_o + dE_o$ within the differential solid angle $d\Omega_1$ is defined as $I_o(E_o) \, dE_o \, d\Omega_1$. These photons strike the surface of the specimen at an incidence angle $\psi_1$.

Table 2.2 Substitutions for Translating Eqs. (2.25) through (2.32) into the Wavelength Coordinate System

| Energy Parameter | Wavelength Parameter | Description |
|---|---|---|
| $E_o$ | $\lambda_o$ | Primary radiation energy or wavelength from the excitation source |
| $E_i$ | $\lambda_i$ | Fluoresced radiation energy or wavelength measured for the ith element |
| $I_o(E_o)\, dE_o$ | $-I_o(\lambda_o)\, d\lambda_o$ | Number of incident photons per second per steradian in energy interval $E_o$ to $E_o + dE_o$ or the wavelength interval $\lambda_o$ to $\lambda_o - d\lambda_o$ |
| $\mu(E_o)$ | $\mu(\lambda_o)$ | Total specimen mass attenuation coefficient for the energy $E_o$ or the wavelength $\lambda_o$ |
| $\mu(E_i)$ | $\mu(\lambda_i)$ | Total specimen mass attenuation coefficient for the fluoresced energy $E_i$ or the fluoresced wavelength $\lambda_i$ |
| $\tau_{Ki}(E_o)$ | $\tau_{Ki}(\lambda_o)$ | Photoelectric mass absorption coefficient for the K shell of element i at the energy $E_o$ or the wavelength $\lambda_o$ |
| $\tau_i(E_o)$ | $\tau_i(\lambda_o)$ | Total photoelectric mass absorption coefficient for element i at the energy $E_o$ or the wavelength $\lambda_o$ |
| $Q_{if}(E_o)$ | $Q_{if}(\lambda_o)$ | Fluorescence probability for measured line from element i |
| $\phi_K$ | $\lambda_{abs} = (12.4)/\phi_K$ | Absorption edge energy or wavelength for the K shell of element i |
| $E_{max}$ | $\lambda_{min}$ | The maximum energy or minimum wavelength in the excitation spectrum |
| $\eta(E_i)$ | $\eta(\lambda_i)$ | The detection efficiency of the x-ray spectrometer for the fluoresced photons at energy $E_i$ or wavelength $\lambda_i$ |
| $I_i(E_i)$ | $I_i(\lambda_i)$ | Number of detected fluoresced photons per second per steradian per steradian for the ith element at energy $E_i$ or wavelength $\lambda_i$ |

Example:  $\displaystyle I_i(E_i) = \frac{\eta(E_i)}{4\pi \sin \psi_1} \int_{E_o = \phi_K}^{E_{max}} \frac{Q_{if}(E_o) I_o(E_o)\, dE_o}{\mu(E_o)\csc\psi_1 + \mu(E_i)\csc\psi_2}$  becomes

$$I_i(\lambda_i) = \frac{\eta(\lambda_i)}{4\pi \sin \psi_1} \int_{\lambda_o = \lambda_{min}}^{\lambda_{abs}} \frac{Q_{if}(\lambda_o) I_o(\lambda_o)\, d\lambda_o}{\mu(\lambda_o)\csc\psi_1 + \mu(\lambda_i)\csc\psi_2}$$

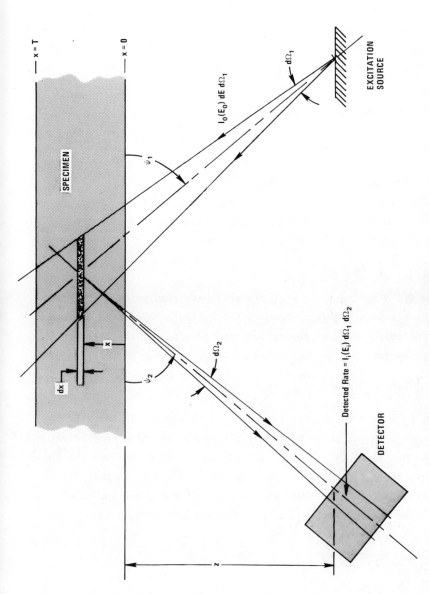

Figure 2.15  Geometry for derivation of the primary fluorescence intensity.  Reprinted by courtesy of EG&G ORTEC.

The number of characteristic x-rays fluoresced in the differential element of thickness dx, located a distance x behind the
specimen surface, is to be calculated.  In order to reach this
volume element, the incident x-rays must pass through an effective
specimen thickness x csc $\psi_1$.  As a result of absorption along this
path length, the rate of photons arriving at the differential volume
element is reduced to

$$I_1 = I_0(E_0) \ dE_0 \ d\Omega_1 \ \exp[-\mu(E_0)\rho x \ \csc \ \psi_1] \qquad (2.13)$$

where $\mu(E_0)$ is the total mass attenuation coefficient in square
centimeters per gram $(cm^2/g)$ for the specimen at the energy $E_0$ and
$\rho$ is the density of the specimen.

In passing through the differential volume element the x-rays
travel through a thickness dx csc $\psi_1$.  Therefore the number of
photoelectric interactions in this element per unit time is given by

$$I_2 = I_1\tau(E_0)\rho \ dx \ \csc \ \psi_1 \qquad (2.14)$$

where $\tau(E_0)$ is the total photoelectric mass absorption coefficient
of the specimen $(cm^2/g)$.  Note that $\tau(E_0)$ is the weighted average of
the photoelectric mass absorption coefficients for the elements
making up the specimen.  That is,

$$\tau(E_0) = \sum_m W_m\tau_m(E_0) \qquad (2.15)$$

where $W_m$ is the weight fraction of the mth element in the specimen
and $\tau_m(E_0)$ is the total photoelectric mass absorption coefficient
of the mth element.  The mth term in Eq. (2.15) determines the number of interactions with the mth element.  Since it is only the excitation of the ith element which is of interest, the number of
photoelectric interactions to be considered is reduced to

$$I_3 = \frac{W_i\tau_i(E_0)}{\tau(E_0)} \ I_2$$

$$= I_1 W_i \tau_i(E_0)\rho \ dx \ \csc \ \psi_1 \qquad (2.16)$$

Furthermore, the term $\tau_i(E_o)$ accounts for ionization in all the
atomic shells of the ith element where the electron-binding energy
$\phi$ is less than the excitation photon energy $E_o$.  Typically it is
only the x-rays caused by electron transitions to *one* of the shells
which is monitored in the fluorescence spectrometer.  For example,
it will be assumed that only the K x-rays of element i are being
measured.  In this case the number of ionizations of the K shell is
reduced to

$$I_4 = \frac{\tau_{Ki}(E_o)}{\tau_i(E_o)} I_3$$

$$= I_1 W_i \tau_{Ki}(E_o) \rho \, dx \, \csc \psi_1 \tag{2.17}$$

where $\tau_{Ki}(E_o)$ is the photoelectric mass absorption coefficient for
only the K shell of element i.

In practice, it is difficult to find accurate tabulated values
of $\tau_{Ki}(E_o)$.  On the other hand, tables of $\tau_i(E_o)$ are readily accessible [6].  Consequently, approximate values of $\tau_{Ki}(E_o)$ are
usually calculated from $\tau_i(E_o)$ in the following manner.

Figure 2.16 shows a plot of the iron total photoelectric mass
absorption coefficient $\tau_i(E)$.  There is an abrupt jump in the coefficient at $E = \phi_K = 7.111$ keV which is the binding energy of an
electron in the K shell.  For photon energies greater than 7.111 keV,
ionization of the K shell is possible, and $\tau_i(E)$ incorporates contributions from all shells including the K shell.  For $E < \phi_K$, the
contribution from the K shell is excluded.  The K edge jump ratio
is defined as

$$r_K = \frac{\tau_+(\phi_K)}{\tau_-(\phi_K)} \tag{2.18}$$

where $\tau_+(\phi_K)$ and $\tau_-(\phi_K)$ are the photoelectric mass absorption coefficients on the high- and low-energy sides, respectively, of the
K edge discontinuity.  For energies close to the absorption edge, it
will be noticed that $\ln \tau_i(E)$ for $E < \phi_K$ is *approximately* parallel
to $\ln \tau_i(E)$ for $E > \phi_K$ when $\ln \tau_i(E)$ is plotted against $\ln E$.  By

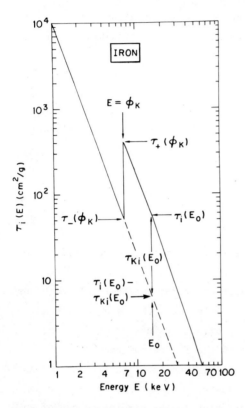

Figure 2.16  Derivation of the approximate value of $\tau_{Ki}(E_o)$. Reprinted by courtesy of EG&G ORTEC.

extrapolating the parallel lower energy portion of the curve as a dashed line in Fig. 2.16, the approximate relationship at energy $E_o$ can be written as

$$\frac{\tau_i(E_o)}{\tau_i(E_o) - \tau_{Ki}(E_o)} \approx \frac{\tau_+(\phi_K)}{\tau_-(\phi_K)} = r_K$$

$$\text{or} \quad \tau_{Ki}(E_o) \approx \frac{r_K - 1}{r_K} \tau_i(E_o)$$

(2.19)

Equation (2.19) represents a convenient *approximate* method of computing $\tau_{Ki}(E_o)$ in Eq. (2.17). However, it is important to be cognizant of its limited accuracy.

Equation (2.17) gives the rate of ionization of the K shell of the ith element.  Following ionization, the vacancies in the K shell are filled by electron transitions from other shells with lower binding energies.  In some of these transitions Auger electrons are emitted.  In the remaining transitions characteristic K x-rays are emitted.  The K-shell fluorescence yield $\omega_K$ describes the fraction of the K-shell ionizations which are followed by emission of characteristic K-series x-rays.  Therefore the rate of emission of K x-rays from the ith element is

$$I_5 = \omega_{Ki} I_4 \tag{2.20}$$

Generally speaking, only one of the lines from the K series will be measured.  The energy of the monitored line will be denoted $E_i$.  If this line represents a fraction f of the total photon rate in the entire K series, then the emitted photon rate in the analyzed line is

$$I_6 = f I_5 \tag{2.21}$$

This photon rate is emitted isotropically in all directions, i.e., into a solid angle of $4\pi$ steradians.  The photon rate emitted toward the detector into the differential solid angle $d\Omega_2$ at a takeoff angle $\psi_2$ is given by

$$I_7 = \frac{d\Omega_2}{4\pi} I_6 \tag{2.22}$$

Before emerging from the specimen the characteristic x-ray can suffer absorption in the path length $x \csc \psi_2$.  This reduces the emitted photon rate to

$$I_8 = I_7 \exp[-\mu(E_i)\rho x \csc \psi_2] \tag{2.23}$$

where $\mu(E_i)$ is the specimen total mass attenuation coefficient for the characteristic x-ray energy $E_i$.  If the detector efficiency for recording a photon of energy $E_i$ is $\eta(E_i)$, then the detected photon rate in the detector is

$$I \ dx \ dE_o = \eta(E_i)I_8 \tag{2.24}$$

This is the contribution of the differential volume element from x to x + dx for the portion of the excitation spectrum between energies $E_o$ and $E_o + dE_o$. Equation (2.24) must be integrated over the range x = 0 to x = T to obtain the contribution from the entire specimen thickness. In performing the integration, it will be assumed that $\psi_2$ is approximately constant. This assumption is reasonable if the distance z is very large compared to x for the portions of the specimen which make a significant contribution to the measured intensity. For most cases in conventional x-ray fluorescence spectrometry, this approximation is justified. Equation (2.24) must also be integrated over the entire range of energies in the excitation spectrum which are capable of K-shell ionization. This range extends from $\phi_K$, the binding energy of the K-shell electrons, up to $E_{max}$, the maximum energy in the excitation spectrum. Substituting the explicit expression for $I_8$ in Eq. (2.24) and performing the double integration yields the detected characteristic photon rate for solid angles $d\Omega_1$ and $d\Omega_2$, i.e.,

$$I_i(E_i)d\Omega_1 \ d\Omega_2 = \int_{E_o=\phi_K}^{E_{max}} \int_{x=0}^{T} I \ dx \ dE_o$$

$$= d\Omega_1 \left(\frac{d\Omega_2}{4\pi}\right) \frac{\eta(E_i)}{\sin \psi_1} \int_{E_o=\phi_K}^{E_{max}} Q_{if}(E_o)$$

$$\times \left(\frac{1 - \exp\{-\rho T[\mu(E_o) \csc \psi_1 + \mu(E_i) \csc \psi_2]\}}{\mu(E_o) \csc \psi_1 + \mu(E_i) \csc \psi_2}\right)$$

$$\times \ I_o(E_o) \ dE_o \tag{2.25}$$

where $Q_{if}(E_o)$ contains all the fundamental parameters associated with fluorescing the characteristic x-rays from element i.

$$Q_{if}(E_o) = W_i \tau_{Ki}(E_o)\omega_{Ki} \ f$$

$$\approx W_i \tau_i(E_o) \frac{r_K - 1}{r_K} \omega_{Ki} f \tag{2.26}$$

Note that the integral over the excitation spectrum can be evaluated only if the shape of the excitation spectrum is known.

In practical spectrometers, large solid angles $\Omega_1$ and $\Omega_2$ are used for the excitation and fluoresced x-rays. In principle, this means that Eq. (2.25) should be integrated over these finite solid angles. Furthermore, the excitation source typically has a finite dimension, and Eq. (2.25) must also be integrated over the source dimensions to be rigorously correct. However, these refinements add a complication which detracts from the clarity of Eq. (2.25) in understanding the fundamental factors affecting the measurement of the fluoresced intensity. For simplicity, Eq. (2.25) will be written as

$$
I_i(E_i) = \frac{\eta(E_i)}{4\pi \sin \psi_1} \int_{E_o=\phi_K}^{E_{max}} Q_{if}(E_o)
$$

$$
\times \left( \frac{1 - \exp\{-\rho T[\mu(E_o)\, \csc\, \psi_1 + \mu(E_i)\, \csc\, \psi_2]\}}{\mu(E_o)\, \csc\, \psi_1 + \mu(E_i)\, \csc\, \psi_2} \right)
$$

$$
\times I_o(E_o)\, dE_o \tag{2.27}
$$

where, strictly speaking, the units of $I_i(E_i)$ are photons per second per steradian per steradian, the units of $I_o(E_o)\, dE_o$ are photons per second per steradian and the formula is understood to apply to the special case of infinitesimal solid angles.

In applying Eq. (2.25) to quantitative analysis by the fundamental parameters method, a further simplifying assumption is often made. It is common practice to assume that the terms in Eq. (2.25) are approximately constant over the entire range defined by the finite geometry. In that case, integration over the finite solid angles amounts to replacing $d\Omega_1$ and $d\Omega_2$ with $\Omega_1$ and $\Omega_2$, respectively, in Eq. (2.25). More frequently, the explicit expression of the solid angles $\Omega_1$ and $\Omega_2$ is deleted and $I_o(E_o)\, dE_o$ and $I_i(E_i)$ are defined as the total incident and detected photon rates. For this finite geometry approximation, $\psi_1$ and $\psi_2$ are defined as the *effective incidence* and *takeoff angles,* respectively, and Eq. (2.27) is

employed.  The justification for using Eq. (2.27) for finite
geometry is entirely pragmatic.

With the limitations of the strict and loose interpretations
of Eq. (2.27) in mind, a number of important effects can be deduced
from Eqs. (2.26) and (2.27).  First of all, the fluoresced intensity
$I_i(E_i)$ is proportional to (a) the weight fraction of element I in
the specimen, (b) the photoelectric mass absorption coefficient for
element i, (c) the fluorescence yield for element i, and (d) the
detection efficiency for the fluoresced line energy.  Since the
photoelectric mass absorption coefficient is largest just above the
absorption edge, the most efficient excitation energies lie just
above the absorption edge energy for element i.

Secondly, the fluoresced intensity is modified by the effects
of primary and secondary absorption in the specimen.  This is a
major source of the so-called matrix effects.  Primary absorption
is defined by the term $\mu(E_o)$ csc $\psi_1$.  It reduces the effectiveness
of the x-rays from the excitation source.  Secondary absorption is
defined by the term $\mu(E_i)$ csc $\psi_2$.  It reduces the intensity of the
desired characteristic x-rays as they leave the specimen.  Note that
$\mu(E_o)$ and $\mu(E_i)$ for the specimen are functions of the specimen composition through the relation

$$\mu(E) = \sum_j W_j \mu_j(E) \qquad\qquad (2.28)$$

where $W_j$ is the weight fraction of element j in the specimen, $\mu_j(E)$
is the total mass attenuation coefficient of element j at energy E,
and the summation over j includes all elements in the specimen such
that $\sum_j W_j = 1$.  Thus, the measured intensity of element i is not
only a function of the concentration of element i but also a function of the concentration of all the other elements in the specimen
as a result of primary and secondary absorption.

If the excitation source is monochromatic (emits only one
energy), Eq. (2.27) simplifies to

$$I_i(E_i) = \frac{\eta(E_i)}{4\pi \sin \psi_1} Q_{if}(E_o)$$

$$\times \left( \frac{1 - \exp\{-\rho T[\mu(E_o) \csc \psi_1 + \mu(E_i) \csc \psi_2]\}}{\mu(E_o) \csc \psi_1 + \mu(E_i) \csc \psi_2} \right)$$

$$\times I_o(E_o) \tag{2.29}$$

If the specimen thickness T is infinite, then the exponential term vanishes in both Eq. (2.27) and Eq. (2.29). That is, for infinitely thick specimens the fluoresced intensity is

$$I_i(E_i) = \frac{\eta(E_i)}{4\pi \sin \psi_1} \int_{E_o=\phi_K}^{E_{max}} \frac{Q_{if}(E_o) I_o(E_o)\ dE_o}{\mu(E_o) \csc \psi_1 + \mu(E_i) \csc \psi_2} \tag{2.30}$$

or for monochromatic excitation,

$$I_i(E_i) = \frac{\eta(E_i)}{4\pi \sin \psi_1} \frac{Q_{if}(E_o) I_o(E_o)}{\mu(E_o) \csc \psi_1 + \mu(E_i) \csc \psi_2} \tag{2.31}$$

The question arises as to how thick a specimen must be to yield virtually all the fluoresced intensity which would be obtained from an infinitely thick specimen. If 99% of the infinite thickness intensity is adequate, then the exponential term in Eq. (2.25) must have a value of 0.01. This defines the critical thickness as

$$T_{99\%} = \frac{-\ln 0.01}{\rho[\mu(E_o) \csc \psi_1 + \mu(E_i) \csc \psi_2]}$$

$$= \frac{-\ln (1 - 0.99)}{\rho[\mu(E_o) \csc \psi_1 + \mu(E_i) \csc \psi_2]} \tag{2.32}$$

Specimens which are thick compared with $T_{99\%}$ can be considered to be infinitely thick.

Formulas similar in form to Eq. (2.31) can be derived for the coherent and incoherent scattered intensities. Instead of $Q_{if}(E_o)$ a term proportional to the scattered intensity described in Sec. 2.5

is inserted and the solid angle effect in Eq. (2.22) is handled
differently.  However, the scattered x-rays suffer primary and
secondary absorption effects similar to the fluoresced x-rays.  An
additional contribution to the fluoresced intensity resulting from
secondary fluorescence will be considered in Sec. 2.11.

At high energies for specimens with low average atomic numbers,
there is some difficulty in choosing the proper value to use for
$\mu(E_o)$ and $\mu(E_i)$ in Eqs. (2.27) to (2.32).  Under these conditions,
the mass scattering coefficients account for a major portion of $\mu$.
Since the scattering interactions do not necessarily prevent the
photon from going on to fluoresce element i, it can be argued that
the proper value to use for $\mu$ lies somewhere between $\tau$ and
$\tau + \sigma_{coh} + \sigma_{inc}$.  It can also be argued that $\tau$ is a better approximation than $\tau + \sigma_{coh} + \sigma_{inc}$.

## 2.10  PRIMARY AND SECONDARY ABSORPTION EFFECTS

In practice, not only the secondary characteristic radiation is pre-
ferentially absorbed by certain matrix elements, but also the primary exciting spectrum.  This gives rise to great complications in
quantitative analysis because whereas the absorption of secondary
characteristic radiation (the secondary absorption effect) is relatively easy to predict and not too difficult to correct for, the
absorption of a broadband primary exciting spectrum (the primary
absorption effect) is a more subtle effect to predict and can be .
much more difficult to correct for.

The secondary absorption effect is due to interaction of the
characteristic x-ray photons excited within the specimen and all
types of atoms making up the specimen.  The total secondary absorption $\mu(\lambda_i)$ csc $\psi_2$ for a given characteristic photon of wavelength
$\lambda_i$ from an analyte element i, is given by Eq. (2.33).

$$\mu(\lambda_i) \ csc \ \psi_2 = \sum_j W_j \mu_j(\lambda_i) \ csc \ \psi_2 \qquad\qquad (2.33)$$

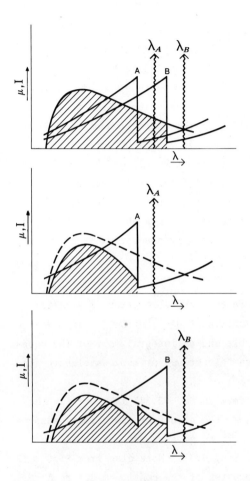

Figure 2.17   Total primary spectrum available for the excitation of elements A and B (the hatched area represents the spectral distribution of photons available for excitation).

$W_j$ is the weight fraction of element j and the index j represents all types of atoms making up the specimen, including the analyte element.   For a given absorbing element j, the value of $\mu_j(\lambda_i)$ is a constant characteristic of the analyte wavelength $\lambda_i$.

The primary absorption effect is probably best illustrated pictorially, and Fig. 2.17 shows two wavelengths $\lambda_A$ and $\lambda_B$ being excited by a band of continous primary radiation. Since both elements A and B are competing for the primary radiation, in effect, only a portion of the continuum will be available for each of the two excited elements. However, since the wavelengths of the absorption edges of A and B do not have the same value, the actual portion of the primary spectrum available for the excitation of one of the elements will be dependent upon the wavelengths of both of the absorption edges. Thus, there is an effect by all of the matrix elements on that portion of the continuum potentially available for the excitation of a given element, and this effect is referred to as the *primary absorption effect*. The total specimen absorption $\alpha_T(\lambda_i)$ is thus made up of both primary and secondary absorption terms.[*]

$$\alpha_T(\lambda_i) = \sum_j W_j [\mu_j(\lambda_o) \csc \psi_1 + \mu_j(\lambda_i) \csc \psi_2] \qquad (2.34)$$

In Eq. (2.34), $\mu_j(\lambda_o)$ represents the absorption of a matrix element j for a primary x-ray photon of wavelength $\lambda_o$. In practice, a range of $\lambda$ values are available and an integral form of the equation will be required unless an "effective" primary wavelength can be assumed (see Sec. 10.7.3).

The secondary absorption term in Eq. (2.34) is generally the dominant term, but in many cases, the effect of the primary absorption is to reduce the variation in total specimen absorption in a typical range of matrices (see Sec. 9.3). Note also from Fig. 2.17 that although the primary absorption of the continuum due to A reduces the more effective primary photon intensity available to excite B, the characteristic line(s) from A are able to excite B (i.e., enhance B). This is not an uncommon situation and primary absorption effects often tend to compensate for enhancement or at least reduce their overall influence on measured characteristic line intensity.

---

[*] The absorption is sometimes written as $\alpha_T(\lambda_i) = \sum_j W_j [\mu_j(\lambda_o) + A\mu_j(\lambda_i)]$, where $A = (\csc \psi_2)/(\csc \psi_1)$. Although this is valid for predicting absorption effects in thick specimens, Eq. (2.34) is the correct form for *both* thick and thin specimens.

## 2.11   ENHANCEMENT AND THIRD ELEMENT EFFECTS

Figure 2.18 shows that the continuum is not the only source of excitation energy causing the fluorescence of element i.  Other wavelengths (energies)  $\lambda_A$  and  $\lambda_B$  arising from matrix elements A and B may also lie on the short-wavelength side (high-energy side) of the absorption edge of element i and, as such, are capable of exciting element i.  Since the most efficient exciting wavelengths (energies) are those closest to the high-energy side of the absorption edge of the excited element, element A enhances element i more efficiently than does element B.

The example is further complicated by the fact that element B can also excite element A, adding to the overall enhancing ability of A.  This effect is called a *third-element effect,* where in this instance, element B is the third element.  An example of enhancing and third-element effects is given in Fig. 2.19, which illustrates the case of the ternary system Ni/Fe/Cr.  Here chromium is the

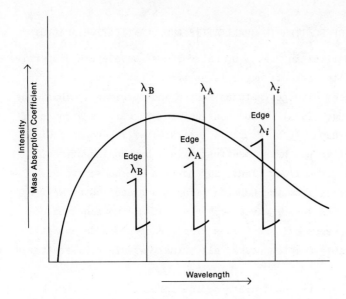

Figure 2.18   Interelement absorption effects.

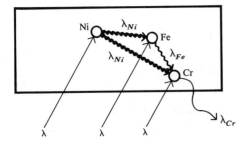

Figure 2.19  Origin of enhancement and third-element effects.

analyte element and nickel and iron are both enhancers of chromium.
Nickel, however, also enhances iron, illustrating the third-element
effect.

The reader who wishes to make a serious theoretical study of
matrix effects should consult the papers of Sherman [7], Shiraiwa
and Fujino [8], and the review article by Sparks [17], where secondary and third-element effects are treated in detail.


2.12  EFFECTS OF SCATTER IN QUALITATIVE AND QUANTITATIVE ANALYSIS

It was shown earlier in Fig. 2.6 that for most wavelength dispersive
spectrometers, the takeoff angle is generally constant but the incident x-ray beam diverges between source and specimen.[*]  Thus, the
value of $\psi$ in Eq. (2.2a) is not constant but rather has a range of
values over perhaps $\pm 10^{\circ}$. Hence, $\lambda_{inc}$ itself will have an equivalent range of values being manifested in a broadening of the incoherently scattered line.  Thus, any portion of the scattered
spectrum resulting from the Compton effect is broadened as well as
shifted toward the longer wavelength region.  One consequence of
this is that whereas emission lines from the specimen do not give
rise to measurable Compton peaks, all primary wavelengths scattered

---

[*]In energy-dispersive spectrometers, the optical path is very broadly
collimated and the detector can receive a wide angular range of
scattering depending upon the receiving solid angle of the detector.

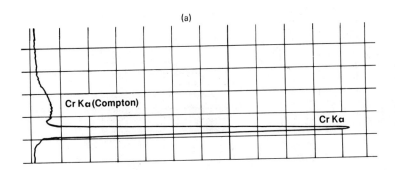

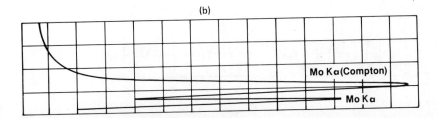

Figure 2.20  Scans over the Cr Kα (a) and Mo Kα (b) scattered x-ray
tube lines showing the change in the relative intensities of the
coherently and Compton-scattered lines.

by the specimen are subject to the Compton effect and this will be
seen as a broadened, shifted, Compton peak displaced toward the
long-wavelength (low-energy) portion of the spectrum.

The effect of this on qualitative and quantitative analysis may
be quite important.  For example, where bremsstrahlung source
excitation is used, all characteristic lines from the source will
appear as both coherently and incoherently scattered lines.  The
measurable amount of scatter will increase with a decrease in the
average atomic number of the specimen, but the ratio of coherent to
incoherent scatter will depend upon the wavelength of the characteristic tube line.  Figures 2.20(a) and (b) were obtained by scattering the tube radiation from a low-atomic-number specimen, the
x-ray tube was changed for each scan and the portions of the spectra
shown are for the Mo Kα (0.709 Å) and the Cr Kα (2.291 Å) x-rays.

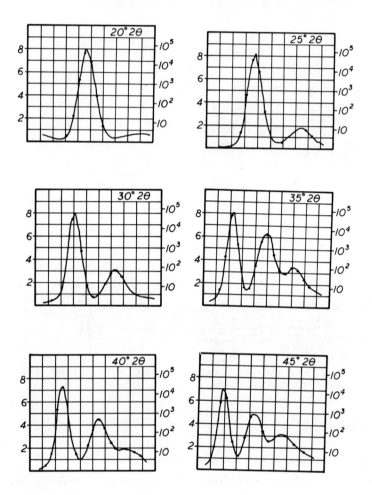

Figure 2.21  Pulse amplitude distributions measured at various diffraction angles from a sample of distilled water using a Chromium anode x-ray tube at 50 kV and 10 mA, scintillation detector, and LiF (200) crystal.

The decrease of the incoherent-to-coherent scattering ratio with increase of wavelength is clearly shown, as is the greater broadening of the incoherently scattered line.

One immediate effect of the appearance of scattered lines along with the emission spectrum of the specimen is the chance of spectral overlap and interference between unwanted scattered lines and the

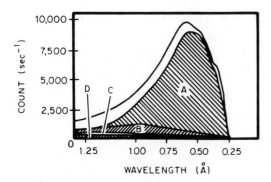

Figure 2.22  The contribution of components to the total background
in a wavelength dispersive spectrometer, as a function of wavelength:
A = first order, B = second order, C = third order, and D = fourth
order.

analytical lines to be measured.  An extreme case of this is the
measurement of a specimen element having the same atomic number as
the x-ray tube anode.

Finally, in the case of wavelength-dispersive spectrometers, it
must be remembered that the measured background which arises from
scatter of the primary tube spectrum, is composed of a series of
harmonic reflections of the scattered spectrum.  That is to say that
the analyzing crystal will diffract first-order scattered spectrum,
second-order, third-order, and so on.  The higher the angle of the
goniometer, the greater will be the chance of higher order harmonics.  Thus, the "order" of the background increases with the
Bragg angle of the spectrometer.  This is illustrated in Fig. 2.21,
which shows a series of pulse height distributions measured from a
sample of distilled water (i.e., an excellent scatterer) at different Bragg angles.  Note that at the higher Bragg angles, several
energies are present in the background, i.e., the first-order wavelength $\lambda_o$ for the Bragg angle, a wavelength $\lambda_o/2$ from the primary
bremsstrahlung, and so on.  Over the typical wavelength range of
the spectrometer, harmonics up to the sixth order may significantly
contribute to the background.

This effect of increasing the order of the background with increasing Bragg angle is the reason why the use of the pulse-height

selector for removal of background becomes more efficient at higher
Bragg angles (in effect at longer wavelengths). This is clearly
illustrated in Fig. 2.22 which shows the contribution of first-,
second-, third-, and fourth-order components to the total scattered
bremsstrahlung background as a function of the diffraction angle
$\psi = 2\theta$.


## 2.13  POLARIZATION OF X-RAYS

Early experimenters with x-radiation [1,9] noted that x-rays could
be polarized like other forms of electromagnetic radiation.  Barkla
[9] showed that the primary bremsstrahlung radiation was partially
polarized.  A later investigator, Ross [10], showed that at the
high-energy limit of the white spectrum, the x-ray photons were very
nearly completely plane polarized.  This is in sharp contrast to
primary characteristic x-rays which are unpolarized.

A scattering process will also result in polarization of x-rays
according to the classical Thompson theory expressed by Eq. (2.1).
For an unpolarized primary beam, scattering at an angle ($2\theta$) of $90^{\circ}$
results in nearly complete plane polarization of the scattered
x-rays.  Thus, crystal diffraction of x-rays produces plane-
polarized radiation.  If the diffraction angle ($2\theta$) approaches $90^{\circ}$,
polarization becomes nearly complete.

These effects have been used by investigators to reduce the
background radiation levels in x-ray spectroscopy.  For Bremsstrahlung radiation, the maximum polarization vector is parallel to the
electron path in the x-ray tube.  In many conventional wavelength
spectrometer designs, the specimen and crystal surfaces will become
parallel when $2\theta$ equals the takeoff angle $\psi_2$ ($\simeq 45^{\circ}$).  In addition,
the x-ray tube axis and the rotational axes of the crystal and de-
tector are parallel.  This so-called parallel optics cannot take
advantage of the polarization of the white spectrum.  If, however,
the surface plane of the crystal and specimen remain perpendicular
to one another and the x-ray tube is positioned so that the maximum

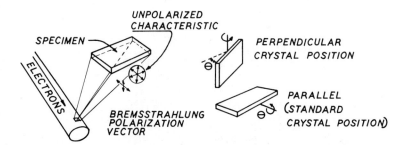

Figure 2.23  X-ray optics geometry to take advantage of the polarization effect on the background.

polarization vector vibrates as shown in Fig. 2.23, background reduction will occur.

Champion and Wittem [11] used this effect and redesigned a conventional wavelength spectrometer with a resulting maximum background reduction of nearly 5:1, under ideal circumstances.

This maximum reduction in scattered bremsstrahlung background
occurs under characteristic lines that are diffracted by the analyzing crystal at or near $90^{\circ}$ $2\theta$.  At other angles, the reduction in
scattered background is lower.

Energy-dispersive systems can also take advantage of polarization effects [12-14] by polarizing the incident excitation flux and
properly positioning the Si(Li) detector with respect to the sample
according to the principles outlined above.  Compton scattering of
tube spectra (at $90^{\circ}$) can provide a weak source of polarized x-rays.
Howell and Pickels [14] have recently investigated the use of the
Borrmann anomolous transmission effect to produce a low-intensity
source of highly polarized x-rays.  Aiginger et al. [15] have used
the novel approach of diffracting the CuKα characteristic line from
the (113) planes of a copper single crystal $2\theta \simeq 90^{\circ}$ thus producing
a high-intensity source of polarized copper Kα photons with a
significant reduction in scattered bremsstrahlung radiation.  Recently, Sparks and Hastings [16] have demonstrated that multiple
scattering effects within the sample severely deteriorate the advantages gained by using a polarized radiation source.

REFERENCES

1.   A. H. Compton and S. K. Allison, *X-rays in Theory and Experiment*, D. Van Nostrand, Princeton, N.J., 1935.
2.   R. D. Evans, *The Atomic Nucleus*, McGraw-Hill, New York, 1955.
3.   L. V. Azaroff, *Principles of X-ray Spectrometry*, McGraw-Hill, New York, 1973.
4.   B. E. Warren, *X-Ray Diffraction*, Addision Wesley, Reading, Mass., 1969.
5.   Y. H. Woo, *Phys. Rev.*, *27*:119 (1926).
6.   W. H. McMaster, N. Kerr Del Grande, J. H. Mallett, and J. H. Hubbell, *UCRL-5-174, Secs. 1,2,3* (1969).
7.   J. Sherman, *Spectrochem. Acta*, *7*:283 (1955).
8.   T. Shiraiwa and N. Fujino, *Jap. J. Appl. Phys.*, *5*:886 (1966).
9.   C. G. Barkla, *Phil. Trans. Roy. Soc.*, *204*:467 (1905).
10.  P. A. Ross, *J. Opt. Soc. Amer.*, *16*:375 (1928).
11.  K. P. Champion and R. N. Wittem, *Nature*, *199*:1086 (1963).
12.  T. G. Dzubay, B. V. Jarrett, and J. M. Jaklevic, *Nucl. Instrum. Methods*, *115*:297 (1974).
13.  R. H. Howell, W. L. Pickels, and J. L. Cate Jr., *Advan. X-ray Anal.*, *18*:265 (1974).
14.  R. H. Howell and W. L. Pickles, *Nuc. Instrum. Methods*, *120*:187 (1974).
15.  H. Aiginger, P. Wobrauschek, and C. Brauner, *Nuc. Instrum. Methods*, *120*:541 (1974).
16.  C. J. Sparks and J. B. Hastings, Personnel communication 1975 (O.R.N.L. Report in press).
17.  C. J. Sparks, *Adv. X-ray Anal.*, *19*:19 (1976).

# 3

# Sources for the Excitation of Characteristic X-rays

## 3.1 INTRODUCTION

Almost any high-energy photon will act as a source for the production of characteristic x-rays provided that it is sufficiently energetic to eject an electron from the appropriate atomic level of the
element to be excited. The minimum energy required is simply the
binding energy $\phi$ of the electron in the appropriate shell. Binding
energies for the different shells and subshells are often denoted
by subscripts such as $\phi_K$, $\phi_L$, $\phi_{L(III)}$, etc. Frequently the binding
energy is identified as the absorption edge energy $E_{abs}$ since it
corresponds to the energy at which the mass absorption coefficient
for the element changes abruptly. It is also referred to as the
*critical excitation potential* since $\phi$ in kiloelectronvolts (keV)
corresponds to the minimum voltage in kilovolts (kV) on the x-ray
tube required to excite the characteristic x-rays. In Table 3.1,
critical excitation potentials are listed for exciting the K$\alpha$, L$\alpha$,
and M$\alpha$ lines of selected elements covering the range from atomic
numbers 9 to 90. Since x-ray spectrometers generally cover the
range from 0.3 to 18 $\overset{o}{A}$ or 0.7 to 40 keV, excitation energies from 1
keV to beyond 50 keV are normally required.

Table 3.1  Variation of Excitation Potential with Atomic Number

| Atomic Number | Element | Kα Position Wavelength (Å) | Kα Energy (keV) | Kα Excitation Potential (keV) | Lα Position Wavelength (Å) | Lα Energy (keV) | Lα Excitation Potential (keV) | Mα Position Wavelength (Å) | Mα Energy (keV) | Mα Excitation Potential (keV) |
|---|---|---|---|---|---|---|---|---|---|---|
| 9  | F  | 18.32 | 0.68  | 0.69  |       |       |       |       |      |      |
| 20 | Ca | 3.36  | 3.69  | 4.04  |       |       |       |       |      |      |
| 30 | Zn | 1.44  | 8.63  | 9.66  | 12.25 | 1.01  | 1.02  |       |      |      |
| 40 | Zr | 0.79  | 15.74 | 18.00 | 6.07  | 2.04  | 2.22  |       |      |      |
| 50 | Sn | 0.49  | 25.19 | 29.19 | 3.60  | 3.44  | 3.93  |       |      |      |
| 60 | Nd | 0.33  | 37.18 | 43.57 | 2.37  | 5.23  | 6.22  | 12.68 | 0.98 | 0.98 |
| 70 | Yb | 0.24  | 52.03 | 61.30 | 1.67  | 7.41  | 8.94  | 8.15  | 1.52 | 1.54 |
| 80 | Hg |       |       |       | 1.24  | 9.98  | 12.29 | 5.65  | 2.20 | 2.32 |
| 90 | Th |       |       |       | 0.96  | 12.95 | 16.30 | 4.14  | 2.99 | 3.34 |

The major requirements of an excitation source are that it be stable, efficient, and sufficiently energetic to excite the elements of interest.  To be efficient, it must yield a high counting rate for each analyte line and also provide a high peak-to-background ratio.

The majority of wavelength-dispersive fluorescence spectrometers utilize a sealed x-ray tube powered by a 3- to 4-kW high-voltage generator as an excitation source.  With energy-dispersive spectrometers, on the other hand, it has been possible to use a variety of excitation methods as a result of the higher detection efficiency.  For specific applications requiring a rather narrow range of performance, x-ray- or γ-ray-emitting radioisotopes are convenient.  However, for a broad range of capability and the more demanding applications, some form of x-ray tube excitation is usually employed.  X-ray tube power dissipation on energy-dispersive instruments varies from a fraction of a watt for direct excitation up to several kilowatts using secondary fluorescers.  Although this book will deal primarily with the sources of photon excitation mentioned above, it should be appreciated that excitation by charged particles such as electrons, protons, and alpha particles is also feasible.

## 3.2  BROADBAND EXCITATION WITH THE SEALED X-RAY TUBE

Probably the most commonly used source of excitation is the radiation obtained directly from the anode of an x-ray tube.  Electrons emitted by a heated cathode are accelerated through a potential V and focused to strike the anode.  Each electron, when it reaches the anode, has acquired a kinetic energy

$$E = V \tag{3.1}$$

The energy E is expressed in electron-volts (eV), or kiloelectron-volts (keV).  If the cathode-to-anode voltage is 50 kV, the electron impinges on the anode with a kinetic energy of 50 keV.  Most of the power transferred to the accelerated electrons is dissipated as heat in the anode, and only a small fraction of the power results in the emission of x-rays. Two important types of x-rays are produced: bremsstrahlung and characteristic x-rays.

If the impinging electron energy is greater than the electron-binding energy $\phi$ of a particular atomic shell for the anode material, then ionization of that shell can take place and characteristic x-rays are emitted by the atom. For example, if the x-ray tube voltage V (in kV) on a molybdenum anode x-ray tube is greater than the ionization potential $\phi_K$ (in keV) for the molybdenum K shell, Mo K series x-rays will be generated. The intensity (counts per second) of the characteristic lines is proportional to the current through the x-ray tube, but is a nonlinear function of the over-voltage $V - \phi$.

The bremsstrahlung continuum is produced by electrons scattered by the atomic nuclei in the anode. In a small fraction of the cases where the electron is deflected by the nucleus, an x-ray photon is emitted [1]. The energy of this photon can vary from zero up to the full energy of the incident electron. Thus, a continuum of x-ray photon energies is generated as electrons bombard the x-ray tube anode. This continuum is referred to as *bremsstrahlung,* a term of German origin which loosely translates into *deceleration radiation.* In a conventional, thick anode x-ray tube, the continuum appears as illustrated in Fig. 3.1(a). The maximum photon energy in the spectrum is

$$E_{max} = V \qquad\qquad\qquad (3.2)$$

where $E_{max}$ is in kiloelectronvolts and V is the x-ray tube voltage in kilovolts. When the spectrum is plotted as a function of wave-length [Fig. 3.1(b)], this maximum energy corresponds to a minimum wavelength.

$$\lambda_{min} = \frac{12.4}{E_{max}} = \frac{12.4}{V} \qquad\qquad\qquad (3.3)$$

where $\lambda_{min}$ is in angstroms. At photon energies approaching $E_{max}$,

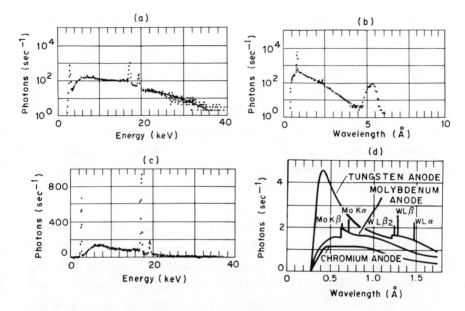

Figure 3.1  Spectra from thick anode x-ray tubes.   (a)  The energy
spectrum from a molybdenum anode x-ray tube measured with a Si(Li)
detector (40 kV tube voltage, 39-μm-thick beryllium x-ray tube window, 90° electron beam incidence angle, 32° x-ray takeoff angle).
Although the low spectral intensity near 40 keV makes accurate
measurement of $E_{max}$ difficult, the spectrum approaches zero intensity near 40 keV.  The Mo K lines occur at 17.4 keV and 19.8 keV,
and the L lines are visible at 2.4 keV.   (b)  The spectrum from (a)
transformed into the wavelength coordinate system.  Note that the
Mo L line at 5.3 Å is broad due to the Si(Li) detector energy resolution.  The value of $\lambda_{max}$ is readily apparent at 0.31 Å.   (c)  The
spectrum from (a) on a linear vertical scale.   (d)  Spectra from
tungsten, molybdenum, and chromium anode x-ray tubes.  The positions
of characteristic lines are marked by vertical lines of arbitrary
height.  Adapted from R. Tertian, *Fluorence X, Theorie et Pratique
de l'Analyse,* Thesis, Université de Paris and reprinted by courtesy
of EG&G ORTEC.

the shape of the continuum is approximately described by Kramers'
formula [2].

$$I(E)\ dE = kiZ\ \frac{(E_{max} - E)\ dE}{E} \tag{3.4}$$

$I(E)\ dE$ is the number of photons between the energies $E$ and $E + dE$,
$i$ is the x-ray tube current, $k$ is an instrumental constant, and $Z$

is the atomic number of the anode. In terms of wavelength, Eq. (3.4) becomes

$$I(\lambda) \; d\lambda = 12.4kiZ \left(\frac{\lambda}{\lambda_{min}} - 1\right) \frac{d\lambda}{\lambda^2} \qquad (3.5)$$

From Eq. (3.5) it is easy to show that the maximum continuum intensity occurs at $\lambda = (3/2)\lambda_{min}$ when plotted as a function of wavelength. At lower x-ray energies (longer wavelengths), absorption of the x-rays as they attempt to emerge from the anode becomes important, and Eqs. (3.4) and (3.5) must be multiplied by absorption correction terms in order to describe the observed spectrum [3,4]. In fact, anode absorption is a major reason for the decrease in intensity at low energies in Fig. 3.1(a). In an x-ray tube with a thick window, window absorption also attenuates the intensity in the low-energy (long-wavelength) region. Attempts to more accurately describe the shape of the bremsstrahlung continuum have been made by several authors. Most of this work has been reviewed by Smith et al. [5]. Measured x-ray tube spectra have been reported by Birks and co-workers [6-8], and more recently by Keith and Loomis [9].

The intensity of the fluoresced line from element i in a thick specimen may be derived as a function of the excitation spectrum (Sec. 2.9):

$$I_i(E_i) = \frac{\eta(E_i)}{4\pi \sin \psi_1} \int_{E=\phi}^{E_{max}} \frac{Q_{if}(E)I_o(E) \; dE}{\mu(E) \csc \psi_1 + \mu(E_i) \csc \psi_2} \qquad (3.6)$$

For broadband excitation with a thick x-ray tube anode, the excitation spectrum is approximately described by

$$I_o(E) \; dE = \left[kiZ\left(\frac{E_{max} - E}{E}\right)f(\chi) + k_j iF_j(V - \phi_j)\right.$$
$$\left. \times \; \delta(E_j - E)\right] \; dE \qquad (3.7)$$

The first term in Eq. (3.7) represents the bremsstrahlung continuum expressed in Eq. (3.4) multiplied by the absorption correction function $f(\chi)$. The second term accounts for the anode characteristic line intensities at energies $E_j$. The term $\delta(E_j - E)$ is zero for $E \neq E_j$ and is equal to unity where a characteristic line exists at $E = E_j$. The absorption-edge energy for the line at energy $E_j$ is given by $\phi$, while $k_j$ combines the instrumental constant and the relative intensity of the jth characteristic line. $F_j(V - \phi_j)$ is a complicated nonlinear function describing the dependence of the characteristic line intensity on the tube voltage V. $F_j(V - \phi_j)$ increases with V for moderate values of $V - \phi_j$; but can be a decreasing function of V at extremely large values of $V - \phi_j$. The probability of fluorescing a particular line of a desired element in a sample is

$$Q_{if}(E) = \tau_i(E)W_i\omega_i f\left(\frac{r_i - 1}{r_i}\right) \qquad (3.8)$$

in which $r_i$ is the absorption jump ratio, $\omega_i$ the fluorescent yield, and f the ratio of the intensity of the measured line to all other lines in the spectral series (see Sec. 2.9).

From the above expression, it will be noted that the energy dependence of $Q_{if}(E)$ is totally contained in the photoelectric mass absorption coefficient $\tau_i(E)$ for element i. $\tau(E)$ varies rapidly with energy as illustrated in Fig. 3.2. An increase in energy by a factor of 2 yields a decrease in $\tau$ by approximately a factor of 7. Thus, for regions where the excitation spectrum $I_o(E)$ dE varies slowly with energy, the excitation efficiency is controlled by $\tau_i(E)$. As Fig. 3.2 reveals, only those photons with energies slightly above the absorption-edge energy are efficient in exciting the analyte lines in the specimen. Consequently, where a wide range of elements is to be analyzed simultaneously, the excitation spectrum must contain a wide range of energies. The bremsstrahlung continuum is useful in this respect. Conversely, characteristic lines from the anode can provide efficient excitation for elements with absorption edges at slightly lower energies. Characteristic anode lines are often used to enhance the sensitivity for specific analyte elements. This technique is carried to the extreme in the monochromatic

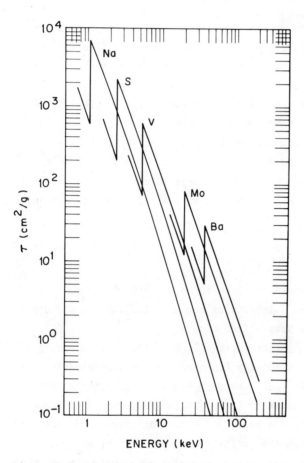

Figure 3.2  The photoelectric mass absorption coefficient for several elements at and above the K absorption edge. Reprinted by courtesy of EG&G ORTEC.

excitation methods where the bremsstrahlung continuum is suppressed and the characteristic lines are used for trace element excitation.

For broadband excitation, Eq. (3.7) shows that the analyte line intensity $I_i$ is proportional to the tube current i. It can also be appreciated that the analyte line intensity is a complicated nonlinear function of the x-ray tube voltage. This relationship is often approximated by the expression

$$I_i = Ki(V - \phi)^n \tag{3.9}$$

where n lies in the range $1 \leq n \leq 2$ and depends on the anode material, on the analyte element, and to some extent on the range of the
tube voltage V.  Here $\phi$ is the absorption-edge energy for the
analyte line in the specimen.  It should be realized that Eq. (3.9)
is a very crude approximation.  However, it is useful in qualitatively demonstrating the important effects of x-ray tube voltage
and current on analyte line intensity.  In this respect, a value
n = 1.6 is often chosen.  Equation (3.9) clearly shows that the
analyte line intensity is very sensitive to the *overvoltage* $V - \phi$.

It can be seen from Eq. (3.7) that the analyte line intensity
excited by the continuum is proportional to the atomic number of the
anode.  Thus, higher atomic number anodes would be expected to yield
higher sensitivities.  Unfortunately this is not always the case.
In high-power, grounded-anode, sealed x-ray tubes, the higher the
atomic number of the anode, the greater is the fraction of incident
electrons which are scattered in the direction of the x-ray tube
window.  This in turn requires a relatively thick window (perhaps
500- to 1000-μm beryllium) to dissipate the heat, with a subsequent
loss in the transmission of longer wavelengths.  On the other hand,
characteristic anode lines are often used to excite the lighter elements, and the ratio of characteristic anode line intensity to continuum intensity is higher on lower atomic number anodes.

A wide variety of x-ray tube anode materials are available.
The more commonly used anodes are chromium, rhodium, tungsten, and
molybdenum.  In addition, aluminum, copper, silver, gold, and
platinum have all been employed as single anodes.  Dual-anode tubes,
combining two different anode materials, are also used to provide
more flexibility in the choice of excitation condition.  Tungsten
with chromium and molybdenum with tungsten are two of the more
common combinations.  Other pairs have also been used to optimize
excitation for a particular application (for example, rhodium combined with tungsten).

If the task is to analyze a broad range of medium- to high-
atomic-number element K lines, then a high-atomic-number anode such

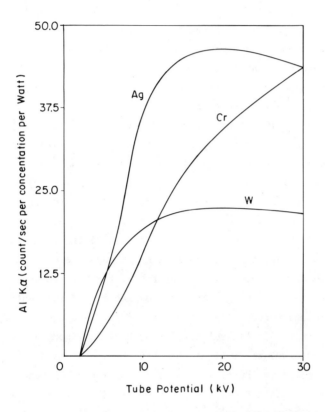

Figure 3.3   Isowatt excitation efficiency curves for the Al Kα line as a function of voltage for Ag, Cr, and W anodes.

as tungsten, gold, or platinum should be chosen for the intense bremsstrahlung spectrum it offers.   For example, the bremsstrahlung intensity from tungsten is approximately three times greater than from chromium.

For light-element excitation, absorption of the longer wave-lengths in the x-ray tube window is an important limitation.   There-fore, it is necessary to reduce the window thickness or completely eliminate it [10-12].   As a specific example, low-atomic-number anodes are useful for reducing the required beryllium window thick-ness on high-power x-ray tubes.   Judicious choice of the anode material for the energies of its characteristic lines can also be

helpful.  The principle is to select a characteristic line energy
which is just above the energy at which the beryllium window attenuation is severe.  On high-power x-ray tubes, the K lines from a
chromium anode are often used.  With low-power tubes having thinner
windows, the L lines from molybdenum, rhodium, or silver anodes
prove useful.  Characteristic line intensities at these longer
wavelengths are quite strong and can considerably enhance light-
element sensitivities, provided they are chosen to lie just to the
short-wavelength side of the beryllium window transmission cutoff.

In all cases of anode selection one must ensure that the coherently and incoherently scattered characteristic lines from the
x-ray tube do not interfere with characteristic lines to be analyzed
from the specimen.  On the other hand, the characteristic anode
lines may also be chosen to enhance the sensitivies for elements of
slightly longer wavelengths.  This technique can be applied for
high-atomic-number elements as well as low-atomic-number elements.

The influence of characteristic lines from the x-ray tube anode
on the overall excitation process can be seen from reference to
Fig. 3.3.  Here the detected counting rate on the aluminum K$\alpha$ line
per watt of tube power and for a 1% aluminum concentration has been
plotted as a function of tube voltage for three different anode
materials:  tungsten, silver, and chromium.  These curves are called
*isowatt curves* since all data is taken at the same tube power or
wattage.  That is, as the tube voltage is increased, the current is
decreased to hold the power dissipation in the tube constant.
Isowatt curves are appropriate for evaluating the effectiveness of
high-power tubes in wavelength-dispersive spectrometers, since the
tube frequently must be operated near its maximum power limit to
achieve optimum sensitivity.  The majority of the excitation of
aluminum in Fig. 3.3 for the silver and chromium anodes comes from
the characteristic tube lines; i.e., the silver L lines at approximately 3 keV and the chromium K lines at approximately 5.4 keV.  The
M lines from tungsten at approximately 1.8 keV are too low in energy
to penetrate the x-ray tube beryllium window.  Consequently,

excitation of aluminum with the tungsten anode is mainly by bremsstrahlung. It will be seen from the figure that the curves for
both silver and tungsten reach a maximum at about 15 kV and then
fall off as the voltage is increased further. The curve for
chromium, however, is still increasing at 30 kV (it too peaks eventually at about 60 kV). In the case of silver and chromium where
excitation is due to the characteristic anode lines, the decrease
in the isowatt curves at extremely high voltages is caused by self-
absorption in the anode. As the tube voltage is increased, the
characteristic anode lines are produced deeper in the anode and are
more heavily absorbed as they attempt to emerge. For tungsten,
where bremsstrahlung excitation predominates, only the lower energy
photons ($\sim$2 to 10 keV) are effective in exciting the aluminum K
lines. Here again, extremely high voltages force the bremsstrahlung
to be produced deep in the anode, and self-absorption causes a decrease in excitation efficiency. Figure 3.3 clearly demonstrates
that the characteristic radiation from the silver and chromium
anodes can be more effective than the bremsstrahlung from the
tungsten anode in exciting light elements such as aluminum, especially since any tungsten M lines are filtered out by the tube window.
If the excitation of aluminum were to have arisen purely from the
bremsstrahlung continuum, the shape of the isowatt curves would
have been approximately the same for all three anodes; although the
overall intensities would have been proportional to the anode
numbers. The fact that low-atomic-number elements (usually of
$Z < 24$ for a chromium anode tube) are excited mainly by the anode
characteristic lines has great significance in quantitative analysis
where primary absorption effects must be considered (see Chaps. 2,
9, and 10). For example, with a chromium anode x-ray tube, the
*effective wavelength* for the excitation of elements of atomic number less than 23 can be taken as 2.291 Å; i.e., the Cr K$\alpha$ wavelength. As will be seen in Chap. 10, this fact greatly simplifies
matrix absorption calculations.

From what has been described above it is obvious that the
unfiltered, thick anode, x-ray tube offers considerable flexibility
in achieving broadband excitation.  Once the appropriate anode material has been selected, the question of the optimum tube voltage
and current settings arises.  Fundamental differences in the characteristics of the energy dispersive and wavelength-dispersive x-ray
spectrometers lead to different criteria for choosing the optimum
tube voltage and current.

### 3.2.1   Choosing Optimum X-ray Tube Settings for Broadband Excitation with the Wavelength-Dispersive Spectrometer

Wavelength-dispersive systems generally require high-power x-ray
tubes (1000 to 4000 W).  These tubes are designed to work over a
range of voltages and currents, with the limitation that the total
power on the tube must not exceed the manufacturer's specified
value.  This upper limit on power dissipation is typically from 2
to 4 kW.  Hence, it is useful to plot the counting rate of an excited analyte line per watt of power versus the tube voltage (isowatt curve).  As the tube voltage is increased to more effectively
excite the analyte, the tube current must be reduced to avoid over-
dissipation in the x-ray tube.  As demonstrated in the previous section, the shape of the isowatt curve for a given element depends on
the anode material.  Figure 3.4 shows the isowatt curves for different elements excited with a chromium anode tube, the elements being
chosen to give $K\alpha$ lines which roughly span the usual operating
range of the x-ray spectrometer.  The curves indicate that the maximum intensities are obtained for excitation voltages between 50
and 100 kV for most wavelengths longer than about 0.7 Å.  It should
be noted, however, that maximum intensity does not in itself guarantee best analytical precision; the peak-to-background ratio must
also be considered.  In practice, the optimum excitation voltage
for wavelength-dispersive spectrometers is about six times the
critical excitation potential, but in most cases should be greater
than 50 kV.  Obviously for the very short wavelengths (< 0.7 Å), it

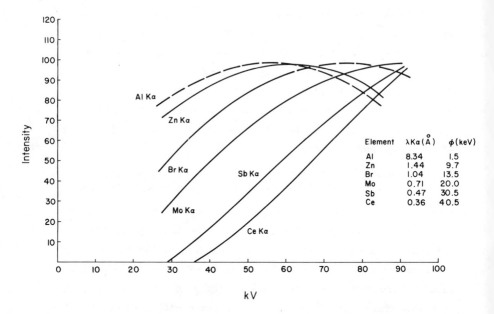

Figure 3.4  Isowatt curves for a range of elements using a chromium anode x-ray tube.

is not possible to reach the optimum excitation voltage since high-voltage generators are generally rated at 60, 80, or 100 kV maximum. It will also be seen from Fig. 3.4 that adjustment of these voltages over the 60- to 100-kV range is desirable where best sensitivity on each of a broad range of elements is required.

### 3.2.2  Optimum X-ray Tube Settings for Broadband Excitation with the Energy-Dispersive Spectrometer

Whereas wavelength-dispersive systems tend to be limited by tube dissipation, energy-dispersive systems employing broadband excitation are count-rate limited.  With a tube power of less than 15 W it is easy to generate an x-ray intensity well beyond the high counting rate limit of the energy spectrometer.  Consequently, there is usually no concern for x-ray tube power dissipation when using direct excitation by the x-ray tube.  The anode material and tube voltage are normally selected to provide optimum excitation of the elements of interest, while the tube current control is regarded as

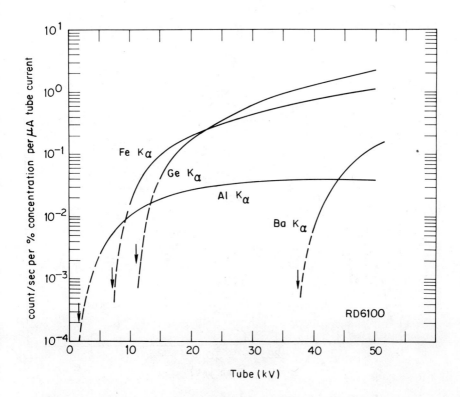

Figure 3.5  Sensitivity curves for an energy dispersive system with
a molybdenum anode.  Arrows mark the absorption-edge energies,
dashed lines represent extrapolation.  Reprinted by courtesy of EG&G
ORTEC.

a secondary adjustment to obtain as high a counting rate as the
energy spectrometer can handle.  Since power dissipation is not of
concern, it is more useful to plot the sensitivity curves as counts
per second per percent of concentration divided by the tube current
versus the x-ray tube voltage.  Figures 3.5 and 3.6 illustrate these
plots for molybdenum and tungsten anodes.  Note the curve for the
Ge Kα line crossing the Fe Kα curve in Fig. 3.5.  This behavior differs from Fig. 3.6 and is caused by the characteristic molybdenum K
lines from the anode selectively exciting the Ge Kα line.

    Even with tube voltages as low as 1.5 times the absorption-
edge energy ϕ, it is possible to achieve adequately high counting

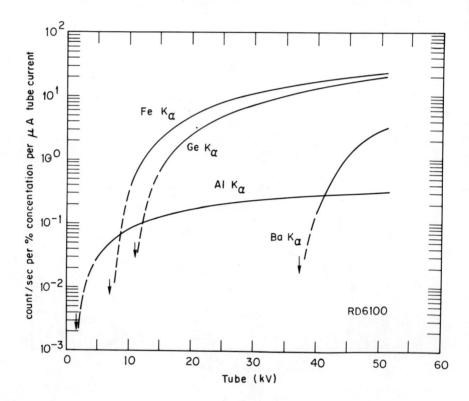

Figure 3.6 Sensitivity curves for an energy dispersive system with a tungsten anode. Arrows mark the absorption-edge energies, dashed lines represent extrapolation. Reprinted by courtesy of EG&G ORTEC.

rates by turning up the current control. In fact, it is not always desirable to use the maximum available tube voltage. For example, if one wishes to excite a 1% concentration of iron in water, it is discovered that the primary tube spectrum scattered from the specimen into the detector represents the major source of counting rate. As the tube voltage is raised above 25 kV, the rapid increase in intensity in the scattered spectrum forces the operator to reduce the tube current. Consequently, the intensity of the iron Kα line suffers, while the peak-to-background ratio is not significantly improved. With energy dispersive systems and direct x-ray tube excitation, the tube voltage is generally run between two and six times the absorption-edge energy of the highest energy lines to be

Table 3.2  Detection Limits for Trace Elements in $Na_2B_4O_7$ Glass on an Energy Dispersive Spectrometer Using Various Broadband Excitations

| | | $C_{MDL}$ (ppm) [a] | | |
|---|---|---|---|---|
| Element/Line | Energy (keV) | W Anode 50 kV, 1 μA | Mo Anode 50 kV, 1 μA | Mo Anode 15 kV, 200 μA |
| Na Kα | 1.04 | 38,000 | 20,000 | 480 |
| Al Kα | 1.50 | | | 42 |
| K  Kα | 3.31 | 54 | 94 | 7.4 |
| Ba Lα | 4.46 | 30 | 46 | 18 |
| Co Kα | 6.92 | 2.8 | 5.6 | 4.2 |
| Ba Kα | 31.9 | 32 | 36 | -- |

[a] For 400 sec livetime.

excited.  In cases where the highest energy lines represent high concentrations and good precision is required on low-concentration low-energy lines, the tube voltage may be reduced to less than twice the absorption-edge energy of the high-energy lines.  This allows higher counting rates to be achieved on the lower energy lines, but will sacrifice intensity stability on the high-energy lines as explained in Sec. 3.4.

The detection limits for trace elements in $Na_2B_4O_7$ glass listed in Table 3.2 demonstrate the effects of choosing the anode material and x-ray tube voltage on specimen excitation.  In the third and fourth columns of Table 3.2 the x-ray tube is operated at 50 kV. This is the typical setting if a broad range of elements is to be detected simultaneously; i.e., elements having lines between 1 and 40 keV can be analyzed.  Note the low tube current which yields the maximum spectrometer counting rate.  The detection limit for sodium is better with the molybdenum anode because of the characteristic Mo L lines at 2.3 keV.  The 1.8 keV tungsten M lines do not pass through the x-ray tube window, and therefore they are ineffective for light-element excitation.  Detection limits for the K Kα, Ba Lα, and Co Kα are better with the tungsten anode because of the tungsten L lines in the range 8.3 to 11.3 keV.  Although not shown in Table 3.2, the molybdenum anode provides better results for elements in

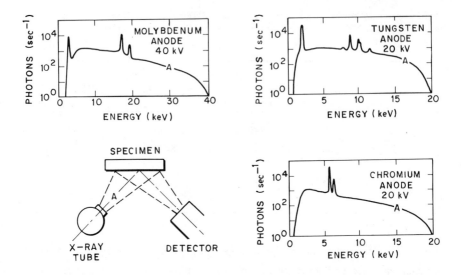

Figure 3.7  The geometry used for broadband excitation.  The spectra depict several commonly used anode materials; the vertical scale is logarithmic.  Reprinted by courtesy of EG&G ORTEC.

the 8.3- to 15-keV energy range due to the characteristic Mo K lines.  For energies above 17 keV, the tungsten anode provides slightly better detection limits as a result of its higher bremsstrahlung intensity.

The fifth column in Table 3.2 shows how light-element detection limits are improved by lowering the x-ray tube voltage and increasing the tube current.  Broadband excitation with tube voltages in the 10- to 15-kV range generally is the most effective excitation method for detecting light elements in trace quantities on an energy-dispersive spectrometer.

## 3.3  MONOCHROMATIC EXCITATION USING X-RAY TUBES

Figure 3.7 illustrates the geometry for broadband x-ray tube excitation on a nondispersive spectrometer, as well as the energy spectra produced by the x-ray tube for various anode materials.  The brem-

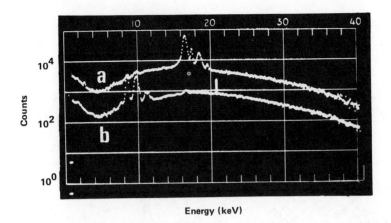

Figure 3.8   X-ray tube spectra scattered from a thick Lucite specimen:   a depicts the Mo anode (spectrum shifted up by one decade):
b depicts the W anode.   Reprinted by courtesy of EG&G ORTEC.

sstrahlung continuum under the anode characteristic lines is useful
in simultaneously exciting a broad range of elements with good
efficiency.   On the other hand, in trace element analysis the x-ray
tube continuum becomes a nuisance.   Often the trace elements are in
a low-atomic-number matrix and the tube spectrum is rather efficiently scattered into the detector by the specimen.   As a result, the
trace element peaks sit on a very high background and tend to be
obscured by the large statistical fluctuations in this background.
Figure 3.8 demonstrates the high background caused by the scattered
x-ray tube spectrum with a low-atomic-number specimen.   The solution
to this problem is to eliminate the contribution of the tube spectrum in regions containing trace-element peaks.   This generally
involves some type of selective filtering or use of a secondary
fluorescer.   Both techniques waste the majority of the x-ray tube
output and require a substantial increase in the tube power.   For
this reason the technique has become popular primarily with energy
dispersive systems where detection efficiencies are already very
high.   In the following sections the more common monochromation
methods are outlined.

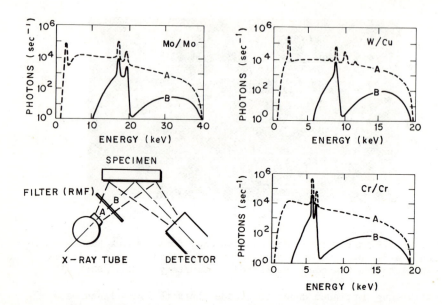

Figure 3.9  Typical geometry for the regenerative monochromator filter method.  The spectra shown will illustrate its use with three different anode-filter combinations.  The dashed line shows the spectrum before filtering, and the solid line is the spectrum obtained after filtering.  W/Cu signifies a tungsten anode followed by a copper filter.  The vertical scales are logarithmic.  Reprinted by courtesy of EG&G ORTEC.

## 3.3.1  The Regenerative Monochromator Filter

Figure 3.9 illustrates the use of a filter between the x-ray tube and the specimen to lower the detectable limits in trace-element analysis.  The concept is best described with reference to the molybdenum anode tube and a molybdenum filter.  Figure 3.10 is a plot of the transmission of a 127-$\mu$m-thick molybdenum filter as a function of x-ray energy.  This is obtained from the equation

$$\frac{I}{I_o} = \exp[-\mu(E)\rho x] \qquad\qquad (3.10)$$

where $I_o$ is the x-ray intensity incident upon the filter, I is the x-ray intensity transmitted through the filter, $\mu(E)$ is the mass attenuation coefficient of the filter (a function of energy), $\rho$ is

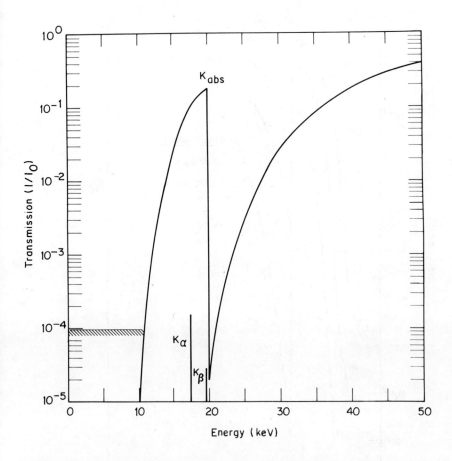

Figure 3.10   The transmission of a 127-μm molybdenum filter as a
function of energy.   The crosshatched line at lower energies represents the limit set by detector background due to incomplete charge
collection.   Reprinted by courtesy of EG&G ORTEC.

the density of the material, and x is the filter thickness.   The
filter has a narrow band, or window, of high transmission in the
vicinity of the characteristic Mo K lines.   When used with a Mo
anode x-ray tube, this filter will pass the Mo K lines from the
anode but will severely attenuate the bremsstrahlung continuum at
lower energies.   Trace elements in the energy range from 6 to 13 keV
will be efficiently excited by the Mo K lines, while the background
beneath these trace element peaks will be extremely low.

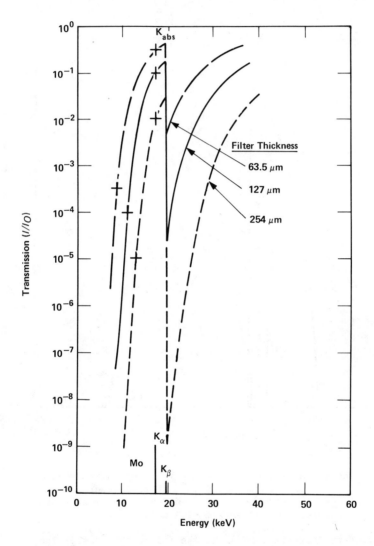

Figure 3.11  The transmission of molybdenum filters of three different thicknesses.  The + signs mark the energies where the transmission has decreased to 1/1000 of the transmission at the Mo Kα
energy.  Reprinted by courtesy of EG&G ORTEC.

Consequently, detectable limits can be lowered by as much as a factor of 4.

As Figs. 3.9 and 3.10 indicate, a low background region is also
produced just above the Mo K absorption edge.  With tube voltages

from 40 to 50 kV, some high-energy bremsstrahlung will pass through
the filter to effectively excite trace elements such as Cd and Ag in
the low background notch just above 20 keV.

The attenuation of the bremsstrahlung just above the Mo K absorption edge is caused by fluorescence of the Mo in the filter.
This effect produces additional Mo K lines to be used in exciting
lower energy elements.  The regeneration of the Mo lines has
prompted the use of the term *regenerative monochromator filter* [13].
However, the intensity of the regenerated Mo K lines on the specimen only becomes significant when the filter is close to the specimen.  The spectrum obtained with the regenerative monochromator
filter is virtually equivalent to that obtained with a transmission
anode tube [14].  On the other hand, the regenerative monochromator
filter provides better flexibility from the standpoint of anode
heat dissipation, choice of filtered as well as unfiltered spectra,
choice of filter material, and adjustment of filter thickness.

The sharpness with which the filter transmission drops on the
low-energy side of the Mo K lines sets the high-energy limit for
lines which can sit on a low background.  This high-energy limit can
be improved by increasing the filter thickness as shown in Fig.
3.11.  The plus signs mark the point on each curve where the transmission has dropped to 1/1000 of the transmission at the Mo K$\alpha$ line.
Clearly, the background falls off at higher energies as the filter
thickness is increased.  There is, of course, a penalty for this
improvement.  The transmission at the Mo K$\alpha$ line also decreases, and
the tube current must be raised proportionally to regain the same
net excitation intensity.

As indicated in Fig. 3.9, a Cr anode with a Cr filter can be
used for light elements, while a Cu filter on a W anode tube handles
the range from 2 to 6 keV quite well.  Figure 3.12 shows how the Cu
filter selectively passes the W L$\alpha$ line.  Background is also low
above the Cu K absorption edge.  Best results are usually obtained
with a tube voltage of 40 to 45 kV for the Mo/Mo combination, and
20 to 25 kV for the W/Cu arrangement.  Although the method can be
applied to single anode tubes, it is normally used in conjunction

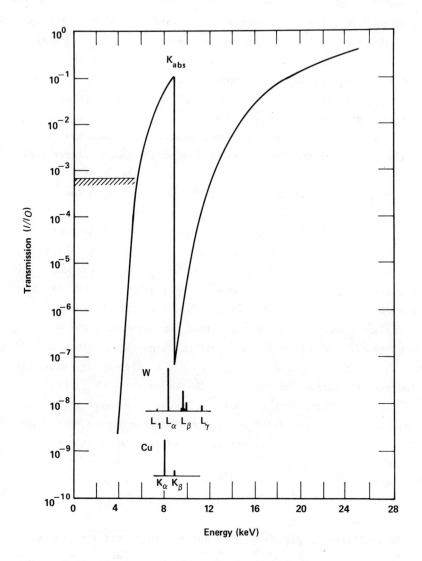

Figure 3.12   The transmission of a 63.5 μm copper filter versus energy.   Reprinted by courtesy of EG&G ORTEC.

with a dual or multiple anode x-ray tube to provide more flexibility in the choice of excitation energies [13,15].

The effect of the filter is demonstrated in Fig. 3.13 for a tungsten anode and a copper filter.   In Fig. 3.14 similar results

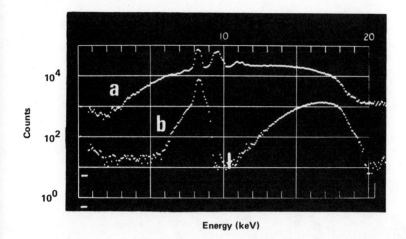

Figure 3.13 The x-ray spectrum of a W anode x-ray tube scattered from a Lucite specimen: a depicts no RMF (spectrum shifted up by one decade); b shows the effect of a 60 μm copper filter on the x-ray tube. Reprinted by courtesy of EG&G ORTEC.

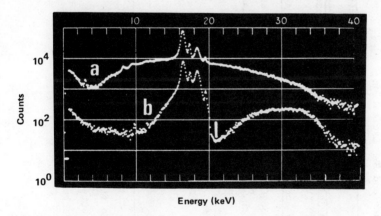

Figure 3.14 a depicts the spectrum from a Mo anode x-ray tube; b shows the results of applying a Mo RMF. The two spectra have been scattered by a Lucite specimen. The attenuation of the Mo Kα intensity by a factor of 5 is not reflected in the figure due to a shift in the vertical scale between a and b. Reprinted by courtesy of EG&G ORTEC.

Table 3.3  Detection Limits for Trace Elements in $Na_2B_4O_7$ Glass for Broadband and RMF Excitation

| | Broadband 50 kV, 1 µA W Anode | Broadband 50 kV, 1 µA Mo Anode | Broadband 15 kV, 200 µA Mo Anode | RMF 25 kV, 200 µA W/Cu[a] | RMF 45 kV, 100 µA Mo/Mo[b] |
|---|---|---|---|---|---|
| | | Detection Limit $C_{MDL}$ (ppm) (for 400 second livetime) | | | |
| Line | | | | | |
| Na Kα | 38,000 | 20,000 | 480 | | |
| Al Kα | c | c | 42 | | |
| K Kα | 54 | 94 | 7.4 | 15 | |
| Ba Lα | 30 | 46 | 18 | 8.7 | |
| Co Kα | 2.8 | 5.6 | 4.2 | 2.1 | |
| Ge Kα | d | d | | | 3.0 |
| Ba Kα | 32 | 36 | | | 0.28 |

[a] W anode/Cu filter.
[b] Mo anode/Mo filter.
[c] Al not detectable at 500 ppm.
[d] Ge not detectable at 3 ppm.

are shown for a molybdenum anode and a molybdenum filter.  Table
3.3 compares detection limits using RMF excitation to those obtained
on the same fluorescence spectrometer with several different types
of broadband excitation.

The regenerative monochromator technique has several advantages.  A tube power of less than 20 W is adequate with energy
dispersive systems.  This makes it easier and less expensive to design a stable x-ray source.  The system efficiently provides good
peak-to-background ratios, particularly on low-energy characteristic
anode lines.  Broadband excitation capability can be selected by
simply removing the filter.  The filter is inexpensive and can be
added to almost any existing instrument.

### 3.3.2  The Secondary Fluorescer Method

The technique for improving trace-element detection limits is to
concentrate all the excitation energy in a narrow band just above
the absorption-edge energy of the highest energy line to be excited.
Excitation radiation spread out over lower energies causes high
background, while at much higher energies it contributes to the
spectrometer counting rate without efficiently exciting the trace
elements.  Radioactive sources such as $^{55}$Fe, $^{109}$Cd, $^{57}$Co, and $^{241}$Am
have been used to achieve the desired monochromatic excitation.
With the exception of the $^{109}$Cd source, the source strengths are
usually too low to achieve detectable limits competitive with optimized x-ray tube excited systems.  The secondary fluorescer method
was developed to emulate the monochromatic excitation provided by
radioisotopes, but with the higher intensities achievable with x-ray
tubes.

Figure 3.15 illustrates the secondary fluorescer technique.  A
high-power x-ray tube (typically 3 kW) is used to excite a secondary fluorescer target.  The tube anode material is generally chosen
to have a high bremsstrahlung output so that a wide range of
secondary fluorescer materials can be efficiently excited.  Tungsten
is a convenient choice for the anode, although a chromium anode can
also be used.  A fraction of the characteristic lines fluoresced in

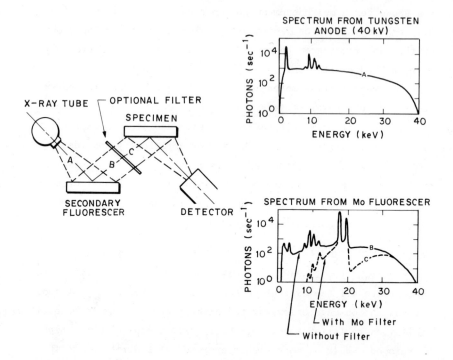

Figure 3.15  Geometry for the secondary fluorescer method.  The top
spectrum is produced by the x-ray tube.  The bottom spectrum is from
the secondary fluorescer.  Reprinted by courtesy of EG&G ORTEC.

the secondary target reach the specimen to excite the desired trace
element lines.  The exciting spectrum incident upon the specimen
looks like the solid curve in the lower right-hand corner of Fig.
3.15, where a Mo secondary fluorescer has been used.  This technique is very inefficient in utilizing the output intensity of the
tube since only a small fraction of the fluoresced radiation from
the secondary radiator is emitted in the correct direction to strike
the specimen.  Consequently, tube power in the 0.7- to 3-kW range is
required.  At the same time, the output of the secondary fluorescer
must be carefully collimated, and the general area of the x-ray
tube and secondary target must be well shielded to prevent the intense primary radiation from reaching the specimen or detector and
causing high background levels.

Table 3.4   Detection Limits for Trace Elements in $Na_2B_4O_7$ Glass Corrected for Differences in Spectrometer Performance

| Line | Fluorescer | Secondary Fluorescer $C'_{MDL}$ (ppm)[a] | | RMF Anode/Filter | RMF $C'_{MDL}$ (ppm)[a] | | Broadband 15 kV $C'_{MDL}$ (ppm)[a] | |
|---|---|---|---|---|---|---|---|---|
| | | A | B | | C | D | C | D |
| 1 | Na Kα | 570[b] | | | | | 410 | 300 |
| 2 | Na Kα | 1100 | 3400[b] | | | | | |
| 3 | Al Kα | 41[b] | | | | | 24 | 27 |
| 4 | Al Kα | 36 | 90[b] | | | | | |
| 5 | K Kα | 1.3 | 4.7[b] | | | | | |
| 6 | K Kα | 4.4 | | W/Cu | 9.9[b] | 9.6 | 7.5 | 4.7 |
| 7 | Ba Lα | 4.8 | | W/Cu | 6.3[b] | 5.6 | 10 | 12 |
| 8 | Co Kα | 0.64 | | W/Cu | | 1.4 | | |
| 9 | Co Kα | 2.8 | 2.2 | Mo/Mo | 3.4 | 1.9 | | 2.7 |
| 10 | Ge Kα | 0.29 | 0.24 | Mo/Mo | 0.25 | 0.18 | | |

[a]For 480 sec livetime.
[b]Limited by maximum tube current.

Table 3.5  X-ray Tube Power Required for $C'_{MDL}$

**Broadband**

| Anode | C | | D | |
|---|---|---|---|---|
|  | kV | µA | kV | µA |
| Mo | 15 | 150 | 15 | 200 |

**RMF**

| Anode/Filter | C | | D | |
|---|---|---|---|---|
|  | kV | µA | kV | µA |
| W/Cu | 20 | 200[a] | 25 | 200 |
| Mo/Mo | 45 | 15 | 45 | 100 |

**Secondary Fluorescers**

| Fluorescer | A | | B | |
|---|---|---|---|---|
|  | kV | mA | kV | µA |
| Cl | 20 | 80[a] | | |
| Ti | 20 | 38 | 22 | 400[a] |
| Cu | 20 | 75 | | |
| Mo | 40 | 23 | 48 | 55 |

[a]Limited to maximum tube current.

Note that the spectrum produced by the simple system is not quite monochromatic. Some of the bremsstrahlung from the x-ray tube is scattered from the fluorescer onto the specimen. At energies below the characteristic Mo lines, the background can be reduced by introducing a thin Mo filter as shown in Fig. 3.15. This addition is identical in function to the regenerative monochromator filter previously described; although a thinner foil would be used in this case to maintain a higher intensity in the characteristic lines.

Table 3.4 compares detection limits with secondary fluorescers to the results with the RMF method and 15-kV broadband excitation [16,17]. Four different fluorescence analyzers were tested (units A, B, C, and D), and the results were corrected for differences in performance for the energy-dispersive spectrometers employed on each unit. Unit A used a chromium anode tube, while unit B used a tungsten anode tube. Unit A was a commercial, general-purpose instrument. Unit B was specifically designed for atmospheric aerosol analysis, where closer coupling between the tube, fluorescer, sample, and detector could be employed with some sacrifice of insensitivity to specimen positioning errors. Table 3.5 lists the x-ray tube operating conditions required for Table 3.4. For medium- to high-atomic-number elements, the secondary fluorescer method provides detection limits equivalent to the RMF element, but requires much higher x-ray tube power. For light elements, secondary fluorescer excitation efficiency is limited by available x-ray tube current. Consequently, light-element detection limits are inferior to those provided by 15-kV broadband excitation with a Mo anode x-ray tube.

The chief advantage of the secondary fluorescer is flexibility in the choice of monochromatic excitation energies. A fairly wide range of materials can be used as secondary fluorescers. It also provides a slightly sharper cutoff in background below the exciting lines than is achieved with the regenerative monochromator filter.

The greatest disadvantage of the scheme is the difficulty it causes in obtaining broadband excitation. Even if the x-ray tube

can be mechanically moved to irradiate the specimen directly, its
output is too intense to be used in direct excitation. Most high-
power tubes are unable to operate stably in the microampere (μA)
current range that is required for direct excitation with energy-
dispersive spectrometers mounted close to the specimen. The alternative is to replace the secondary fluorescer with an efficient
scatterer such as carbon or some form of hydrocarbon. This scatters
the x-ray tube spectrum onto the specimen. Unfortunately, absorption in the scatterer severely attenuates the low-energy segment of
the spectrum, rendering the method ineffective for light-element
excitation.

### 3.3.3  Benefits and Limitations of Monochromatic Excitation

Although monochromatic excitation can improve the minimum detection
limits over a restricted range of elements, it compromises performance in other areas. Most obvious is the need for higher intensity
in the primary radiation source.

   A more severe limitation is the restricted range of elements
for which good sensitivity can be obtained. Figure 3.16 demonstrates
the principle. A rock specimen containing high concentrations of
aluminum, silicon, potassium, calcium, titanium, manganese, and
iron; and trace amounts of nickel, copper, zinc, gallium, lead,
rubidium, strontium, and yttrium has been analyzed with broadband
as well as monochromatic excitation. With broadband excitation,
good sensitivities are obtained on the light elements as well as
the heavier elements. However, it is difficult to analyze the trace
elements, with nickel and lead, in particular, being completely obscured by the scattered x-ray tube background. Monochromatic excitation using a Mo anode with a Mo RMF significantly improves the
visibility on the trace elements. Unfortunately, this improvement
is gained at the expense of sensitivity on the light elements
(aluminum, silicon, potassium, calcium, and titanium). Inspection
of Eqs. (2.26) and (2.31), and Fig. 3.2 reveals that excitation
efficiency is poor with monochromatic excitation when the

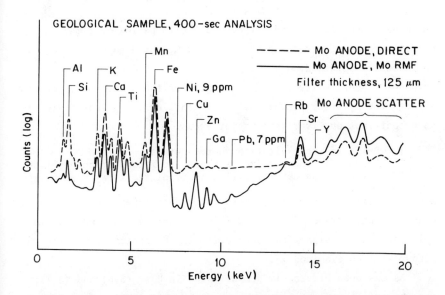

Figure 3.16  A comparison of broadband and monochromatic excitation (RMF) of the same specimen.  The dashed spectrum was acquired using an unfiltered Mo anode with 45 kV.  The solid spectrum was taken with the RMF method employing a Mo anode and a Mo filter.  Reprinted by courtesy of EG&G ORTEC.

excitation energy is high relative to the absorption-edge energy of the element of interest.  This means that monochromatic excitation is only useful for elements lying close to the excitation energy. More will be said about this characteristic in Chapter 11 which deals with trace analysis, and in the discussion on excitation with radioisotopes.

As shown in Chap. 2 [Eq. (2.29)], monochromatic excitation eliminates the integration over energy and considerably simplifies theoretical calculations of fluoresced intensities.  For this reason monochromatic excitation is advantageous where it is desirable to apply the *fundamental parameters* quantitative model (see Chap. 10).

## 3.4  SIGNIFICANCE OF DRIFT IN THE X-RAY GENERATOR

Many of the quantitative analyses performed with x-ray fluorescence spectrometers require relative accuracies of 1% or better.  Since

the measurement is carried out over a finite period of time, and
normally requires comparing an intensity measured on the unknown to
an intensity obtained on a standard specimen, any change in the instrument's sensitivity for the element over the measurement period
affects the accuracy of the analysis.  This drift in sensitivity can
have contributions from three major areas:

1.  Drift in the excitation source output
2.  Changes in the specimen's shape, position, or composition
3.  Drift in the detection efficiency of the x-ray spectrometer

A more detailed treatment of these major categories can be found in
Chap. 11, and further discussion is also included in Chap. 5.  In
this section the significance of drift in the excitation source
output will be examined.

As has been previously demonstrated in Eqs. (3.6) and (3.7);
the sensitivity of the instrument for the analyte line is proportional to the intensity of the excitation source and is also a
function of the spectral distribution of the source.  Anything which
changes the intensity or spectral distribution of the excitation
source will cause a change in the analyte sensitivity.  If the
changes are uncontrolled or unpredictable, then analytical errors
due to drift are encountered.  Some of the sources of drift are inherent to the instrument design and cannot be controlled by the
operator.  On the other hand, there are techniques which can minimize the importance of the drifts inherent in the instrument.

Anything which causes a change in the distance between the
focal spot on the x-ray tube anode and the specimen is a source of
drift due to the change in solid angle subtended by the specimen.
Items falling into this category include thermal expansion of the
parts of the instrument, specimen positioning errors, and focal-
spot wander in the x-ray tube.  Focal-spot wander can be caused by
slight changes in the voltages controlling the x-ray tube optics or
may be a result of external, varying magnetic fields deflecting the
electrons as they travel from the cathode toward the anode.  Aging
effects in the tube may result in a change of the shape or emission

of the heated cathode.  This also affects the focal-spot size and
position on the anode.  In addition, residual gases in the tube can
lead to a poorly controlled emission spectrum.  Except, perhaps, for
keeping magnetic fields  away from the instrument and replacing
worn-out components, the operator has very little control over the
sources of drift listed in this paragraph.

Normally the stability of the instrument is sensitive to room
temperature, and the range and purity of the ac line voltage used
to power the instrument.  These are parameters that the operator can
and should control to within the tolerances specified by the manufacturer.  In this respect there are many other environmental factors which can cause faulty performance.  The operator wishing to
maintain good stability should avoid a dusty or dirty environment,
and may find it important to eliminate excessive room noise or
severe floor vibrations.  High-voltage generators are usually designed to delivery constant potential and current within certain
specified stability limits, but with two conditions:  that variations in the input voltage be within about ±10% and that any
fluctuations be smooth, i.e., take place over a period of several
seconds.  Figure 3.17 shows a typical output trace from a recording
voltmeter which has been set to monitor fluctuations in the
(nominal) 220-V main supply.  The full length of the trace shown
represents about 2.5 h.  It will be seen that the average voltage
level is about 223 V with the majority of fluctuation falling
within 1.0 V.  It will also be seen, however, that larger ripples
are also present where the voltage may suddenly change by up to 3 V.

Although in this particular instance the output from the high-
voltage generator would probably not be  affected by the indicated
ripple, should the sudden change in voltage become much worse, the
stabilization circuits in the generator would be unable to cope
with the change.  The worst case of this is where rapid and large
input-voltage changes (transients) occur.  Smoothing is generally
achieved by using a feedback system to a variable transformer which
is placed before the high-voltage transformer.  The feedback system

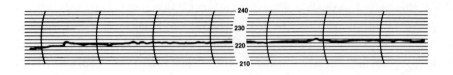

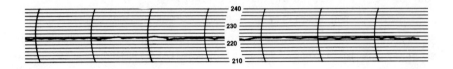

Figure 3.17 Normal 220-V ac power line voltage stability. The two successive strips cover a 2.5-h period.

essentially compensates for fluctuations in the mains voltage. If a motor driven variable transformer is used, several tenths of seconds may elapse before the transformer is able to compensate for sudden changes in line voltage. Where solid-state feedback and compensation circuits are employed, correction for line voltage drift is achieved in about 50 ms. Thus, even in the case of solid-state stabilization circuitry, transients of less than 50 ms duration may not be removed. Clearly, it is important to provide a well-regulated and transient-free line voltage to power the instrument.

A good qualitative appreciation for the importance of proper x-ray tube voltage selection can be gained by examing Eq. (3.9). Taking the logarithm and differentiating leads to

$$\ln I_i = \ln K + \ln i + n \ln (V - \phi) \qquad (3.11)$$

$$\frac{dI_i}{I_i} = \frac{di}{i} + \frac{nV}{V - \phi} \frac{dV}{V} \qquad (3.12)$$

Equation (3.12) relates the fractional change in the analyte line intensity $dI_i/I_i$ to the fractional changes in tube current $di/i$ and tube voltage $dV/V$. Note that a 0.1% drift in tube current results in a 0.1% change in analyte line intensity. However, a 0.1% drift in tube voltage causes a much larger change in analyte line intensity. The voltage drift is multiplied by the factor $nV/(V - \phi)$.

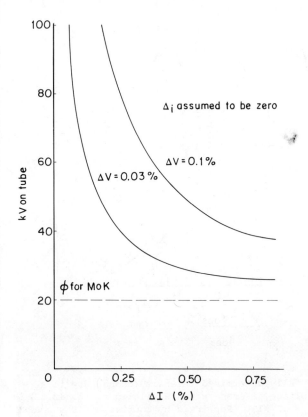

Figure 3.18  The effect of overvoltage on analyte line intensity stability.  Parameters are related to Eq. (3.12) as follows: $\Delta I (\%) = 100\% \times dI_i/I_i$; $\Delta V = 100\% \times dV/V$; $\Delta i = 100\% \times di/i$; $n = 1.6$. The relationship has been plotted for excitation of the Mo K$\alpha$ line.

If the tube voltage V is set very close to the analyte line absorption-edge energy $\phi$, the multiplying factor becomes extremely large and the analyte line intensity is very sensitive to slight drifts in the tube voltage.  This situation should be avoided wherever possible.  Equation (3.12) shows that the x-ray tube voltage should be large compared to $\phi$ in order to achieve good stability.  Even in this case the value of n, which lies between 1 and 2, multiplies the effect of voltage drift.  Obviously the x-ray fluorescence spectrometer is much more sensitive to voltage drift than tube current drift.  Equation (3.12) is illustrated in Fig. 3.18 for a 0.03%

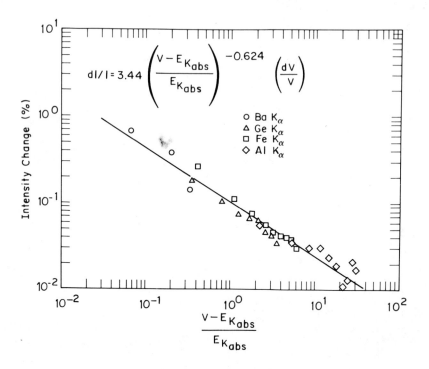

Figure 3.19 Experimentally determined intensity change due to a 0.03% drift in tube voltage. Parameters are related to Eq. (3.12) via $E_{K(abs)} = \phi$; $dI/I = dI_i/I_i$. Reprinted by courtesy of EG&G ORTEC.

and a 0.1% instability in x-ray tube voltage, where the Mo $K\alpha$ line is being excited. Figure 3.19 shows experimental data derived from Fig. 3.6 for a wide range of data. The straight line and equation in the figure represent a pragmatic fit to the data. Figures 3.18 and 3.19 demonstrate the need to set the tube voltage higher than twice the absorption-edge energy of the highest energy line of interest. In the case of monochromatic excitation, $\phi$ should be taken as the absorption edge energy corresponding to the monochromatic excitation line.

Drift in the analyte line sensitivity due to changes in the x-ray source output can be categorized as either short-term or long-term drift. It should be appreciated that these are rather

vague terms.  What short-term drift is to one analyst may be considered long-term drift by another.  Furthermore, the means of
specifying the magnitude of the drift varies.  In some cases drift
is specified as a standard deviation in intensity over a period of
time, while in other situations it is listed as the maximum positive
and negative excursions.  One should also be aware that a drift
specification may be stated only for one or two of the many factors
which can cause drift.  For example, a stability specification of
0.05% for line voltages between 198 and 242 V ac may only imply
stability against line voltage changes.  It does not necessarily
include drifts caused by room temperature variations.  Drift speci-
fications may include changes in x-ray spectrometer sensitivity as
well as the x-ray generator or may simply apply to the stability
of regulation of the x-ray tube voltage and current.

   With the lack of standardization in the definition of drift,
it becomes expedient to adopt an operational definition of the
difference between short- and long-term drift.  *Short-term drift* is
that component of drift which is significant during the time taken
to make a quantitative measurement of one unknown specimen by reference to an instrument calibration standard (see Chap. 10).  This,
in fact, involves the measurement of two specimens:  the unknown and
the  instrument standard.  The purpose of the instrument standard
is to test for drift in the instrument's sensitivity.  By ratioing
the intensity measured on the unknown to that measured on the standard, the effects of long-term drift are approximately canceled
(see Sec. 11.2.5).  Thus, *long-term drift* is defined as the component or components of drift which are significant over a period
of time which is long compared to the time taken to make a quantitative measurement of one unknown specimen by reference to an instrument calibration standard.  By definition, the ratio method cannot
correct for short-term drift.  Consequently, long-term drift can be
compensated by operator technique, while short-term drift cannot.

   The short- and long-term drift characteristics vary from instrument to instrument, and the relative importance of the various

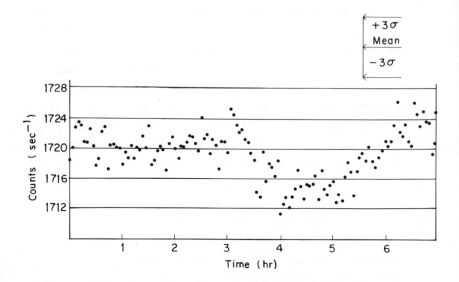

Figure 3.20  Drift curve for an x-ray spectrometer.

causes of drift is a complex function of instrument design and
maintenance.  Where the drift characteristics are thought to be
important in an analytical determination, the only safe procedure
is to measure the drift contribution using the specimen to be
analyzed.

Normally, short-term drift reflects the sensitivity of the
generator to abrupt changes in line voltages or to changes in the
selection of x-ray tube current and voltage.  Most of the long-term-
drift effects show up over periods of several hours and are rather
readily suppressed by the ratio method.  Figure 3.20 illustrates a
series of count readings, each lasting 100 s, recorded using a
spectrometer set on one wavelength during a time in which no parameter on the spectrometer was changed.  The spectrometer was temperature stabilized and a scintillation counter was employed; thus,
the fluctuations above those caused by counting statistics should
be primarily due to the x-ray source.  The short-term instability
of the high-voltage generator was specified as a standard deviation
$\sigma_g$ = 0.06%.  The standard deviation from counting statistics is

calculated as $\sigma_s$ = 100/(1720 counts/s $\times$ 100 s)$^{1/2}$ = 0.077%.  Thus, the total expected short-term drift is

$$\sigma_T = (\sigma_g^2 + \sigma_s^2)^{1/2}$$
$$= [(0.06)^2 + (0.077)^2]^{1/2}$$
$$= 0.1\%$$

The indicated short-term error limits in Fig. 3.20 correspond to $\pm 3\sigma_T$ or $\pm 0.3\%$.  It will be seen from study of the data that over almost any 30-min period, the data lie within the predicted short-term drift limits.  Over a period of hours, however, an additional cyclic drift is present.  This is a typical example of long-term drift.  Long-term drift is commonly several times larger than short-term drift.  Other examples of drift analyses are given in Sec. 5.6 and Sec. 11.2.5.


## 3.5  EXCITATION WITH RADIOISOTOPES

Although x-ray tubes provide a very versatile source of excitation, together with their power supplies they are bulky and consume a great deal of power.  This is particularly true of the 4-kW generators.  Radioisotope sources, on the other hand, are extremely compact and require no electrical power.  Consequently, there are a number of applications where radioisotopes are the preferred source of excitation.  Most obvious is the general class of portable fluorescence analyzers where weight and power limitations are stringent.  Many routine process control applications are simple analytical problems which do not demand a sophisticated instrument. In such cases the compactness and higher reliability of the radioisotope source are attractive features.  The conditions which frequently lead to the selection of radioisotope excitation are the following:

1. Minimum size, weight, and power consumption are extremely important.

2.  The analytical problem is simple and encompasses a limited
    range of elements.
3.  The ultimate in sensitivity or detection limits is not
    required.
4.  The resolution of a wavelength-dispersive spectrometer is
    not required, so that the high detection efficiency of a
    proportional counter, an NaI(Tl) detector, or a Si(Li)
    detector can be utilized.
5.  System simplicity, ruggedness, reliability, and long-term
    stability are important.

Since radioisotopes provide lower output intensity and less flexibility in the selection of the excitation spectrum, x-ray tubes are
still the preferred source for the more complex and demanding
analytical problems.

The excitation radiation from a radioisotope is produced when
an unstable nuclear isotope decays into a different isotope. This
second isotope, or daughter nucleus, may also be unstable and decay
into yet another isotope. The radioactive decay scheme may involve
only one parent-daughter relationship or can encompass a long chain
of decay sequences. The radioisotopes commonly used in x-ray
fluorescence spectrometry emit photons in the form of γ-rays from
the nucleus, or characteristic x-rays from the atomic shells.
Sources emitting charged particles are more rarely employed. Most
γ-ray and x-ray emitting sources are monochromatic, or nearly
monochromatic, in that they emit photons having only one or a few
discrete energies. Broadband x-ray sources have been constructed
using β⁻ emitters and a suitable low-atomic-number absorber [18].
The electrons (β⁻ particles) emitted by the radioisotope decelerate
in the absorber to produce bremsstrahlung. The maximum energy of
the x-ray continuum so produced corresponds to the maximum β-ray
energy allowed by the radioactive decay.

Radioactive source technology is extensive. In this section
only the more important concepts will be summarized. For more details on the design of radioisotope excited fluorescence spectrometers, the excellent review article comprising Ref. 18 is recommended. Radioisotopes can be a significant health hazard if im-
properly used or handled. Consequently, anyone using or planning

to use radioisotopes is advised to receive proper training in the
use of radioactive sources.  References 19 to 21 are recommended
as a starting point.

## 3.5.1 Source Activity

The strength of a source is measured in terms of its *activity* in
curies, where 1 curie $\equiv$ 1 Ci $\equiv$ 3.7 $\times$ $10^{10}$ transformations/s.  A
transformation refers to the decay of a single radioactive nucleus
into its daughter nucleus.  Whereas radioactive sources are normally
described by their activity for manufacturing, licencing, and regulation purposes, this description of source strength is not
particularly useful for the x-ray spectroscopist.  In designing a
radioisotope excited fluorescence spectrometer for a specific
application, it is the flux density in photons per square centimeter per second [20] at the specimen surface which is of interest
for each photon energy emitted by the source.  To compute the flux
density from the activity one must know the details of the decay
scheme [22], the geometry of the source and collimator system, the
self-absorption characteristics of the source, and the filtering
provided between the source and the specimen (including the source
container).  For sources having a complicated decay scheme and/or
emitting several photon energies, this can be a complex task
[1,22].  For the more common applications the source activities
usually lie within the 0.5- to 100-mCi (millicurie) range.

## 3.5.2 Half-life and the Law of Radioactive Decay

Because each transformation reduces the number of parent radioactive nuclei, and the probability of a transformation occurring per
unit time is proportional to the number of radioactive nuclei remaining in the source, the source activity decreases with time.
For a simple decay scheme involving the transformation of a radioactive parent nucleus into a stable daughter nucleus, the decrease
of activity with time is given by the relation

$$A_2 = A_1 \exp\left(\frac{-(t_2 - t_1)\ \ln 2}{T_{1/2}}\right) \qquad\qquad (3.13)$$

The activity measured at the time $t_1$ is $A_1$, and $A_2$ is the activity
observed at the later time $t_2$. The half-life $T_{1/2}$ is characteristic of the radioactive parent nucleus and defines the time required
for the activity to decrease to one-half of its value. This can be
checked in Eq. (3.13) by noting that $A_2 = (1/2)A_1$ when $t_2 - t_1 =$
$T_{1/2}$. Since the decay law is exponential, the time taken to decay
to one-half of the measured activity is always $T_{1/2}$, no matter
when the activity is measured.

Equation (3.13) is important because the fluoresced analyte
line intensity in a radioisotope excited fluorescence spectrometer
is proportional to the activity yielding the excitation lines. If
the spectrometer is operated over a time period significant with
respect to the source half-life, the concentration-versus-counting-
rate calibration curves must be corrected for the decrease in source
activity. If the half-life of the source is known, Eq. (3.13) can
be applied for the correction. Some radioactive sources involve a
series of decays where each daughter nucleus is also unstable.
This leads to further decays, finally ending in a stable nucleus.
Since each transformation has its own half-life, the equation describing the activity as a function of time is much more complicated.
Fortunately, in most cases of interest the longest half-life in
the series predominates and the simpler Eq. (3.13) can be applied.
Table 3.6 lists half-lives for some of the commonly employed
radioisotopes. The longer half-lives are more desirable from the
standpoint of replacing sources when their activity has decreased
below the lowest usable value, and correcting intensities for decaying source activity. Even though radioisotopes have a diminish-
ing excitation intensity, they offer better long-term stability
than x-ray tubes since the decay in activity is very predictable.

3.5.3  Geometry and Commonly Used Radioisotopes

A variety of geometries are feasible for radioisotope excited
fluorescence spectrometers. Figure 3.21 depicts the three major
categories.

Table 3.6  Characteristics of Radioisotopes Commonly Used for X-ray Fluorescence Spectrometers

| Isotope | Primary Decay Mode | Half-life $T_{1/2}$ (years) | Useful Photon Energies Emitted | % Theoretical Yield (photons per 100 decay transformations) | Typical Activity (mCi) |
|---|---|---|---|---|---|
| $^{55}$Fe | Electron capture | 2.7 | 5.9- and 6.4- keV Mn K x-rays | 28.5 | 5 to 100 |
| $^{109}$Cd | Electron capture | 1.3 | 22.2- to 25.5- keV Ag K x-rays | 107 | 0.5 to 100 |
| | | | 88.2- keV γ-ray | 4 | |
| $^{241}$Am | Alpha | 458 | 14- to 21-keV $N_p$ L x-rays | 37 | 1 to 50 |
| | | | 59.6-keV γ-ray | 36 | |
| $^{57}$Co | Electron capture | 0.74 | 6.4- and 7.1-keV Fe K x-rays | 51 | 1 |
| | | | 14.4-keV γ-ray | 8.2 | |
| | | | 122-keV γ-ray | 88.9 | |
| | | | 136-keV γ-ray | 8.8 | |
| $^{3}$H | Beta | 12.3 | Bremsstrahlung source: endpoint at 18.6 keV | | 3000 to 5000 |
| $^{147}$Pm | Beta | 2.6 | Bremsstrahlung source: endpoint at 225 keV | | 500 |

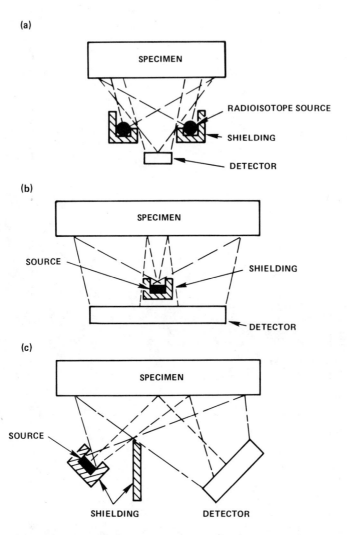

Figure 3.21 The three geometry categories for radioisotope-excited fluorescence spectrometers: (a) annular source, (b) central source, and (c) side source. Reprinted by courtesy of EG&G ORTEC.

The annular source is ideal for small-area detectors such as the Si(Li) detector. A large number of individual sources can be placed in a ring around the detector window to form a very high activity composite source. Shielding must be added to prevent the

detector from directly viewing the source. This is usually accomplished with a *graded* shield. Materials having a high mass
absorption coefficient such as lead or tantalum are used to attenuate the source output with minimum thickness. To eliminate the
characteristic x-rays generated in the shielding, the shielding is
surrounded with a thin layer of lower atomic number material. Frequently a third layer of even lower atomic number is employed to
eliminate the characteristic lines of the second material. Of
course, the entire chamber containing the source, detector, and
specimen must also be shielded to reduce exposure to operating
personnel to less than 0.25 mR/h.

The central source is efficient for large-area detectors such
as NaI(Tl) detectors, where extremely high activity is not required. Note that the source and its shield shadow the central
area of the detector, slightly reducing its detection efficiency.
The side source geometry is effective for medium- to large-area
detectors when the source area is comparable to the detector area.

In all three cases a wide range of incidence and takeoff
angles is included. This must be kept in mind when applying
theoretical formulas for the fluoresced and scattered intensity.
As with x-ray tube excitation, the excitation intensity is not
uniform across the specimen surface. Thus, it is important to ensure specimen homogeneity. Sensitivity of fluoresced intensity to
specimen position is a function of the geometry and source design
in addition to specimen composition. Hence, it must be experimentally determined for the intended application.

Table 3.6 summarizes the characteristics of some of the
commonly used radioisotopes. $^{55}$Fe is useful for exciting the
light-element K lines from sodium to titanium. The silver K lines
from the $^{109}$Cd source are efficient for exciting the medium-atomic-
number element K lines from about chromium to niobium. The 88.2-keV
$\gamma$-ray from $^{109}$Cd is extremely effective for exciting the K lines
from heavy elements such as platinum, gold, mercury, and lead.
With $^{241}$Am, the neptunium L lines are frequently filtered out and

the source is used for exciting the K lines of medium- to high-atomic-number elements. Tritium ($^{3}$H) and $^{147}$Pm can be used as bremsstrahlung sources for broadband excitation.

Often the availability of discrete photon energies from radioisotopes can be used to an advantage in selectively exciting particular elements in the specimen. In some cases the radioisotope can be chosen to have a line energy which is below the absorption-edge energy of an interfering major concentration element but above the absorption edge of the trace element to be measured. In other situations the excitation of a heavier element can be enhanced relative to a lighter element by choosing a radioisotope with a photon energy just above the absorption edge of the heavier element. A more flexible choice of excitation energies can be obtained by using a radioisotope of much higher activity to excite a secondary fluorescer which in turn excites the specimen. This technique is similar to the secondary fluorescer method for x-ray tubes described in Sec. 3.3.2. It has the disadvantage of requiring a rather high source activity. Excitation with radioisotopes having discrete photon energies usually allows simplification of matrix correction models since only a few excitation energies are involved.

## 3.6 ELECTRON EXCITATION

The most widespread use of electron excitation for x-ray spectrometry is in scanning electron microscopes (SEM) and electron-beam microprobes, where a finely focused beam of electrons is used to analyze the microcomposition of the specimen surface in regions as small as a few micrometers in diameter. The energy of the bombarding electrons is typically in the range of 5 to 50 keV.

Electron excitation can also be utilized for bulk analysis by x-ray spectrometry. The electron source is typically a heated tungsten wire which is held at a potential that is negative relative to the specimen target. This structure is very similar to the conventional x-ray tube, with the specimen replacing the normal

pure element anode. The discussion in Sec. 3.2 applies, except
that the anode is a multielement specimen. The electron source
gives a high excitation efficiency for the long-wavelength characteristic lines, and also causes a high bremsstrahlung background as
illustrated in Fig. 3.1. The stability of the electron source is
mainly dependent upon the electron emission from the filament and
the applied potential between filament and sample. The electron
emission is, in turn, related to the current passing through the
filament and the vacuum surrounding the filament and the specimen.
A stable electron source requires, therefore, a stabilized filament current and high-voltage supply, plus a good ($10^{-7}$ torr)
vacuum. Great care must also be taken to render the specimen conductive since this might otherwise become charged and lead to instability of the incident electron beam. The major advantage offered by the electron source is its high excitation efficiency over
the whole analytical x-ray region. It is particularly effective
for light-element excitation.


3.7  EXCITATION BY PROTONS

It has been known for many years that characteristic x-rays can be
excited by charged-particle bombardment. The first reported work
in this area was published as long ago as 1912 [23]. Recent interest in methods of excitation by particles, in practice mainly by
protons, has stemmed primarily from the current need for ultratrace
analysis of the microspecimens involved with air pollution and
environmental problems. This need has come at a time when quite a
large number of Van de Graff generators and cyclotrons are available, resulting in a considerable upsurge in interest. This is
reflected in the large number of recently published papers in this
field [24-26].

     Relative to the previously discussed excitation sources, the
main advantages to be offered by the use of charged particles is
the very low background obtained from the source itself and the

98                                  Sources for X-ray Excitation

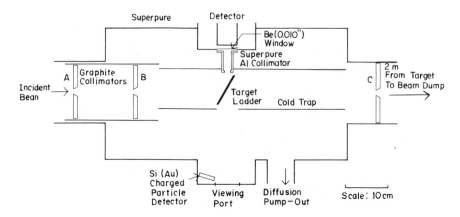

Figure 3.22  Typical target chamber geometry for charged particle
excitation on a nuclear accelerator.

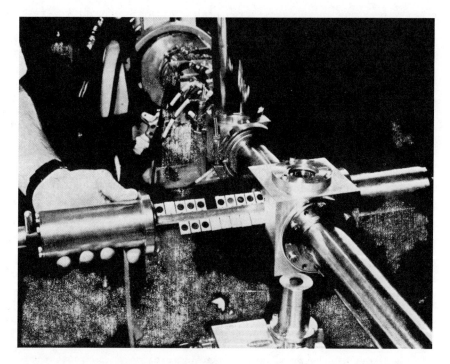

Figure 3.23  Specimen changer for the apparatus in Fig. 3.22.

almost limitless flux available.  In electron excitation, the
majority of the background comes from the continuum generated by
the decelerating electrons; and in photon excitation most of the
background arises from the scatter of this same continuum.  Since
protons, for example, are not scattered by the electrons of the
sample atoms, the scattered source background is always relatively
low.  In practice, the major factor that limits the lowest detectable quantities is usually the material of the sample support [27].
In cases where the backing material can be dispersed entirely,
extremely low concentrations can be detected.  For example, 27
elements at concentration levels ranging from 0.1 to 100 μg/g have
been detected in single strands of human hair stretched onto an
aluminum frame [28].

Figure 3.22 shows a schematic diagram of a typical target
chamber of a proton-source spectrometer.  The energy-dispersive
spectrometer is particularly well suited to this type of system and
is generally used as the characteristic line selection device.  It
is also common practice to incorporate a multiple sample handling
system into the spectrometer; the target chamber assembly shown in
Fig. 3.23 is fairly typical.  Rather high excitation energies are
required for proton excitation.  However, the 0.1- to 10-MeV ac-
celerating potentials which are needed are available on nuclear
particle accelerators.  Proton excitation is only cost effective
when the nuclear accelerator is already installed and justified on
the basis of other work.

REFERENCES

1.   Robley D. Evans, *The Atomic Nucleus,* McGraw-Hill, New York,
     1955.
2.   H. A. Kramers, *Phil. Mag., 46:*836 (1923).
3.   N. G. Ware and S. J. B. Reed, *J. Phys. E. Sci. Instrum., 6:*286
     (1973).
4.   S. J. B. Reed, *X-ray Spectrom., 4:*14 (1974).
5.   D. G. W. Smith, C. M. Gold, and D. A. Tomlinson, *X-ray
     Spectrom., 4:*149 (1975).

6.   L. S. Birks, *X-Ray Spectrochemical Analysis*, Wiley Interscience, New York, 1969.

7.   J. V. Gilfrich and L. S. Birks, *Anal. Chem.*, *40*:1077 (1968).

8.   J. V. Gilfrich, P. G. Burkhalter, R. R. Whitlock, E. S. Warden, and L. S. Birks, *Anal. Chem.*, *43*:934 (1971).

9.   H. D. Keith and T. C. Loomis, *X-ray Spectrom.*, *5*:93 (1976).

10.  R. Jenkins, *An Introduction to X-Ray Spectrometry*, Heyden, London, 1974.

11.  C. J. Toussaint, *Anal. Chim. Acta*, *55*:373 (1971).

12.  B. L. Henke, R. L. Elgin, R. E. Lent, and R. B. Ledingham, *Norelco Reporter*, *XIV*:112 (1967).

13.  G. R. Dyer, D. A. Gedcke, and T. R. Harris, *Advan. X-ray Anal.*, *15*:228 (1972).

14.  J. M. Jaklevic, R. D. Giauque, D. F. Malone, and W. L. Searles, *Advan. X-ray Anal.*, *15*:266 (1972).

15.  A. J. Hebert and Kenneth Street, Jr., *Anal. Chem.*, *46*:203 (1974).

16.  D. A. Gedcke, E. Elad, and P. B. Denee, *An Intercomparison of Trace Element Excitation Methods for Energy Dispersive Fluorescence Analyzers*, presented at the Twenty-Fourth Annual Conference on Applications of X-Ray Analysis, Denver (August 1975).

17.  D. A. Gedcke, E. Elad, and P. B. Denee, *X-ray Spectrom.*, *6*:21 (1977).

18.  J. R. Rhodes, *Energy Dispersion X-Ray Analysis: X-Ray and Electron Probe Analysis*, ASTM STP 485, American Society for Testing and Materials, Philadelphia, 1971, p. 243.

19.  *General Safety Standard for Installations Using Non-Medical X-Ray and Sealed Gamma-Ray Sources, Energies up to 10 MeV*, NBS Handbook 114, U.S. Government Printing Office, Washington, D.C., 1975.

20.  K. Z. Morgan and J. E. Turner, *Principles of Radiation Protection*, Wiley, New York, 1967.

21.  *Radiological Health Handbook*, U.S. Government Printing Office, Washington, D.C., January 1970.

22.  C. M. Lederer, J. M. Hollander, and I. Perlman, *Table of Isotopes*, Wiley, New York, 1967.

23.  J. Chadwick, *Phil. Mag.*, *24*:594 (1912).

24.  T. B. Johansson, R. Akselsson, and S. A. E. Johansson, *Nucl. Instrum. Methods*, *84*:141 (1970).

25.  J. L. Duggan, W. L. Beck, L. Albrecht, L. Munz, and J. D. Spaulding, *Advan. X-ray Anal.*, *15*:407 (1971).

26.  J. A. Cooper, *Nucl. Instrum. Methods*, *106*:525 (1973).

27.  A. W. Herman, L. A. McNelles, and J. L. Campbell, *Nucl. Instrum. Methods*, *109*:429 (1973).

28.  V. Valkovic, *Nature*, *243*:543 (1973).

29.  E. J. Feldl and C. J. Umbarger, *Nucl. Instrum. Methods*, *103*:341 (1972).

# 4

## Instrumentation

4.1  X-RAY FLUORESCENCE SPECTROMETERS AND THEIR MAJOR COMPONENTS

The x-ray fluorescence spectrometer consists of three main parts:
the excitation source, the specimen presentation apparatus, and the
x-ray spectrometer.  The function of the excitation source is to
excite the characteristic x-rays in the specimen via the x-ray
fluorescence process.  The specimen presentation apparatus holds the
specimen in a precisely defined position during analysis and provides for introduction and removal of the specimen from the excitation position.  The x-ray spectrometer is responsible for separating
and counting the x-rays of various wavelengths or energies emitted
by the specimen.  In this book the term *x-ray spectrometer* denotes
the collection of components used to disperse, detect, count, and
display the spectrum of x-ray photons emitted by the specimen.  When
referring to the entire instrument, including excitation source,
sample presentation apparatus, and x-ray spectrometer, the term
*x-ray fluorescence spectrometer* will be used.  In this latter sense
the term *x-ray fluorescence analyzer* is sometimes encountered in
the literature.

As described in Chap. 3, a variety of excitation sources can
be employed.  With the wavelength-dispersive spectrometer where high
source intensity is required, a high-power x-ray tube normally
excites the specimen directly.  Both the characteristic lines and
the bremsstrahlung continuum from the x-ray tube are utilized.  This
same method is also commonly employed with the energy-dispersive
spectrometers, with the exception that the x-ray tube is operated
at much lower power.  A variety of quasi-monochromatic excitation
techniques based on x-ray tubes are frequently used with energy-
dispersive spectrometers.  These include the secondary fluorescer
method and the regenerative monochromator filter technique.  In all
of the systems which use an x-ray tube, the nominal incidence angle
for the excitation radiation on the specimen surface (the angle
between the specimen surface and the incident radiation direction)
is normally somewhere in the range of 30 to $50^{\circ}$.  In most instruments the incidence angle is poorly defined as a result of the large
divergence angle designed into the source of radiation (see Figs.
3.7, 3.9, and 3.15).  The source divergence angle is necessarily
large to deliver as much of the source radiation to the specimen
as possible.  This fact should be appreciated when applying any of
the theoretical models which include the incidence angle.  With
energy-dispersive spectrometers, radioisotopes are often used for
excitation.  Such systems require no power supply for the excitation
source.  Excitation geometries for radioisotope-excited systems can
be quite diverse as illustrated in Fig. 3.21.  For the most part,
this book deals with fluorescence spectrometers based on the above
mentioned photon excitation methods, excluding systems using excitation by electrons or heavy charged particles.

    With x-ray tube sources, a high-voltage power supply is needed
to provide the high voltage for the x-ray tube and to control the
current through it.  Voltages in the range of 10 to 100 kV are
required, with x-ray tube currents ranging from 1 $\mu$A to 5 mA for
energy dispersive spectrometers, and from 5 to 100 mA on wavelength-
dispersive spectrometers or energy-dispersive spectrometers

employing the secondary fluorescer method. Generally, the power supply must provide up to 250 W of output power for low-power x-ray tubes, and up to 4 kW of output power for high-power tubes. The low-power generators can operate from conventional power lines, but special 220-V entry lines and cooling water are commonly required for the 4-kW supplies.

In the older x-ray generators, the output voltage to the x-ray tube was of the *full-wave rectified* type. That is, the voltage varies from zero up to the preselected value and back to zero at twice the ac line frequency. This type of supply makes it extremely difficult to control and stabilize the excitation conditions. In more modern instruments, the power supply is of the *constant potential* type. That is, a large capacitor is added across the high-voltage output to smooth the output and provide a dc voltage with very little residual line frequency ripple. The equations describing the x-ray tube output spectrum in Chap. 3 and the counting statistics equations used throughout this book, apply only to the constant potential type of supply. They are not valid for half-wave or full-wave rectified supplies.

Although constant potential supplies make it easier to stabilize the voltage and current applied to the x-ray tube, not all supplies take full advantage of this capability. Most x-ray generators control the current through the x-ray tube by measuring the electron current arriving at the anode, or leaving the cathode, and comparing it to the value selected on the generator control panel. If the actual current does not match the selected value, a correction signal is applied to the current control circuit. This may involve adjusting the voltage on the cathode heater filament or changing the voltage on a cathode emission control grid to achieve the correct current through the tube. This is a negative feedback servo system. Good stability of the tube current depends on the sensitivity of this servo system to ambient temperature, line voltage, and to other "difficult-to-identify" sources of long-term drift. The high voltage applied across the x-ray tube may be

controlled in a similar fashion.  The actual applied voltage is compared to the value selected on the x-ray generator control panel.
Any mismatch is converted into an error signal which adjusts the
generator output to obtain the correct voltage.  Here again, the
stability of the high voltage is determined by the sensitivity of
this servo loop to ambient temperature, line voltage, and other
difficult-to-identify sources of long-term drift.  Some x-ray generators do not control the high voltage by regulating the output
voltage to the x-ray tube (secondary regulation).  Instead, they
regulate the much lower voltage driving the input of the step-up
transformer in the high-voltage supply (primary regulation).  This
makes the electronic design much easier since the controller can
measure voltages in the 50- to 220-V range instead of the 20- to
100-kV range.  However, the stability of the voltage applied to the
x-ray tube now depends upon the stability of (a) the high-voltage
step-up transformer, (b) the high-voltage recitifiers, (c) the high-
voltage smoothing capacitor, and (d) the x-ray tube loading.  Since
these parameters are not well controlled, one cannot expect good
stability or repeatability from a supply having only primary regulation.  Even with a constant potential supply having good secondary regulation, excellent excitation source stability is not
guaranteed.  Various factors such as changes in the x-ray tube
anode condition, interfering magnetic fields, thermal expansion,
and excessive gas in the x-ray tube can lead to poor excitation
spectrum intensity stability.

Most x-ray tubes are designed to operate within certain limits
on current, voltage, and power.  Usually the maximum limit on power
is significantly less than the product of the maximum current and
the maximum voltage.  Thus, most x-ray generators can be operated
under conditions which will exceed the maximum allowable x-ray tube
dissipation.  If the generator is not equipped with automatic overload prevention features, the operator must ensure that destructive
operating conditions are not selected.

A variety of designs have been utilized for the specimen presentation apparatus.  With any of these designs it is most important
that the specimen positioning be precise, repeatable, and reliable.
Typically, 500-μm variations in the elevation of the analyzed
specimen surface can cause errors in the quantitative measurement
of the order of 0.5%.  For this reason, the specimen presentation
mechanism must be kept free of dust and dirt and should be maintained in good working order.  Air paths can be used for analyzing
the shorter wavelengths, but light-element analysis requires a
vacuum path.  Most instruments provide for evacuating the specimen
chamber or introducing helium into the x-ray paths for light-element
analysis.  The helium path is useful where the specimen cannot withstand vacuum.  With energy-dispersive spectrometers, helium should
be used with caution, since it may penetrate the thin beryllium
window on the detector end cap and cause deterioration of the detector cryostat vacuum.  General caution should be exercised regarding introducing specimens which contain materials which are
corrosive to the instrument.  In this respect, the beryllium windows used on x-ray tubes, NaI(Tl) detectors and Si(Li) detectors,
are particularly sensitive to the vapors from salt solutions, bases,
and acids.

Major categorical differences between x-ray fluorescence
spectrometers occur in the x-ray spectrometer section.  The x-ray
spectrometer is generally one of two types.  A wavelength-dispersive spectrometer utilizes the diffracting property of a single
analyzing crystal to separate or disperse the polychromatic beam of
radiation from the specimen.  An energy dispersive spectrometer uses
the proportional characteristics of a suitable detector to produce
a distribution of voltage pulses proportional to the spectrum of
photon energies from the specimen.  A multichannel pulse height
analyzer measures the pulse height distribution and subsequently
displays the spectrum.

Wavelength dispersive x-ray fluorescence spectrometers may be
built around a single sequential x-ray spectrometer or may be of the
simultaneous multichannel type.  In the sequential spectrometer, the
entire range of wavelengths is covered by adjusting the diffracting
angle of the analyzing crystal.  Although several types of crystals
can be interchanged to efficiently cover the range from 0.2 to 20
$\overset{o}{A}$, they all use the same goniometer to select the diffraction angle.
In other words, only one wavelength can be analyzed at a time.  If
several wavelengths are to be counted, they must be dealt with
sequentially.  Sequential spectrometers are flexible and can be
used for a variety of analytical problems.  Multichannel simultaneous spectrometers are most efficient where a predetermined suite
of elements is to be repetitively measured on a dedicated basis
over a long period of time.  Usually, anywhere from 7 to 28 fixed
spectrometers are located around the specimen.  Each spectrometer
contains the appropriate crystal set to the proper angle for the
wavelength it is to measure.  Although most of the spectrometers
will be set on characteristic lines corresponding to the elements
of interest, some are also used to sample the background.  These
spectrometers have a very narrow range of adjustment sufficient to
fine tune for the optimum counting rate on the line to be monitored.
Because the data can be acquired for all elements of interest
simultaneously, the sacrifice in flexibility results in much
shorter measurement times.  This can be important where a large
number of specimens must be analyzed in a short time.  Most multichannel wavelength spectrometers offer a sequential spectrometer
channel as an accessory to regain flexibility and provide for
studying unanticipated contaminant elements.  A few instruments
also are available with a Si(Li) energy dispersive spectrometer as
an attachment.

A variety of energy-dispersive x-ray fluorescence spectrometers are available ranging from portable, battery-powered units
which are designed to analyze for a single element up to the most
sophisticated Si(Li) energy dispersive systems which provide a wide

flexibility for analyzing all elements from sodium to uranium in
the periodic table.  Four detectors are appropriate for energy
dispersive systems.  In order of improving energy resolution, they
are the NaI(Tl) detector, the proportional counter, the Si(Li) detector, and the Ge(Li) detector.  The NaI(Tl) detector and the
proportional counter are often used in portable, single-element
fluorescence spectrometers.  In this respect, they are frequently
combined with a balanced filter technique to compensate for their
poor energy resolution.  Both the Si(Li) detector and the Ge(Li)
detector have adequate energy resolution to resolve the K$\alpha$ lines
from adjacent elements for atomic numbers greater than 10.  The
Si(Li) detector is optimum over the 1- to 40-keV energy range,
while the Ge(Li) detector is better suited to the 6- to 200-keV
range due to its higher atomic number.  Both solid-state detectors
are commonly used with a multichannel pulse-height analyzer for
multielement analysis, but can be used with a single-channel pulse-
height analyzer (pulse height selector) for analysis of one up to a
few preselected elements.


4.2  DISPERSION, DETECTION, AND COUNTING WITH A
     WAVELENGTH-DISPERSIVE SPECTROMETER

Figure 4.1 illustrates the typical flat-crystal spectrometer
geometry in a wavelength-dispersive spectrometer.  The x-ray tube
is closely coupled to the specimen and illuminates a large area on
the specimen surface.  The primary collimator selects the x-rays
emitted by the specimen which will be allowed to pass on to the
analyzing crystal.  A series of thin, high-atomic-number parallel
plates form the collimator, and limit the divergence of the x-ray
beam which strikes the analyzing crystal.  For wavelengths which
satisfy Bragg's law, the x-rays are diffracted from the crystal
through the angle $2\theta$ into the detectors.  A parallel-plate secondary collimator on the entrance to the gas-flow proportional counter
further restricts the divergence of the x-ray beam as it leaves the

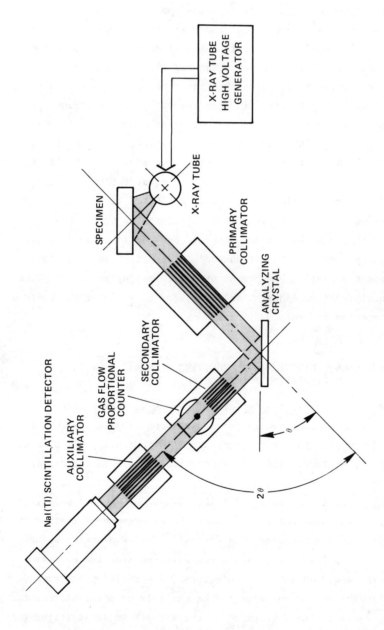

Figure 4.1  The flat crystal spectrometer geometry.  Reprinted by courtesy of EG&G ORTEC.

analyzing crystal.  An auxiliary collimator performs a similar
function in front of the NaI(Tl) scintillation detector.  In some
fluorescence spectrometers, either the gas-flow proportional
counter or the scintillation detector can be selected to detect the
diffracted x-rays.  In other spectrometers the two detectors can be
operated simultaneously, in tandem.  In this instance, the scintillation detector counts the x-rays which pass through the proportional counter.  The proportional counter is most effective for
long-wavelength x-rays (low energies) since it has a thin window
and good pulse-height (energy) resolution.  At shorter wavelengths
where the gas-flow proportional counter's detection efficiency is
low, the high efficiency of the scintillation detector becomes
useful in spite of its poorer energy resolution

The wavelength dispersing or selecting device is the analyzing
crystal.  Bragg's law

$$n\lambda = 2d \sin \theta \qquad\qquad n = 1, 2, 3, \ldots \qquad\qquad (4.1)$$

relates the wavelengths $\lambda$ which are reflected through the angle $2\theta$
to the crystal lattice spacing d.  To scan a range of wavelengths,
the detectors are swept through the angle $2\theta$, while the crystal
rotates at half the rate through the angle $\theta$.

Detection of the x-rays is provided by the flow proportional
counter or the scintillation detector.  Both types of detectors
convert each detected x-ray photon into a pulse of electrical
charge.  The magnitude of the charge pulse is proportional to the
energy of the x-ray photon and inversely proportional to the
wavelength.

Two general classes of signal processing electronics have been
used following the detectors:  (a) the integrating electrometer
and (b) the single-photon counting system.  Figure 4.2 is a simplified illustration of the integrating electrometer principle.  A
network of switches permits the integrating capacitor to be connected to the detector output, isolated from the detector, or dis-
charged to ground potential.  An electronic voltmeter having a very

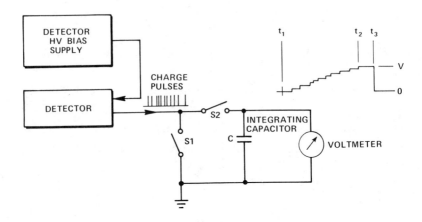

Figure 4.2   The integrating electrometer signal-processing electronics.   Reprinted by courtesy of EG&G ORTEC.

high input impedance monitors the voltage on the capacitor.   Initially, switches S1 and S2 are both closed and the voltage on the
capacitor is zero.   At the time $t_1$ data acquisition commences when
the switch S1 is opened and charge pulses from the detector are allowed to accumulate on the capacitor.   Each pulse increments the
capacitor voltage by an amount

$$\Delta V = \frac{q}{C} \tag{4.2}$$

where q is the charge in each pulse and C is the capacitance of the
integrating capacitor.   If N pulses or x-rays have been detected
by the end of the counting interval at time $t_2$, the voltage on the
capacitor will be

$$V = \frac{Nq}{C} \tag{4.3}$$

which is proportional to the number of x-rays detected.   At time $t_2$
the switch S2 is opened, the voltage on the capacitor is read, and
switch S1 is closed.   At time $t_3$ switch S2 closes to discharge the
capacitor in preparation for the next counting interval.   Thus, the
number of x-rays detected in the time interval $t = t_2 - t_1$ is read
as a voltage on the capacitor, and the x-ray intensity is

proportional to the voltage reading divided by the time interval.
This type of system is simple and free of deadtime losses.  However,
other effects which cause nonlinearity of the relationship between
x-ray intensity and the voltmeter reading can be significant.  The
output reading is proportional to the gain of the detector, which
sets the charge per pulse q.  Since it is difficult to eliminate
gain changes of the order of 1% or less, instability of the intensity measurement calibration can be a significant source of drift.
In addition, the method sacrifices the ability to select or reject
n-th order diffraction harmonics from the analyzing crystal.  Although it is simple and therefore attractive for multichannel systems, the integrating electrometer has enough drawbacks that it is
seldom used on the more modern spectrometers.

The typical single-photon-counting signal-processing electronics
is shown in Fig. 4.3.  In this case the charge pulse from each detected x-ray photon is counted individually.  The total number of
pulses counted over a time interval t divided by t measures the
x-ray photon counting rate.  In this case minor drift in the detector gain does not significantly alter the number of pulses
counted, and the stability of the intensity calibration is substantially improved.  On the other hand, the deadtime arising from the
separate counting of each charge pulse leads to a nonlinear relationship between the observed counting rate and the true x-ray
photon rate.  This nonlinearity is referred to as *deadtime losses*
and becomes important when the duration of each pulse is a significant fraction of the mean time between pulses; i.e., at high
counting rates.

Figure 4.3 is typical of the systems employed in modern wavelength dispersive x-ray fluorescence spectrometers.  The detector
may be either a NaI(Tl) scintillation detector or a gas-flow proportional counter.  In either case a high-voltage bias in the range
of 500 to 3000 V is required to make the detector operate.  This
voltage is provided by the detector high-voltage bias supply.  The
function of the preamplifier is to collect the charge pulse from

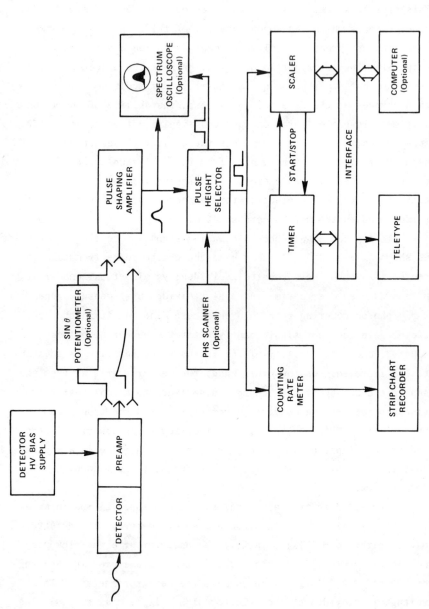

Figure 4.3   The typical signal-processing electronics for single-photon counting in a wavelength-dispersive spectrometer.   Reprinted by courtesy of EG&G ORTEC.

the detector and provide the low driving impedance necessary to
pass the signal through a coaxial cable to the main amplifier, which
is usually located some distance away in a control panel.  The
charge pulse from the detector is collected in the preamplifier by
integrating it on a small capacitor to produce a voltage pulse.
The height of this voltage pulse is proportional to the energy of
the x-ray.

There are three problems associated with the pulse at the preamplifier output:

1. The pulse amplitude is extremely small
2. The pulse duration is too long
3. There is generally an unacceptable noise level superimposed
   on the signal by the preamplifier, which would degrade the
   pulse height (energy) resolution.

Consequently, the pulse-shaping amplifier serves three purposes.
First, it amplifies the signal to make it lie in the 0- to 10-V
pulse height range.  Second, suitable pulse-shaping filters are
incorporated to yield a shorter pulse duration so that high counting rates can be employed with minimal deadtime losses.  Third, the
filters are selected to minimize the noise contribution from the
preamplifier.  The result is a pulse at the amplifier output which
rises from the baseline at ground potential, has a pulse height
which is proportional to the detected x-ray photon energy, and has
a duration near the baseline somewhere in the range of 1 to 9 μs
depending on amplifier design and detector type.

Most of the pulses observed at the amplifier output have an
amplitude which corresponds to the desired wavelength selected by
the analyzing crystal.  Unfortunately, unwanted pulses also occur.
These pulses generally have a different pulse height and can result
from second- or higher order diffraction [n = 2, 3, 4, in Eq. (4.1)],
fluorescence of the analyzing crystal, or nonideal interactions in
the detector.  Consequently, the performance of the x-ray spectrometer can be improved if only those pulses of the correct amplitude
are counted.  This is the function of the pulse-height selector
(PHS).  For every amplifier pulse which falls within a preselected

narrow range of pulse heights, the pulse-height selector generates
a short, standard, digital logic pulse at its output.  The preselected range of pulse heights is referred to as the pulse height
*window*.  Thus, the counting rate of the digital logic pulses at the
pulse height selector output is identical to the counting rate of
the *desired* pulse amplitudes at the output of the main amplifier.
Occasionally, the pulse-height selector is referred to as a *single-*
*channel pulse height analyzer* or *single-channel analyzer* (SCA).
This nomenclature comes from the field of nuclear spectrometry and
is intended to provide contrast with the multichannel analyzer
(MCA) commonly used for energy-dispersive spectrum analysis.

In addition to the *window mode* of operation, the pulse height
selector can be used in the *integral mode*.  In this case all pulse
heights which exceed a preselected value generate a digital logic
pulse at the output.  The preselected level is referred to as the
*integral discriminator threshold*.  This mode is commonly used when
it is sufficient to count all pulses above the amplifier output
noise level.  Where the sin $\theta$ potentiometer feature is not available, the integral mode becomes necessary for wavelength scans.

Two methods are available for recording the counting rate at
the pulse-height-selector output:  one analog and the other digital.
An analog measurement is provided by the counting ratemeter.  This
device converts the average pulse rate at its input into a voltage
which drives a front panel meter or the y-axis input of a strip-
chart recorder.  The meter is calibrated to indicate the average
counting rate.  A digital measurement is made with the timer and
scaler combination.  A scaler is an electronic device which counts
the pulses at its input by means of a digital register.  The contents of the register are erased or set to zero before each counting interval.  The duration of the counting interval is controlled
by the timer.  When the timer reaches its preset limit, it stops
the scaler from counting and the content of the scaler register is
exhibited on a digital front panel display.  If the number of pulses
counted was N, and the preset counting time interval was t, then the

counting rate is measured as N/t.  Data from the scaler and timer
can usually be printed out on a hard-copy device such as a teletype.
In some cases, an interface to a digital computer provides greater
flexibility in data acquisition control, readout, and data analysis.
The timer/scaler mode is commonly used when quantitative measurements are being made at specific wavelengths.  The counting ratemeter is useful when a section of the wavelength spectrum must be
scanned for qualitative analysis.  The strip-chart recorder advances as the crystal spectrometer scans through a range of $2\theta$
angles.  Consequently, a graph of counting rate versus $2\theta$ angle is
recorded, which defines the spectrum of intensity versus wavelength.

The counting ratemeter can also be used in conjunction with
the pulse-height-selector scanner and strip-chart recorder to plot
the pulse height spectrum produced by the detector at a fixed angle
setting of the crystal spectrometer.  The pulse-height-selector
scanner slowly moves the center of the pulse-height-selector window
over the entire range of pulse heights from maximum down to minimum
as the strip-chart recorder advances.  In this way, the counting
rate is plotted as a function of pulse height to yield the detector
pulse-height spectrum.  This mode can be helpful in studying escape peaks and higher order diffraction problems so that optimum
window selection can be achieved.  It can also be used as a service
test for degraded detector resolution.

At  moderate to high counting rates, the spectrum oscilloscope
can be helpful in monitoring the detector pulse height spectrum
and in setting the position and width of the pulse-height-selector
window.  This monitor is a rather novel analog device and is described more fully in Ref. 1.  The sin $\theta$ potentiometer is used to
simplify the setting of the pulse-height-selector window during
wavelength scanning.  This principle will be discussed in Sec.
4.2.8.

4.2.1  The Bragg Crystal Spectrometer

The Bragg crystal spectrometer may consist of a flat, curved
(Johann), curved and ground (Johannson), or logarithmically curved

crystal.  Flat-crystal spectrometers employ parallel plate collimators between the source and the crystal (the primary collimator),
and between the crystal and the detector (the secondary collimator).
Curved-crystal spectrometers employ slits which lie on the focusing
circle of the spectrometer, again between the source and the crystal
(the source slit) and between the crystal and the detector (the
detector slit).  In this type of geometry the effective surface of
the crystal also lies on the focusing circle.  In each of the above
geometries the diffraction angle $\theta$ of a given wavelength $\lambda$ is fixed
by the 2d spacing of the crystal

$$n\lambda = 2d \sin \theta \qquad\qquad n = 1, 2, 3, \ldots \qquad\qquad (4.1)$$

In all types of crystal spectrometers the reflection efficiency of
the crystal and the resolution of the spectrometer $d\lambda/\lambda$ are determined by the solid angle defined by the collimators or slit system
and by the integral reflection coefficient R of the analyzing
crystal.  The integral reflection coefficient is given by

$$R = \int_{\theta_1}^{\theta_2} I \, d\theta \qquad\qquad\qquad\qquad (4.4)$$

where $\theta_1$ and $\theta_2$ represent the angular extremes to which the diffraction profile extends.  The peak width at half maximum (FWHM) of
the diffraction profile is obtained by convoluting the $(FWHM)_c$ of
the crystal with that of the collimators.  Thus

$$FWHM = \sqrt{(FWHM)_c^2 + (FWHM)_{primary\ collimator}^2 + (FWHM)_{secondary\ collimator}^2}$$

Both the R and FWHM values of a crystal can be varied by making the
surface of the crystal microcrystalline (mosaic), by abrasion, or
similar process.  The angular dispersion $d\theta/d\lambda$ of a crystal spectrometer is related to the d spacing of the analyzing crystal and the
order n of the reflection, i.e.,

$$\frac{d\theta}{d\lambda} = \frac{n}{2d \cos \theta} \qquad\qquad\qquad\qquad (4.5)$$

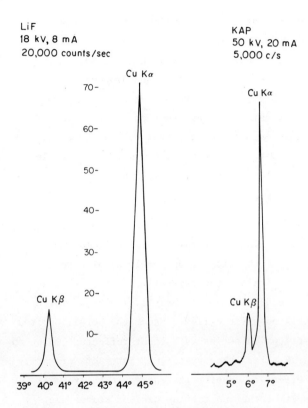

Figure 4.4   A comparison of angular dispersion between two analyzing crystals.   The left hand side of the figure is the Cu K line spectrum taken with a LiF(200) crystal.   The right-hand copper spectrum was acquired with a KAP crystal.

where θ is the diffraction angle.

It will be seen from Eq. (4.5) that the angular dispersion is inversely related to the 2d spacing of the analyzing crystal.   This is illustrated in Fig. 4.4 which shows the copper K spectrum recorded with a LiF(200) crystal, 2d = 4.028 Å; and a KAP crystal, 2d = 26.4 Å.   As would be predicted from Eq. (4.5), the angular separation of the Kα and Kβ lines is more than six times greater with the LiF(200) spectrum.   Note also the greater reflecting power of LiF(200) which for Cu Kα is almost 30 times better than KAP.

The maximum attainable diffraction angle in most crystal spectrometers is about $75^{\circ}$. Hence, from Eq. (4.1), the maximum wavelength measurable with a given crystal is about 1.8d, where d is the interplanar spacing of the crystal. Thus, to cover the full wavelength range of 0.2 to 20 $\overset{\circ}{A}$, a crystal of at least 22 $\overset{\circ}{A}$ would be required, but this crystal would have very poor angular dispersion for the shorter wavelengths. Hence, most crystal spectrometers have several analyzing crystals which are easily interchangeable for covering different parts of the wavelength range.

## 4.2.2  The NaI(Tl) Scintillation Detector

Figure 4.5 schematically illustrates the construction of the NaI(Tl) scintillation detector. A single crystal of sodium iodide doped with a low concentration of thallium is optically coupled to a photomultiplier tube. Typically, the crystal is 2.54 cm in diameter by 1 mm thick. The 0.127-mm-thick beryllium window provides a light-tight entrance window for x-rays. Under the beryllium is a 1-μm-thick aluminum foil which serves as a light reflector. When an x-ray photon of energy E enters the crystal, it interacts primarily by the photoelectric effect to deposit its energy in the form of ionization. As a result of the ionization, excited states are formed in the crystal. As these states decay, a scintillation or flash of light is emitted. The intensity of the scintillation is proportional to the x-ray energy E and decays exponentially with a 250-ns time constant. As this light falls on the photocathode it causes the emission of photoelectrons. The number of electrons emitted is proportional to the light intensity. As shown in Fig. 4.5, a series of dynodes is arranged between the photocathode and the anode of the tube, with each dynode at a successively higher voltage. Each electron emitted by the photocathode is attracted to the first dynode and knocks out several electrons. These secondary electrons are attracted to the second dynode where the multiplication process is repeated. The final current reaching the anode is therefore a replica of the current pulse leaving the photocathode,

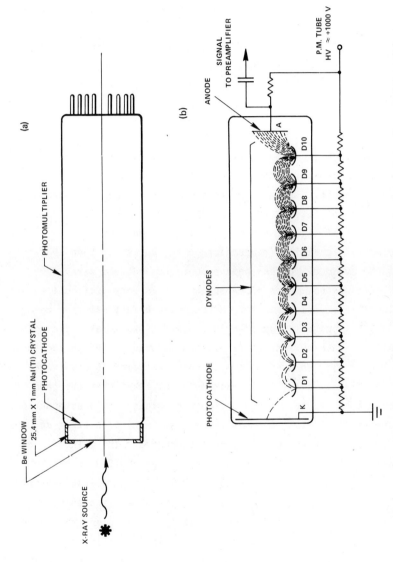

Figure 4.5  The NaI(Tl) scintillation detector:  (a) the integral assembly;  (b) a schematic representation of the photomultiplier tube and associated circuitry. Reprinted by courtesy of EG&G ORTEC.

but is amplified by a very large factor (between $10^5$ and $10^8$).
Hence, the total charge in the anode pulse is proportional to the
x-ray energy E. As discussed in Sec. 4.2, the preamplifier inte-
grates this charge to produce a pulse whose amplitude is proportional to the original x-ray energy.

The NaI(Tl) detector suffers from poor resolution due to the
inefficiencies in the several processes which convert the x-ray
energy to an electrical signal. The full-width-at-half-maximum
(FWHM) energy resolution $\Gamma$ is given by

$$\Gamma = k \sqrt{E} \qquad\qquad\qquad\qquad\qquad\qquad (4.6)$$

This resolution is measured by accumulating a peak in the energy
spectrum and measuring the full width of the peak at one-half its
maximum intensity. For a very good counter assembly, k can be as
low as 2.46 in units of $(keV)^{1/2}$. Worse resolutions are common,
particularly if the counter has aged, has been damaged by a rapid
temperature transition (temperature changing faster than $10^{\circ}C/h$
should be avoided), or has been exposed to a direct x-ray tube
beam. Because of the poor resolution, it is very difficult to set
the lower level discriminator in the pulse height selector to accept all the pulses produced by a 5.9-keV Mn K$\alpha$ x-ray while rejecting all the dark current noise pulses produced by the photocathode
(see Fig. 4.6). When working with energies below 10 keV (wavelengths above 1.24 $\overset{\circ}{A}$), it is extremely important that the gain of
the entire spectrometer does not change, since a gain attenuation
would alter the relative lower level discriminator setting and reduce the fraction of the x-ray pulses counted. The gain of the
photomultiplier is determined by the voltage applied to the tube.
Therefore, the high-voltage supply and the dynode voltages provided
by the resistor string must be extremely stable. The shaping
amplifier should contain a good baseline restorer, and the pulse-
height selector should be dc coupled for good baseline stability.
Even though the energy resolution and dark current noise are not
strong functions of the photomultiplier high voltage, the
manufacturer's recommended voltage setting should be used. Al-

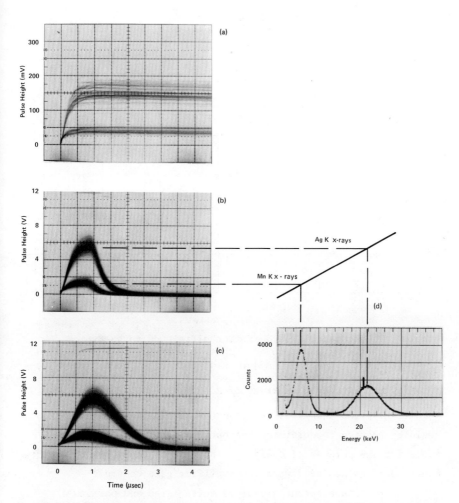

Figure 4.6  Pulse shapes encountered with a NaI(Tl) detector and spectrometer electronics.  The detector is responding to the 22 keV silver K lines from a radioactive $^{109}$Cd source and the 6 keV Mn K lines from a radioactive $^{55}$Fe source.  (a) The preamplifier output pulses.  (b) Output pulses from a delay-line clipped pulse-shaping amplifier with a 1 μsec delay line clip and a 1/4 μsec integration time constant.  (c) Output pulses from the alternative semigaussian shaping amplifier with a 0.5 μsec time constant.  (a), (b), and (c) are multiple traces on an oscilloscope.  (d) The energy (pulse-height) spectrum obtained by analyzing the delay line amplifier output on a multichannel pulse-height analyzer.  Reprinted by courtesy of EG&G ORTEC.

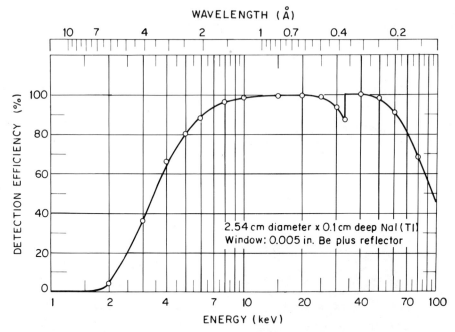

Figure 4.7  The calculated detection efficiency for a typical
NaI(Tl) scintillation detector.  The notch at 33.2 keV corresponds
to the iodine K absorption edge.  Reduction of efficiency as a result of the iodine escape peak above 33.2 keV has not been incorporated.  Reprinted by courtesy of EG&G ORTEC.

ternatively, the change of noise and resolution with high voltage
could be plotted to determine the optimum setting.  Gain adjustments for the purpose of energy calibration or energy range adjustment should be achieved with the amplifier gain control.

The phototube output signal is quite large and permits a
rather crude preamplifier design to be used.  However, both the
phototube and preamplifier must be capable of handling high counting
rates without gain attenuation.  Usually gain attenuation in the
photomultiplier becomes a limitation at high counting rates.  Gain
attenuations of less than 1% are possible up to 10,000 counts/s,
where deadtime losses of the order of 3% are encountered.  Since
the number of photoelectrons produced per x-ray photon is low, it
is important to utilize all of the signal provided.  This leads to
amplifier pulse shapes with a time from start to peak amplitude of

about 1 μs and widths at baseline between 2 and 4 μs, as illustrated
in Fig. 4.6.   In modern systems the amplifier pulse width becomes
the dominant system deadtime and is primarily of the paralyzable
type.

Figure 4.7 illustrates the full energy peak detection efficiency for the NaI(Tl) detector.   At low energies, x-rays are absorbed
in the entrance window (0.127 mm beryllium plus 1 μm aluminum) and
the efficiency is low.   At very high energies the crystal becomes
transparent to x-rays and the detection efficiency is reduced.   The
NaI(Tl) detector is useful for energies above 6 keV (wavelengths
shorter than 2.1 Å).   Although the beryllium window is not nearly
as fragile as the thin beryllium windows forming the atmosphere-
vacuum barrier on Si(Li) detectors, it is sensitive to corrosion.
For this reason one should avoid touching the window and should
avoid a corrosive environment.   Corrosion of the beryllium window
can cause a light leak and ruin the detector performance.

Since one seldom observes the energy spectrum from the NaI(Tl)
detector in routine x-ray spectroscopy, it is possible for a fault
to develop in the spectrometer, resulting in data distortion without the flaw being obvious.   Periodic checks of the energy resolution, energy calibration (at low and high count rate), and the
extent of the dark current noise spectrum should catch any malfunc-
tion.   These parameters can be measured manually by setting a narrow
window on the pulse-height selector and recording x-ray intensity
(count rate) versus lower level discriminator setting.   This same
pulse height spectrum can be recorded automatically using an electronic sweep generator or pulse-height-selector scanner to move the
pulse-height-selector window while plotting the ratemeter output on
a strip-chart recorder.

Further information on the scintillation detector and
associated electronics can be found in Ref. 2.

4.2.3   The Gas-Flow Proportional Counter

Figures 4.8(a) and 4.9 illustrate the mechanical structure of the
gas-flow proportional counter.   The detector consists of a cylinder
operated at ground potential with a thin, coaxial center wire.

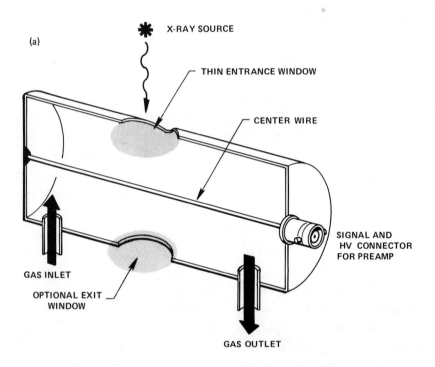

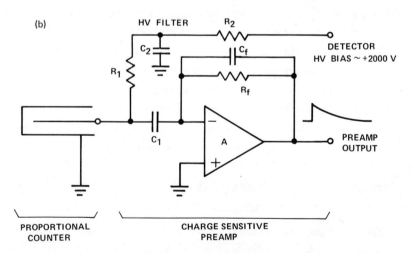

Figure 4.8 The gas-flow proportional counter: (a) a simplified representation of the detector structure; (b) the preamplifier-detector interface. Reprinted by courtesy of EG&G ORTEC.

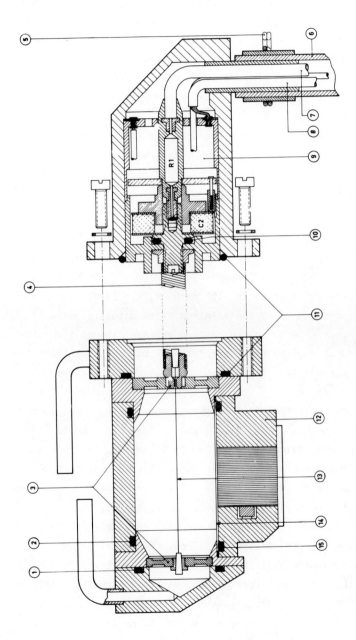

Figure 4.9 A sectional drawing of the typical flow proportional counter used in an x-ray fluorescence spectrometer: 1, 2, 10, 11, and 15 = "0" ring seals; 3 = anode wire supports; 4 = connector spring; 5 = hose clip; 6 = cable assembly; 7 and 8 = signal and high-voltage leads; 9 = preamplifier; 12 = collimator assembly; 13 = anode wire; and 14 = entrance window. Reprinted with permission from Philips Electronic Instruments, Mahwah, New Jersey.

The center wire is insulated from the cylinder and operated at positive high voltage (1000 to 3000 V). A gas atmosphere of controlled
temperature and pressure is provided within the detector via inlet
and exhaust tubes. A constant flow of gas is necessary for proportional counters with very thin entrance windows for soft x-ray
spectroscopy since these windows usually are not completely leak
tight.

An x-ray of energy E enters the detector through the thin
entrance window and interacts with the gas inside via the photoelectric process to produce a cloud of ionization. The average number n of ions produced is given by

$$n = \frac{E}{\varepsilon} \qquad\qquad\qquad (4.7)$$

where $\varepsilon$ is the average energy required to produce one ion. Values
for $\varepsilon$ range from 20 to 28 eV depending on the gas used. For the
typical mixture of 90% argon and 10% methane, $\varepsilon \approx 27$ eV. As shown
in Fig. 4.10, the number of ion-electron pairs actually collected
from the detector depends on the applied voltage. Up to the voltage designated as $V_i$, the field strength is too weak to collect the
ion-electron pairs before they recombine. Above $V_i$ lies the
ionization chamber region. Here the applied voltage causes the
electrons to drift to the center wire and the ions to the cylinder
wall such that the number of ion-electron pairs collected is independent of detector voltage. Equation (4.7), multiplied by the
charge of a single electron, defines the collected charge in this
case. Between the voltages $V_p$ and $V_{1p}$ lies the proportional region.
Here the applied voltage is high enough to accelerate the electrons
on the way to the center wire such that they collide with other
atoms and cause further ionization. Thus, the signal is amplified
in a process not too dissimilar from the scheme employed in a
photomultiplier. The amplitude of the resulting signal is still
proportional to the incident x-ray photon energy. The range from
$V_p$ to $V_{1p}$ is where the proportional counter is normally operated.
Above $V_{1p}$ the strict proportionality is lost; first in the limited
proportional region where x-rays of different energies still pro-

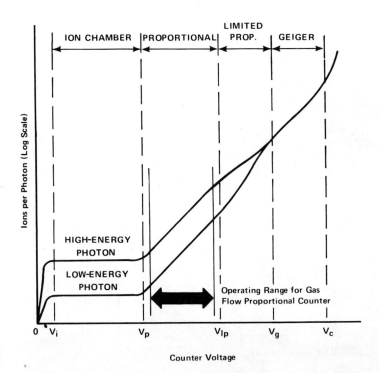

Figure 4.10  The effect of counter voltage on the number of ions collected.  A gas-flow proportional counter is operated in the proportional region between $V_p$ and $V_{lp}$.  This graph should not be confused with the operating plateau of a Geiger counter, which is a plot of counting rate against detector voltage.  Adapted from Ref. 3 and reprinted by courtesy of EG&G ORTEC.

duce different output pulse heights, and then in the Geiger region where all x-rays yield the same pulse height.  Above the Geiger region, a sustained discharge occurs.

In the proportional region, the average number of ion-electron pairs collected is given by

$$n' = An = A \frac{E}{\epsilon} \qquad (4.8)$$

where A is the average gas gain.  The electrons are collected at the center wire in a very short time, typically of the order of a few hundred nanoseconds, while the positive ions take several hundreds of microseconds to drift to the outer wall.  Only a minor portion of the signal from the proportional counter results from

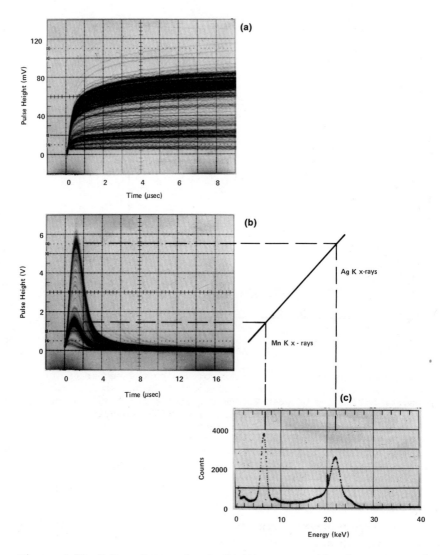

Figure 4.11  Pulse shapes observed with a proportional counter and
spectrometer electronics.  The detector is responding to Ag and
Mn K lines from radioisotopes.  (a) The charge-sensitive preamplifier output.  (b) The output from a semigaussian shaping amplifier
with a 1/2 μsec shaping time constant.  (a) and (b) are multiple
traces from an oscilloscope.  (c) The energy (pulse-height) spectrum
obtained by analyzing the amplifier output on a multichannel pulse
height analyzer.  Reprinted by courtesy of EG&G ORTEC.

collection of the electrons.  The major portion of the output pulse
is formed by the motion of the positive ions.  Consequently, the
integrated charge pulse rises rapidly at first as the positive
charge moves in the strong field near the center wire, then more
and more slowly as the charge drifts into the weaker field regions
near the wall [see Fig.  4.11(a)].  To avoid extremely long pulse
widths, the filters in the shaping amplifier are designed to select
only the fast rising portion of the positive ion pulse.  To minimize electronic noise contribution, a low-noise charge-sensitive
preamplifier collects the induced charge from the center wire con-
verting the signal into a voltage pulse on a small capacitor (see
Fig. 4.13).

Contributions to the detector FWHM energy resolution $\Gamma$ are
defined by Eq. (4.9).

$$\Gamma = [(2.35 \ \sqrt{\epsilon FE})^2 + \Gamma_A^2 + \Gamma_{noise}^2 + \Gamma_p^2]^{1/2} \qquad (4.9)$$

The first term is the broadening due to ionization statistics.  The
Fano factor F lies between 0.09 to 0.23 depending on the detector
gas.  Statistical fluctuations in the gas gain contribute through
the second term $\Gamma_A$.  Electronic noise from the preamplifier is
accounted for in the term $\Gamma_{noise}$, and the term $\Gamma_p$ covers fluctuations due to variations in the counter design parameters over the
dimensions of the counter.  For cases where $\Gamma_{noise}$ can be neglected,
the energy resolution is typically described by

$$\Gamma = kE^{1/2} \qquad (4.10)$$

with $k = 0.35 \ (keV)^{1/2}$ in a good detector.  Resolution can be sensitive to detector voltage [3,4].  Therefore, a plot of $\Gamma$ versus
applied voltage should be generated to aid in choosing the best
compromise in high-voltage setting.  Figure 4.11 illustrates the
proportional counter energy resolution.

A commonly used filter for the shaping amplifier is the semigaussian filter with a 0.5 µs time constant.  If positive ion collection times were infinitesimal, this filter would produce a

pulse approximately 4 µs wide near the baseline. Unfortunately, the long positive ion collection times degrade this width to about 9 µs by causing a slowly decaying tail on the pulse (see Fig. 4.11). Shorter shaping time constants result in a less than proportionate decrease in the length of this tail, and can degenerate the resolution by cutting into the rising portion of the detector pulse while simultaneously increasing the preamplifier noise contribution. Deadtimes in the proportional counter system range from 1 to 9 µs, depending on amplifier design and the pulse-height-selector setting, and are primarily of the paralyzable type.

The chief limitation on counting rate for the proportional counter is the attenuation of the gas gain that shows up at high counting rates [3,5-7]. This phenomenon is most noticeable as a reduction in output pulse height at high counting rates. Fundamentally, it is caused by the electrostatic shielding provided by the positive ions as they slowly drift to the wall of the counter chamber. If the positive ion density is high, it reduces the effective voltage near the center wire, thus reducing the gas gain [5]. A dirty center wire or a center wire with a rough surface can produce similar, and often more pronounced, results [3,6,7]. Consequently, it is important to maintain a clean and smooth center wire through the use of clean gas and periodic service [6]. If the preamplifier is noisy, and high gas gains must be used to minimize the noise contribution; then the gain of the system will be observed to attenuate even at relatively low counting rates. A simple solution to this problem is to employ a low-noise preamplifier to permit utilization of lower gas gains [3,5]. With this approach, counting rates up to 70,000 counts per second produce negligible gain attenuation, and deadtime losses become the dominant limitation. Deadtime losses can be as high as 10% at rates of 50,000 counts per second. Here again, a good baseline restorer is required in the amplifier, along with dc coupling to the pulse height selector to preserve stability of the pulse-height-selector window calibration under varying counting rates. Choice of the best detector high-voltage setting involves studying the system charac-

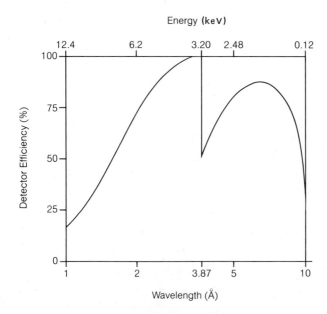

Figure 4.12  The proportional counter detection efficiency.

teristics with respect to energy resolution, preamplifier noise
contribution, and gain sensitivity to counting rate, all as a function of high voltage.  The operating voltage should be chosen to
yield optimum energy resolution while minimizing gas gain attenuation with counting rate.  Once this voltage is determined for the
energy range of interest, further gain calibration adjustments
should be made with the shaping amplifier gain controls.  It is
useful to check the above parameters periodically to verify that
the spectrometer performance is still dependable.  Other parameters
that affect the gain stability, and to some extent the detection
efficiency, are the temperature and pressure of the counter gas.
Gas-density stabilizers are often used to minimize this problem.

     Figure 4.12 shows typical detection efficiency as a function
of wavelength.  Efficiency attenuation at long wavelengths is controlled by entrance window materials and varies depending on window
design.  At short wavelengths, gas type, gas density, and thickness
of the sensitive volume determine the transparency of the detector
to x-rays.

Several review articles on the x-ray proportional counter have been written [3,6,8]. They are highly recommended for further reading. Reference 2 may be consulted for further details on the associated electronic instrumentation.

### 4.2.4 Escape Peaks

Although escape peaks do not appear directly in the wavelength ($2\theta$) spectrum with wavelength-dispersive spectrometers, their presence in the pulse height spectra produced by the proportional counter and the scintillation detector can cause some rather subtle, but significant, distortions of the analytical data.

Consider the case of a proportional counter containing a 90% argon and 10% methane mixture. X-ray photons having a wavelength shorter than 3.871 Å are detected primarily by ionization of the K shell of an argon atom. The critical wavelength 3.871 Å corresponds to the binding energy $\phi$ of the electron in the argon K shell; i.e., $\phi = 3.203$ keV. In this photoelectric interaction the initial x-ray photon disappears and the ejected K-shell electron carries off the energy in excess of the K-shell binding energy. That is, the energy of the ejected photoelectron is

$$E_e = E - \phi \tag{4.11}$$

where E is the initial x-ray photon energy. As the photoelectron moves through the gas it loses its energy by creating further ionization, mainly by interactions with valence electrons. On the other hand, the vacancy in the K shell of the initially excited argon atom is filled by outer-shell electrons accompanied by the emission of either K x-rays or Auger electrons. These emissions account for the remaining energy $\phi$. If they are totally absorbed within the active detector volume, then the entire energy of the initial x-ray photon is converted to a collected charge

$$Q = A \frac{E}{\varepsilon} q_e = A \frac{E_e + \phi}{\varepsilon} q_e \tag{4.12}$$

where $q_e = 1.6 \times 10^{-19}$ C is the charge on an electron.  The resulting detector output pulse occurs at the proper pulse height for the
x-ray of energy E.

In some situations the entire photon energy E is not absorbed
within the active detector volume.  Where the initial photoelectric
interaction takes place near the detector entrance window there is
a high probability that the K x-ray emitted by the excited argon
atom will escape the sensitive detector volume.  Consequently, the
energy deposited in the counter will be

$$E' = E - E_K \tag{4.13}$$

where $E_K \approx 2.96$ keV is the energy of the argon K x-ray.  This event
will yield a peak in the pulse-height spectrum 2.96 keV lower in
energy than the true detected photon energy.  This lower energy peak
is termed the *escape peak*.  Two methods of naming the peak are in
common use.  The first identifies the escape peak by the lost x-ray;
i.e., *argon escape peak*.  The second method identifies the escape
peak by the parent peak of energy E from which it was derived; i.e.,
*Mn Kα escape peak*.  The method in use is usually obvious from the
text.

An identical process occurs in the NaI(Tl) detector, except
that it is the escape of the iodine x-ray which is important.  This
is most noticeable with the iodine K x-rays.  The K absorption edge
occurs at $\phi = 33.164$ keV.  Consequently, x-ray photons with wavelengths less than 0.3738 Å will yield an escape peak in the pulse
height spectrum approximately 29 keV lower in energy than the true
photon energy.

The probability of escape is greater when the decaying atom
is close to the detector entrance window.  Therefore, the intensity
of the escape peak relative to the parent peak is highest for detected photons whose energies are just above the argon K
absorption-edge energy in the proportional counter, and the iodine
K absorption-edge energy in the NaI(Tl) detector.  Higher energy
photons penetrate more deeply into the detector resulting in weaker
escape peaks.

Escape peak positions play an important role in setting the pulse-height-selector window, as discussed in Sec. 4.2.7. There is also a rather special case where escape peak interference can occur. Consider a situation where the crystal spectrometer is set for a wavelength $\lambda$, which is within 10% of the wavelength of the escaping x-ray (i.e., Ar K$\alpha$ or I K$\alpha$ x-ray). If there is a characteristic x-ray emitted by the specimen at half this wavelength, $\lambda/2$, the spectrometer will transmit it to the detector. The escape peak from this higher energy x-ray will lie at nearly the same energy as the first-order diffraction peak from the desired line. Consequently, the first-order peak intensity will be distorted. It is not possible to remove this type of interference with the pulse height selector. For example, suppose the wavelength 4.204 $\overset{\circ}{A}$, corresponding to an energy of 2.949 keV, has been chosen as a background measurement point to estimate the background under a weak potassium K$\alpha$ line in a specimen containing a high concentration of manganese. Second-order diffraction for this spectrometer setting occurs at a wavelength of 2.102 $\overset{\circ}{A}$ or an energy of 5.898 keV. This is exactly the wavelength of the Mn K$\alpha_1$ line. Consequently, a strong second-order peak from the Mn K$\alpha_1$ line will be observed in the proportional counter pulse-height spectrum. The escape peak from the Mn K$\alpha_1$ line will occur at a pulse height corresponding to 2.94 keV or 4.22 $\overset{\circ}{A}$, and the proportional counter will be unable to resolve it from the first-order diffraction peak at 2.949 keV. Consequently, the background measured for the K K$\alpha$ line will be in error and will show a strong correlation with the Mn concentration in the specimen.

## 4.2.5  Types of Preamplifiers

Figure 4.13 illustrates the two general classes of preamplifiers used with proportional counters and scintillation detectors [2]. With both the voltage-sensitive and the charge-sensitive preamplifiers, the positive detector high-voltage bias is introduced

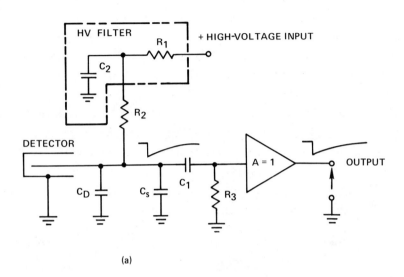

(a)

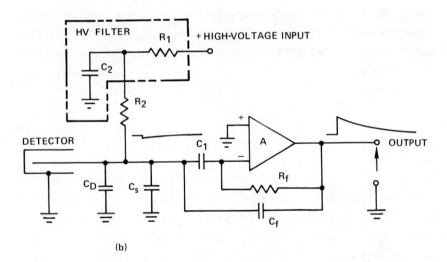

(b)

Figure 4.13  Types of preamplifiers used with proportional counters and scintillation detectors:  (a) the voltage-sensitive preamplifier; (b) the charge-sensitive preamplifier.  Reprinted by courtesy of EG&G ORTEC.

through a filter network to remove power supply ripple and any interference picked up on the cable between the preamplifier and high-voltage supply.  The filter consists of resistor $R_1$ and capacitor $C_2$, with resistor $R_2$ buffering the detector output from the filter network.  Capacitor $C_1$ blocks the detector high-voltage bias from entering the preamplifier.

The voltage-sensitive preamplifier consists of a simple buffer amplifier with a gain A = 1.  The amplifier has a very high input impedance and a low output impedance suitable for driving the co-axial cable to the shaping amplifier.  The charge pulse produced by the detector at time t = 0 is collected on the parallel combination of $C_D$ and $C_S$ to produce a voltage pulse of height

$$V_1 = \frac{Q}{C_D + C_S} \qquad\qquad\qquad\qquad (4.14)$$

where Q is the charge in the pulse, $C_D$ is the detector capacitance, and $C_S$ is the stray capacitance associated with the preamplifier input and the interconnection between the detector and preamplifier. The decay back to baseline for this negative pulse is approximately described by

$$V_0 = \frac{Q}{C_D + C_S} \exp\left[\frac{-t}{R_3}(C_D + C_S)\right] \qquad t \geq 0 \qquad (4.15)$$

for the typical case where $R_2$ is large compared to $R_3$, and $C_1$ is large compared with either $C_S$ or $C_D$.  Since the amplifier gain is unity, Eq. (4.15) represents the shape of the output pulse when the detector charge collection time is insignificant.  In practice, the output pulse will have a noticeable risetime due to the charge collection time of the detector.  Note that the output pulse height is inversely proportional to ($C_D + C_S$).  Any variation in $C_S$ will cause a change in the output pulse height.  Thus, the output pulse height is sensitive to movement of the interconnecting cable between the detector and preamplifier.  If $C_D + C_S$ is large, a small output signal will result.  In older equipment the A = 1 amplifier is typically a vacuum tube connected in the cathode follower

configuration, and its gain is usually slightly less than one. In
transistorized versions, a more sophisticated circuit containing
two or more transistors is commonly used.

The charge-sensitive preamplifier is often employed to eliminate the sensitivity to changes in the stray capacity, and to achieve
larger output pulse heights. It employs an inverting amplifier
with a large open loop gain A. The open loop gain is typically
greater than $10^4$. By virtue of the large open loop gain, almost
all of the detector charge pulse is collected on the feedback
capacitor $C_f$ to produce an output pulse approximately described by

$$V_0 = \frac{Q}{C_f} \exp\left(\frac{-t}{R_f C_f}\right) \qquad \text{for } t \geq 0 \qquad (4.16)$$

where the detector charge collection time is negligible. The
height at $t = 0$ is $Q/C_f$. By choosing $C_f$ to be small compared to
$C_D + C_S$, a much larger output pulse height is achieved than with
the voltage-sensitive preamplifier. In addition, the output pulse
height is made insensitive to $C_D$ and $C_S$ by choosing $AC_f$ to be large
compared to $C_D + C_S$. In essence, this ensures that a negligible
fraction of the charge Q is collected on $C_D + C_S$. The output impedance of the charge-sensitive preamplifier is also very low, and
suitable for driving the cable to the shaping amplifier. Note that
the charge-sensitive preamplifier produces a positive polarity output, whereas a negative pulse is derived from the voltage-sensitive
preamplifier. A more accurate description of the rising portion of
the output pulse shape is given in Figs. 4.6 and 4.11 where the
finite detector charge collection times are included.

4.2.6 The Shaping Amplifier

The purpose of the shaping amplifier is to

1. Increase the amplitude of the preamplifier output signal
2. Limit the duration of the pulse resulting from each detected x-ray photon so that high counting rates can be
   tolerated
3. Optimize the pulse height resolution.

The second and third functions are achieved by including pulse-shaping filters in the amplifier.

Two main classes of amplifiers are encountered in x-ray spectrometry:  the delay-line clipped amplifier and the semigaussian amplifier.  Figure 4.14 illustrates the pulse shaping in the delay-line clipped amplifier for an ideal pulse of zero risetime from the preamplifier.  The pulse is limited in duration or *clipped* by delaying the preamplifier signal by an interval $\tau_d$ and subtracting it from the origional signal.  This is shown in Figure 4.14 at the points (2) the original signal, (3) the delayed signal, and (4) the difference between the prompt and delayed signals.  Since the pre-amplifier signal decays exponentially toward zero pulse height, it is necessary to attenuate the delayed signal to prevent an under-shoot in the clipped signal at point (4).  Thus, the undershoot adjustment potentiometer is used to fine tune the pulse shaping for exact return to baseline on the trailing edge of the clipped pulse. This adjustment is similar in function to the pole-zero cancella-tion network in the semigaussian shaping amplifier.  The delay-line clip, which is sometimes called a *delay-line differentiator,* is effective in suppressing low-frequency noise from the preamplifier. To minimize high-frequency noise contribution from the preampli-fier, a simple RC integrator is often added.  This network slows down the rise and fall times of the pulse, giving the leading and trailing edges exponential shapes,with a characteristic time con-stant $\tau_{int}$ = RC [see point (5), Fig. 4.14].  Depending on the detector application, the integrator time constant is chosen within the range $0 < \tau_{int} \leq \tau_D$.  Several stages of adjustable gain are also added to provide the desired amplification of pulse height. A baseline restorer is added near the output.  Its function is to examine the baseline between pulses and hold the baseline at ground potential.  At very high counting rates the baseline re-storer loses its effectiveness and the baseline shifts below ground potential [9].  This condition is to be avoided since it can shift the desired x-ray pulses out of the pulse-height-selector window.

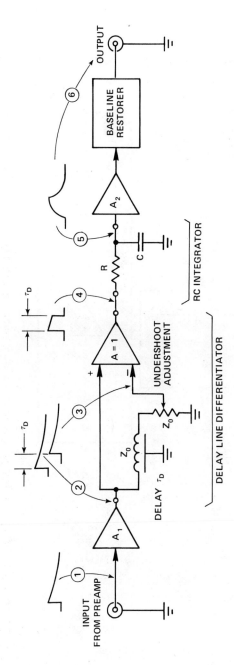

Figure 4.14  Pulse shaping in the delay-line clipped amplifier.  Reprinted by courtesy of EG&G ORTEC.

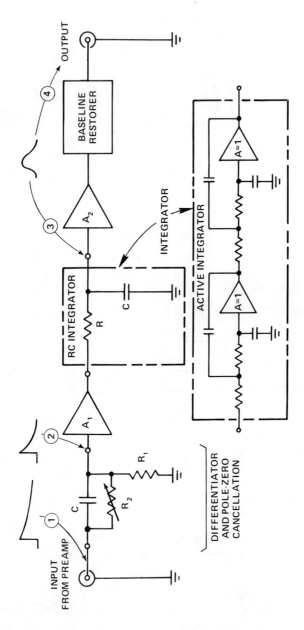

Figure 4.15  Pulse shaping in the semigaussian shaping amplifier.  Reprinted by courtesy of EG&G ORTEC.

Pulse amplitudes at the amplifier output are typically in the 0-
to 10-V range on transistor amplifiers.

The semigaussian shaping amplifier is illustrated in Fig. 4.15.
The CR differentiator at the input is used to reject low-frequency
noise. For preamplifier pulses with negligible risetime it results
in an exponentially decaying pulse [point (2), Fig. 4.15] with a
characteristic exponential decay time constant $\tau = R_1 R_2 C / (R_1 + R_2) \approx$
$R_1 C$. The addition of adjustable resistor $R_2$ provides for removal
of the pulse undershoot below baseline which otherwise would result
from the finite exponential decay of the preamplifier pulse. This
adjustment is normally accessible from the amplifier front panel
and is labeled *pole-zero cancellation adjustment*. Anytime either
the preamplifier or shaping amplifier is replaced or serviced, this
adjustment must be tuned to minimize the time taken for the amplifier output pulse to return to baseline without undershoot. Proper
pole-zero cancellation adjustment is critical to good amplifier performance at high counting rates. As in the delay-line clipped
amplifier, high-frequency noise is suppressed with an integrator
filter, which results in a pulse with a slow risetime [point (3),
Fig. 4.15]. In Fig. 4.15 this filter has been illustrated by a
simple RC network with exponential time constant $\tau = RC$. However,
in practical amplifiers this simple network is replaced with a
sophisticated and more effective active integrator filter as illustrated in the inset in Fig. 4.15. The time constant of the integrator filter is normally set equal to the differentiator time
constant, and switch selection of the common time constant is sometimes provided. As in the delay-line clipped amplifier, several
stages of selectable gain and a good quality baseline restorer
complete the amplifier.

The semigaussian shaping amplifier is the better choice where
the optimum signal-to-noise ratio is necessary. Where minimum
pulse width is more important than minimum noise, the delay-line
clipped amplifier can be used to advantage. Consequently, the
delay line clipped amplifier is often used on wavelength-dispersive
spectrometers where the signal-to-noise ratio can be controlled to

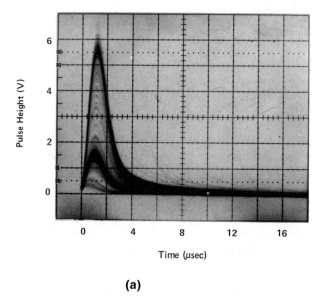

**(a)**

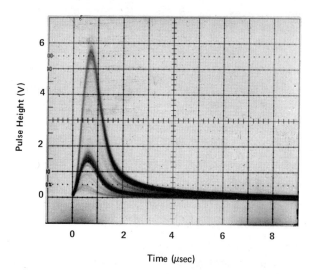

**(b)**

Figure 4.16  Amplifier output pulse shapes with a proportional
counter and semigaussian shaping amplifier.  The x-rays are the Ag
and Mn K lines from radioisotopes.  (a) 1/2 µsec shaping time constant.  (b) 1/4 µsec shaping time constant.  Reprinted by courtesy
of EG&G ORTEC.

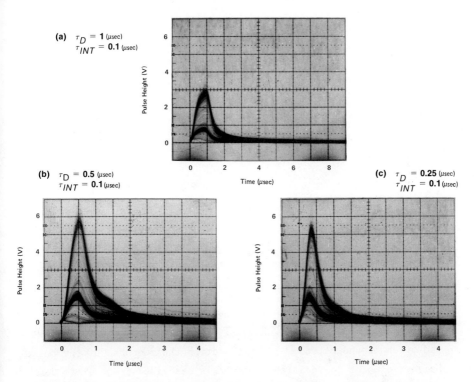

Figure 4.17  Amplifier output pulse shapes with a proportional counter and a delay-line clipped shaping amplifier.  [109]Cd and [55]Fe radioisotopes are the sources of the Ag and Mn K x-rays, respectively.  Delay-line clipping times are (a) 1 μsec, (b) 1/2 μsec, and (c) 1/4 μsec.  The integration time constant is 0.1 μsec in all cases.  Reprinted by courtesy of EG&G ORTEC.

a large extent by the internal gains provided by the scintillation detector and proportional counter.

Figures 4.16 and 4.17 show the pulse shapes obtained for silver and manganese K x-rays with a proportional counter and various choices of shaping time constants with the two types of amplifiers. Note the persistance of the long charge collection tail even with short shaping time constants.

### 4.2.7  The Pulse Height Selector

Although the basic function of the pulse height selector is simple to comprehend, proper techniques for its use are much more subtle

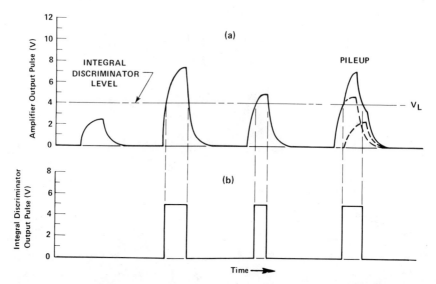

Figure 4.18  Pulse responses with the pulse-height selector in the integral discriminator mode:  (a) pulses from the delay-line amplifier output; (b) the output logic pulses from the pulse-height selector.  Reprinted by courtesy of EG&G ORTEC.

and often misunderstood.  In this section the response of the pulse-height selector to idealized spectra is described in order to demonstrate the various methods of use.

Figure 4.18(b) shows the output pulses from the pulse-height selector in response to some typical amplifier pulses at its input [Fig. 4.18(a)].  The pulse-height selector is being operated in the integral discriminator mode.  The integral discriminator threshold has been chosen at a pulse amplitude of $V_L$ = 4 V.  All amplifier pulses which do not exceed this 4-V level produce no output pulse.  For amplifier pulses which do exceed the 4-V threshold, an output logic pulse is generated.  This logic pulse is typically 5 V in amplitude and lasts for the duration of the amplifier pulse above the integral discriminator threshold.  Although not shown here, the logic pulse is usually regenerated to a standard width of 0.5 μs by an additional circuit before appearing at the pulse-height-selector output.  Thus, the integral discriminator can be used to suppress the counting of lower pulse heights in favor of some higher pulse height.  In fact, the most common use is to prevent counting of the low-amplitude fluctuations caused by preamplifier noise. The first three pulses in Fig. 4.18(a) result from single x-ray photons arriving at the detector. The last pulse is a pileup event where two x-ray photons arrive at the detector at almost the same time (dashed lines) and sum to produce a composite pulse (solid line). Note that the integral discriminator counts only one of the two x-ray photons. A further complication can arise when both pulses have individual pulse heights below the discriminator threshold, but pileup and sum to a composite pulse amplitude which exceeds the discriminator level. Separately they would not be counted, but when they arrive together the discriminator counts them as a valid event satisfying the amplitude test. Such pileup events play a significant role in the deadtime losses which will be examined in a later section.

The question arises as to how to set the integral discriminator to count the desired pulses and reject the unwanted pulses. The answer may not be obvious if the relationship between the amplifier pulse height spectrum and the integral discriminator dial reading is not known. The standard solution to this problem is demonstrated in Fig. 4.19 with an idealized amplifier output pulse-height spectrum containing preamplifier noise at low pulse heights, and two x-ray peaks superimposed on a flat background [Fig. 4.19(a)]. Without an oscilloscope or multichannel pulse-height analyzer this spectrum would not be observed by the instrument operator. However, it can be deduced from the data measured in Fig. 4.19(b). Figure 4.19(b) is generated by measuring the output counting rate of the pulse-height selector in the integral discriminator mode as a function of the integral discriminator setting. The discriminator setting, in volts, can be read from the lower level discriminator control dial. The pulse-height-selector output counting rate is measured with the counting ratemeter or via the scaler/timer combination as shown in Fig. 4.3.

For any discriminator setting $V_L$, the output counting rate of the pulse height selector is equal to the total counting rate of all pulses in the spectrum whose amplitudes exceed $V_L$. For example,

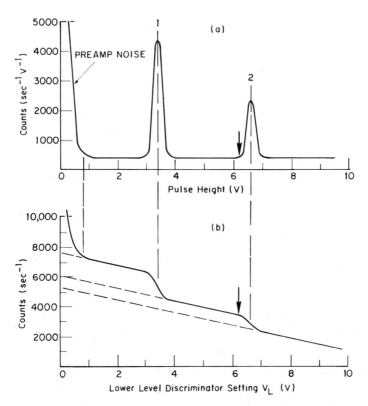

Figure 4.19  The result of scanning the integral discriminator
through a pulse-height spectrum:   (a) the pulse-height spectrum at
the amplifier output; (b) the counting rate at the integral discriminator output as a function of the discriminator level voltage.
Reprinted by courtesy of EG&G ORTEC.

with $V_L$ = 0 V, all pulses and the portion of the preamplifier noise
above 0 V exceed the discriminator threshold and are counted.  This
counting rate will be extremely high mainly as a result of the high-
frequency content of the preamplifier noise.  As the threshold is
raised to about 1 V, the counting rate drops abruptly due to the
discriminator level being raised above the amplitude of the preamplifier noise.  This is usually the most dramatic change in counting rate observed as the discriminator is swept through the spectrum.  When the discriminator level is moved from 1 to 3 V, the
drop in counting rate follows a rather gentle slope since portions

of the flat background are being rejected.  In moving from 3 to
3.8 V, the counting rate drops abruptly as the discriminator is
raised above the pulse heights in peak number 1.  Between peaks 1
and 2 the decrease in counting rate again follows a gentle slope
characteristic of the flat background region.  An abrupt decrease
in counting rate is observed once more as the discriminator passes
through peak number 2.  Above peak number 2 the gentle slope
caused by the flat background is repeated.  Given the data measured
in Fig. 4.19(b), the spectrum in Fig. 4.19(a) can be computed by
taking the mathematical derivative of the curve in Fig. 4.19(b)
and multiplying by minus 1.  This relationship is the reason why
the curve in Fig. 4.19(b) is often called the *integral spectrum* and
hence, the operating mode is called the *integral discriminator mode*.

Once this relationship is understood it is no longer necessary
to actually compute the spectrum of Fig. 4.19(a).  The characteristic features of Fig. 4.19(b) can be directly interpreted to correspond to peaks, slowly varying background, or preamplifier noise.
Furthermore, correct choice of the discriminator setting for elemental analysis can be made directly from the integral spectrum.
For example, suppose it is desirable to count only peak number 2
while rejecting preamplifier noise, peak number 1, and as much of
the background as possible.  A discriminator setting of 6.2 V, as
indicated by the arrow, is the highest setting which accepts all
of the pulses in peak number 2.  If one is concerned that the spectrum may shift to lower pulse heights at high counting rates, a
safer discriminator setting would be approximately 5.5 V.  This
margin of safety against gain attenuation at high counting rates
would be achieved at the expense of including more unwanted background counts.

The more common use of the integral discriminator mode is to
count all pulse heights above the preamplifier noise level while
performing a wavelength scan.  In this case, Fig. 4.19(b) indicates
that the proper discriminator setting is approximately 1.0 V, because below this level the counting rate rises rapidly as a result

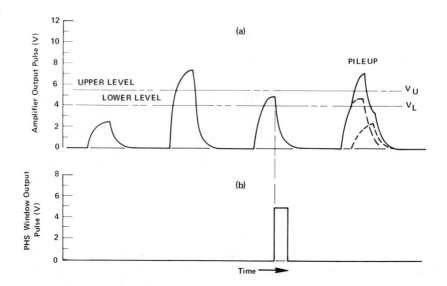

Figure 4.20   The pulse-height selector response in the window mode:
(a) typical pulses from the delay-line amplifier output; (b) the
corresponding pulse-height selector logic-pulse output.   Reprinted
by courtesy of EG&G ORTEC.

of preamplifier noise.

Figure 4.19(b) can be generated by manually moving the lower
level discriminator setting and reading the counting rate at each
setting.   In systems which include a pulse-height-selector scanner
(Fig. 4.3), the discriminator scan can be executed electronically
while a strip-chart recorder plots the integral spectrum from the
ratemeter output.

Figure 4.20 illustrates the window mode of operation.   This
mode is sometimes called the *differential mode* because scanning the
pulse-height spectrum with a very narrow window width yields a good
approximation to the mathematical derivative of the integral
spectrum.

A second discriminator level $V_U$ has been added to the lower
level discriminator $V_L$ normally used in the integral mode.   In
the window mode the pulse-height selector produces a standard output logic pulse only if the amplifier pulse amplitude exceeds the

lower level $V_L$ and does not exceed the upper level $V_U$. Only the
third pulse in Fig. 4.20(a) satisfies this condition. As described
previously, the fourth pulse is a composite pileup pulse. Note
that the earlier pulse (dashed line) contained in the composite
pulse would have fallen within the window if it occurred alone.
However, the pileup caused by the smaller, later pulse (dashed
line) forces the composite pulse height above the upper level, and
neither pulse is counted. Consequently, pulse pileup can cause a
more severe deadtime loss in the window mode than in the integral
mode. In the integral mode the first pulse of the pileup was
counted and the second pulse was missed. However, in the window
mode, pileup caused both pulses to be missed.

Once the amplifier pulse exceeds the lower level, the pulse
height selector must determine when to test the upper level discriminator to see whether it has been exceeded. For this purpose
the upper level discriminator drives a two-state memory which remem-
bers when the upper level has been crossed. Two methods are commonly used to determine the proper logic testing time. The
simplest method is to wait until the signal begins to fall back
through the lower level. At this point the upper level memory is
tested and a standard 0.5 µs wide, 5-V high output logic pulse is
generated if the upper level has not been crossed. Subsequently,
the memory is cleared in preparation for the next pulse. A method
which results in somewhat less deadtime is to employ an additional
circuit which senses the time of peak amplitude of the amplifier
pulse. This is referred to as a *peak detector*. The conditions of
the upper level memory and lower level discriminator are checked
at the peak detection time, and an output logic pulse is generated
if the lower level has been exceeded but the upper level has not.
With either method, the output logic pulse can contribute to the
deadtime, depending on whether its length is long compared to the
duration of the amplifier pulse's trailing edge. Reference 10
elaborates on the types of deadtimes encountered in proportional
counter systems using pulse height selectors.

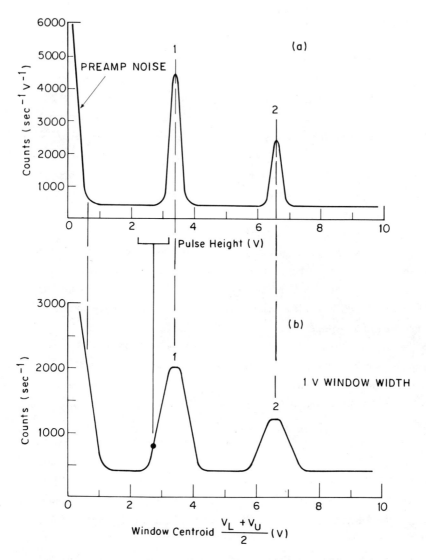

Figure 4.21  The result of scanning the pulse-height selector
through the pulse-height spectrum while in the window mode:
(a) the pulse-height spectrum at the amplifier output; (b) the
counting rate at the pulse-height selector output as a function of
the window centroid voltage.  The window width is 1 V.  Reprinted
by courtesy of EG&G ORTEC.

Figure 4.21(b) shows the result of scanning the pulse height selector through the amplifier pulse-height spectrum shown in Fig. 4.21(a) while using a moderately wide (1 V) window. The output counting rate of the pulse height selector is plotted against the window centroid; i.e., $(V_L + V_U)/2$. Note how the moderately wide window broadens the spectral features and tends to flatten the tops of the peaks. The source of the broadening is most readily described by choosing a 2.7-V setting of the window centroid. In the original spectrum this pulse amplitude lies in a flat background region. However, the window extends from 2.2 to 3.2 V and counts all pulses in that range. This includes part of peak number 1. Consequently, the counting rate of peak number 1 is spread out into adjacent background regions in Fig. 4.21(b).

Frequently, it is desirable to use the window mode in conjunction with a pulse-height-selector scanner in order to measure the spectrum of Fig. 4.21(a) on a strip-chart recorder. If the resolution of Fig. 4.21(a) is to be faithfully reproduced in Fig. 4.21(b), the window width must be small compared to the peak width in Fig. 4.21(a). That is, the window width must be small compared to the pulse-height resolution inherent in the amplifier pulse-height spectrum. A good practical choice is to make the window width less than one-fourth of the width of the narrowest peak in the amplifier pulse-height spectrum. The width being taken as the width of the peak at half the peak amplitude above background.

The second use of the window mode is to select one peak for counting while rejecting all other content of the spectrum. This technique is used to obtain optimum statistical precision when measuring the intensity of a single element. When the peak-to-background ratio is low in the pulse height spectrum and it becomes important to minimize the statistical error arising from the background contribution, the window should be centered on the peak with a window width equal to 1.17 times the full width of the peak at half its height above background. This setting provides the best statistical precision as shown in the discussion for energy

dispersive systems, but it is sensitive to peak shifts caused by
count rate dependent gain changes.  Therefore, the more common
practice is to use a window width which encompasses all of the peak
and often some of the adjacent background.  This choice minimizes
sensitivity to gain changes and sacrifices statistical precision
only in the rate case where the peak-to-background ratio is low in
the pulse-height spectrum.  Where the escape peak is not completely
resolved from the low-energy side of the main peak of interest, the
window must be widened and set asymmetrically to include both the
main peak and its associated escape peak.

The use of the pulse-height-selector window mode to isolate a
single peak in the pulse-height spectrum is most effective in eliminating higher order diffraction interferences as demonstrated in
Fig. 4.22.  The spectrum is a wavelength scan in the vicinity of
the Si Kα line at 7.125 $\overset{\circ}{A}$ (1.740 keV) for a specimen containing
silicon, iron, and zinc.  In the upper spectrum, the pulse-height
selector has been operated in the integral mode with the threshold
set just above the noise level.  In the neighborhood of the Si Kα
peak, higher order diffraction peaks from the Zn Kα and the Fe Kβ
lines are observed.  Fifth-order diffraction of the 1.437 $\overset{\circ}{A}$
(8.630 keV) Zn Kα line occurs where first-order diffraction from a
7.183-$\overset{\circ}{A}$ x-ray would be observed.  Similarly, fourth-order diffraction of the Fe Kβ line ($\lambda$ = 1.757 $\overset{\circ}{A}$, E = 7.057 keV) occurs at a
wavelength corresponding to first-order diffraction of a 6.605-$\overset{\circ}{A}$
x-ray.  These two higher order diffraction peaks interfere with the
Si Kα peak.  Fortunately, the pulse heights produced by the three
peaks are distinctly different.  The Si Kα occurs at 1.740 keV, the
Zn Kα at 8.630 keV, and the Fe Kβ at 7.057 keV.  By choosing the
window mode and setting the pulse-height-selector window to count
only the Si Kα at 1.740 keV, the iron and zinc x-rays are rejected.
The result is the removal of the higher order diffraction interferences as demonstrated in the lower spectrum of Fig. 4.22.  Thus,
the pulse-height-selector window mode can be used to avoid higher

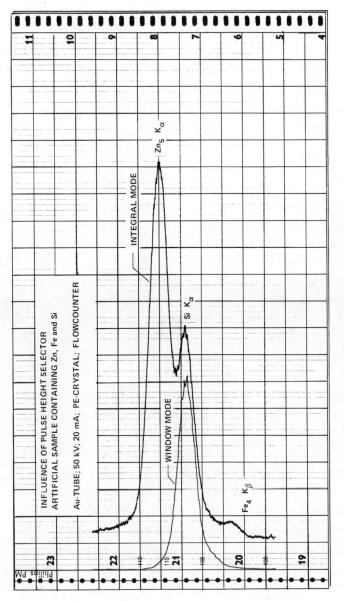

Figure 4.22 The use of the pulse-height selector window mode to reject interference from higher-order diffraction. The upper curve is measured in the integral mode. The lower curve is from the window mode, with the window centered on the Si Kα pulse-height distribution.

order diffraction interferences where the proportional counter
resolution is adequate to separate the different x-ray energies.

Occasionally, maintenance service is performed on the components which affect the gain calibration of the detector, preamplifier, and amplifier system, and the operator must reestablish the
normal gain calibration. The easiest method of adjusting the gain
to restore calibration is to set the crystal spectrometer on a line
of known wavelength and observe the amplifier output on an oscilloscope. The gain is adjusted until the proper amplifier output pulse
height is obtained for the chosen x-ray line. In many laboratories
an appropriate oscilloscope for this task is not available. Consequently, it becomes necessary to use the pulse height selector
and ratemeter for this adjustment.

The method follows the illustrations in Fig. 4.23, where it
has been assumed that the chosen x-ray line is to be centered at a
4-V pulse height. Figure 4.23(d) is plotted by monitoring the
counting ratemeter reading as a function of the gain setting, with
the integral mode selected, and the lower level discriminator set
at 4 V. In Fig. 4.23(a), (b), and (c) the amplifier output pulse
height spectrum is shown for three different gain settings. For
gains from 25 to 85 the counting rate increases smoothly as more
background counts are extended above the 4-V threshold [per Fig.
4.23(a)]. As the gain is increased past 100, the x-ray peak of
interest passes through the 4-V level [see Fig. 4.23(b)], and the
pulse-height selector output counting rate rises rather abruptly.
At higher gains the counting rate increases smoothly and more
slowly as a result of more of the flat background distribution
surpassing the 4-V threshold [Fig. 4.23(c)]. Finally, at gains
approaching 500, the preamplifier noise begins to cross the integral discriminator threshold, and the counting rate rises abruptly.
To center the peak of interest at a 4-V pulse height, the midpoint
of the abrupt rise in counting rate near the gain of 100 should be
chosen. This will provide a good first approximation to the
proper setting, which may be totally adequate in many cases. To
achieve a more accurate setting, the pulse-height selector is

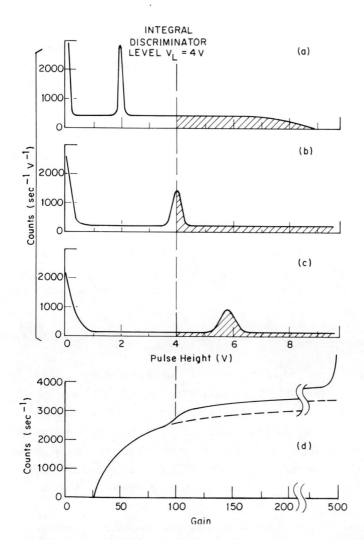

Figure 4.23  Calibration of the detector and amplifier gain using the pulse-height-selector.  The amplifier output pulse-height spectra are shown at gains of (a) 50, (b) 100, and (c) 150.  The ratemeter counting rate as a function of gain setting is shown in (d) for a 4-V integral discriminator threshold.  Reprinted by courtesy of EG&G ORTEC.

subsequently set to the window mode.  The window width is chosen to be less than the x-ray peak width, and the window is centered at 4-V.  Fine adjustments of the gain are made about the gain of 100

until the ratemeter reading is maximized.  This centers the peak on
the window and provides an accurate calibration.  The calibration
can be confirmed by carrying out the spectrum scan per the method
of Fig. 4.21 with a window width which is narrow compared to the
peak width.

### 4.2.8  The Sin θ Potentiometer

As discussed in Sec. 4.2.7, use of the pulse-height-selector window
mode can be helpful in eliminating higher order diffraction interferences, and in improving the background rejection.  The removal
of interferences is most appreciated in a wavelength scan where it
simplifies the problem of spectrum interpretation.

For a fixed gain setting, the amplifier output pulse height V,
observed for an x-ray of wavelength $\lambda$, is proportional to the x-ray
energy E.  This relationship can be expressed in terms of Bragg's
law

$$V = \frac{12.4 \ n \ K_1 \ K_2}{2d \ \sin \theta} \tag{4.17}$$

where $K_1$ and $K_2$ are redundant gain calibration constants.  This
shows that the pulse height during a wavelength scan varies inversely with sin θ.  In the absence of some special method for
following this change in pulse height, the operator must choose the
integral mode on the pulse-height selector and a lower level threshold which is below the lowest pulse height to be encountered in
the scan.  Thus, higher order diffraction interferences must be
tolerated.

A method which alleviates this limitation involves inserting
a gain attenuating potentiometer between the preamplifier output
and the amplifier input.  The potentiometer is driven by the crystal
spectrometer goniometer such that it varies the gain constant $K_2$
according to the relationship

$$K_2 = \sin \theta \tag{4.18}$$

during the wavelength scan.  The resulting amplifier output pulse
height

$$V = \frac{12.4n \ K_1 \ \sin \theta}{2d \ \sin \theta}$$

$$= \frac{12.4n \ K_1}{2d} \tag{4.19}$$

is constant.  Thus, a pulse-height-selector window can be set
around this pulse height and left there for the wavelength scan.
This use of the sin $\theta$ potentiometer allows suppression of higher
order diffraction interferences during a wavelength scan.

It should be noted that the relative pulse height resolution
is proportional to $E^{-1/2}$.  Therefore, the window width must be set
wide enough to accommodate the resolution of the lowest energy
(longest wavelength) x-rays covered in the scan.  This setting will
be suboptimal for the shorter wavelengths.  Furthermore, for very
soft x-rays, the relative pulse height resolution of the detector
may be so poor that there is little to be gained by using the window mode.  The method is most useful at medium to short wavelengths.

### 4.2.9  The Counting Ratemeter

Analog counting ratemeters are often used with wavelength dispersive
x-ray spectrometers to record the x-ray spectrum.  The ratemeter
output is plotted on a moving strip-chart recorder as the spectrometer scans through a predetermined range of $2\theta$.  Ratemeters are
also useful as x-ray intensity monitors in on-line process control
fluorescence spectrometers.  In both applications two ratemeter
response characteristics are important:  (1) the response time
and (2) the fluctuations in the output reading due to the random
arrival of x-rays.  The response time determines the ability of
the ratemeter to follow a quick change in counting rate.  Output
reading fluctuations determine the statistical error in measuring a
counting rate.

Figure 4.24 is a simplified diagram of the linear ratemeter.
Digital input pulses provided by the pulse-height-selector trigger
a "one-shot" circuit.  For each input pulse, the one-shot delivers
a short charge pulse containing a fixed charge Q to the capacitor C.

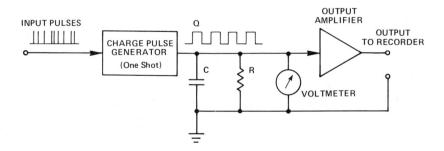

Figure 4.24  A functional diagram of the linear ratemeter.
Reprinted by courtesy of EG&G ORTEC.

Each pulse causes an increase in the voltage across the capacitor,
which leaks off exponentially through resistor R, i.e.,

$$V(t) = \frac{Q}{C} \exp\left(\frac{-t}{RC}\right)$$  (4.20)

where $V(t)$ is the voltage pulse observed on the capacitor at time $t$
for a single input pulse at time $t = 0$.  For a mean, steady-state
counting rate of $I$ counts per second, the average voltage across
the resistor is

$$V = IQR$$  (4.21)

This voltage is read on the voltmeter as a calibrated counting rate
and delivered to a strip-chart recorder through an output buffer
amplifier.  The function of the capacitor is to smooth the signal
developed across the resistor R.  The smoothing effect is described
by the time constant $\tau = RC$.  The larger the time constant, the
smoother the output voltage will be.  This effect is readily observed on the meter readout as the time constant selector switch
is moved from short to long time constants.  At short time constants
the meter needle position will fluctuate with rather large amplitudes.  As the time constant is increased, the amplitude of the
fluctuations will decrease.  Needle jitter is caused by the random
fluctuations in arrival time of the x-ray photons.  By increasing
the ratemeter time constant, the measurement of counting rate is
averaged over a longer time period and the statistical errors are

reduced.  Unfortunately, long time constants not only reduce the
statistical error, but also increase the instrument response time.
It is important to understand the relationship between these two
effects so that spectrometer scanning times and ratemeter time constants can be chosen to produce precise and undistorted spectra.
Evans [11] gives a detailed derivation of the pertinent formulas
which are only summarized here.  He also deals with several other
ratemeter formulas of interest to the x-ray spectroscopist.  The
question of spectrum distortion has been treated quantitatively by
Edwards [12].

The fractional standard deviation of a single instantaneous
reading of the counting ratemeter can be shown to be [11]

$$\frac{\sigma_V}{V} = \frac{1}{\sqrt{2I\tau}}$$                    (4.22)

It is interesting to note that this is the same error as is expected
with a scaler counting for a time interval $2\tau$.  The standard laws
for the reduction of the standard deviation by averaging n independent readings apply only if the readings are separated by a time
large compared to $\tau$ = RC.  For intervals of the order of $\tau$ or less,
the readings are not strictly uncorrelated due to the exponential
decay, and this memory effect must be accounted for [11].

Two response times can be defined:  (1) the risetime $t_r$ and
(2) the equilibrium time $t_e$.  For an initial counting rate $I_1$ and a
final counting rate $I_2$, the *risetime* is defined as the time for the
meter reading to rise from 10% of the distance to the new rate to
90% of the distance to the new rate.  The rise is exponential;
hence, the rise time is given by

$$t_r = 2.2\tau$$                    (4.23)

The risetime is independent of $I_1$ and $I_2$.  This measure of response
time is useful in determining whether the selected time constant
will cause significant smearing of an abrupt change in the counting
rate which represents a feature of interest in the spectrum [12].

Generally speaking, if the change in counting rate takes place over
a time interval t' where t' > 5t$_r$, the distortion will be small.
In choosing scanning rates on the dispersive spectrometer, a good
rule of thumb is to select the scanning rate and ratemeter time
constant so that the time constant is less than 1/10 of the time it
will take to scan across the FWHM of the narrowest peak in the 2θ
spectrum.

The equilibrium time t$_e$ is useful in determining how long to
wait before reading a new steady-state counting rate so that the
indicated value has approached the true value within the error
determined by counting statistics.  The time taken for the ratemeter
output to move from an initial rate I$_1$ to within k standard deviations of the final rate I$_2$ is [11]

$$t_e = \tau \ln \left[ \frac{\sqrt{2I_2\tau}}{k} \frac{(I_2 - I_1)}{I_2} \right] \qquad (4.24)$$

If one standard deviation is chosen as the criterion k = 1.  For a
limit of one probable error, k = 0.6745 [11].

Logarithmic ratemeters are useful for recording spectra where
both large and small peaks are of interest.  Up to five decades of
counting rates can be plotted on the same graph, permitting adequate resolution of peak heights in all decades.  The logarithmic
mode is also useful for routine monitoring since the range from 10
to $10^6$ counts per second is covered without having to touch a counting rate range switch.  Although the capacitor still functions as a
smoothing device, it is no longer possible to identify a unique
time constant.  The resistor in Fig. 4.24 is replaced by a logarithmic device (usually one of the diode junctions in a bipolar transistor) whose resistance is proportional to the logarithm of the
counting rate.  Hence, the time constant now becomes a function of
the measured instantaneous counting rate.  The time constant is
long for low counting rates and short for high counting rates.
Formulas for the response time of the logarithmic ratemeter are
complex.  Details are available in Refs. 13 and 14.  On the other

hand, the fractional standard deviation is approximately constant over the entire range of counting rates [14].  This means that the heaviest smoothing is applied at low counting rates where it is most needed to make a precise counting rate measurement.  Sometimes two responses are offered, a long one for precision (i.e., 5% fractional standard deviation over the entire range), and a short response for quick observations (i.e., 15% fractional standard deviation) [14].

Practical ratemeters deviate from the ideal device performance in several respects.  These include a deadtime caused by the charge pulse generator, nonlinearity in the plot of analog output voltage versus input rate, errors in the meter reading, and sensitivity of the output signal to temperature variations.  All of these deviations are controllable to some extent by the choice of design.

## 4.3   THE ENERGY-DISPERSIVE X-RAY SPECTROMETER

Figure 4.25 illustrates the typical energy-dispersive x-ray spectrometer.  The analysis of the spectrum of x-rays from the specimen is totally accomplished by the Si(Li) detector and its associated electronics.  In this respect the spectrometer functions in a manner similar to the proportional counter system, except with far better energy resolution.  In fact in simple systems, where good energy resolution is not required, a proportional counter sometimes can be substituted for the Si(Li) detector.

The Si(Li) detector diode serves as a solid-state version of the gas-ionization chamber, which is the operating mode of the gas-flow proportional counter when the gas gain is unity (see Fig. 4.10).  When an x-ray photon is stopped in the detector diode a cloud of ionization is generated in the form of electron-hole pairs.  The number of electron-hole pairs created, or in other words, the total electric charge released, is proportional to the energy of the detected photon.  This charge is swept from the detector diode by the high voltage applied across it.  The preamplifier is

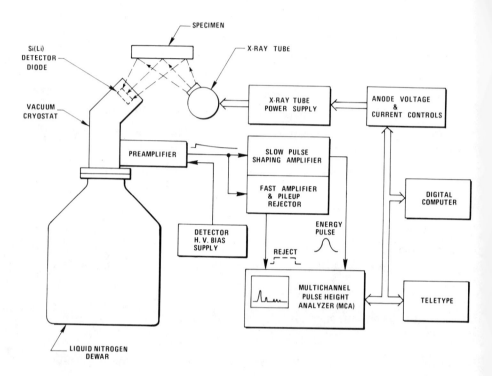

Figure 4.25  A block diagram of the typical energy-dispersive x-ray
fluorescence spectrometer.  The specimen presentation apparatus and
its controls have been omitted from the figure.  Reprinted by
courtesy of EG&G ORTEC.

responsible for collecting this charge on a feedback capacitor to
produce an output pulse whose voltage amplitude is proportional to
the original x-ray photon energy.  To minimize the electronic noise
added to the signal during this process, the Si(Li) detector, along
with the first stage and feedback elements of the preamplifier, is
mounted in a light tight, vacuum cryostat and operated at the
liquid nitrogen boiling temperature (77 K).  As shown in Fig. 4.26,
the x-rays enter the cryostat through a thin beryllium window to
reach the Si(Li) detector.  The beryllium window is typically 7.6
to 13 μm thick.

The signal from the preamplifier is small and has a low
signal-to-noise ratio.  Consequently, the slow pulse-shaping

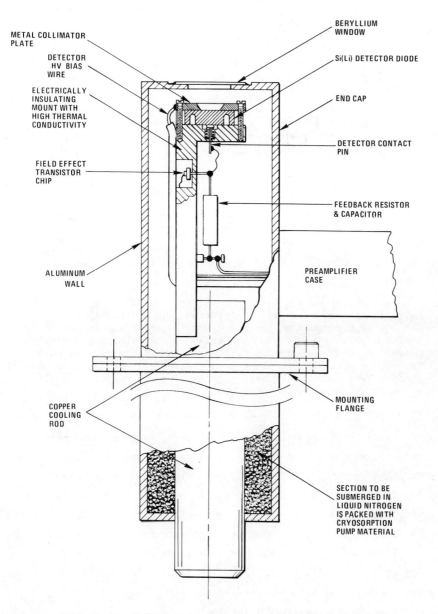

Figure 4.26   A symplified representation of the Si(Li) detector
cryogenic mount for a resistive feedback system.   The length of the
cryostat has been foreshortened in the figure; cryostat length is
typically on the order of 70 cm.   The section above the mounting
flange is the end cap and may vary from a 2.5- to a 6-cm outside
diameter.   Reprinted by courtesy of EG&G ORTEC.

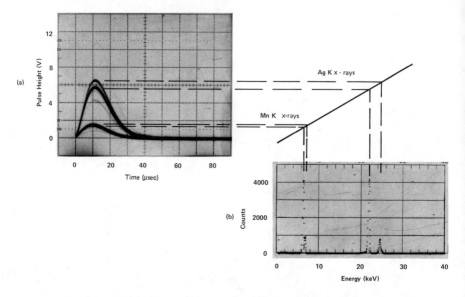

Figure 4.27  The analysis of the shaping amplifier output pulse
heights into the x-ray energy spectrum.  (a)  A multiple-trace
oscilloscope picture of the pulse-shaping amplifier output for a
spectrum containing Mn and Ag K x-rays.  (b)  The analyzed pulse-
height spectrum as viewed on the multichannel analyzer display.
Each dot represents one channel in the analyzer memory.  The channel
numbers have been calibrated in terms of x-ray photon energy.
Reprinted by courtesy of EG&G ORTEC.

amplifier must serve two main functions.  First, it must amplify
the pulse into the 0- to 10-V pulse height range so that pulse
height analysis can be performed.  Second, it applies filters which
suppress extremely low and extremely high frequencies, where the
signal-to-noise ratio is poorest, in order to obtain improved energy
resolution.  As a result the output pulse from the amplifier is a
rather wide pulse (of the order of 40 μs), with a slow rise to peak
amplitude (13 μs) and a slow decay back to ground potential.
Figure 4.27(a) illustrates the pulse shape.  Fortunately, the voltage amplitude of the pulse at the amplifier output is still proportional to the detected x-ray photon energy.

Up to the input of the multichannel pulse-height analyzer the
system has an analog response.  That is, a continuum of pulse amplitudes is possible at the amplifier output.  The purpose of the

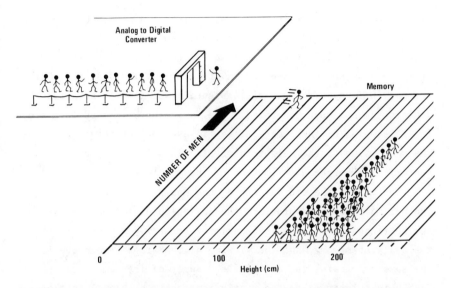

Figure 4.28   An analogy representing the pulse-height sorting function in the multichannel pulse-height analyzer.   Reprinted by courtesy of EG&G ORTEC.

multichannel pulse-height analyzer is to measure the height of each amplifier output pulse and represent this amplitude by an integer number.   This is an analog-to-digital conversion process.   The number of times a pulse of each height has been detected is accumulated in the analyzer memory to form the spectrum of pulse heights.   Subsequently, this information can be displayed as a picture of the analyzed energy spectrum.

The process can be understood more easily by reference to the analogy in Fig. 4.28.   Here, the men in the line arriving at the doorway in the upper left-hand corner of the figure represent the pulses arriving at the multichannel analyzer input.   The task is to measure the spectrum which represents their height distribution. The large area adjacent to the doorway has been laid out in a square.   Viewed from above, the square is a graph, with the height of a man plotted along the x axis, and the number of men in each 10-cm height interval plotted on the y axis.   The job of the man standing by the doorway is to measure the height of each individual as he passes through the doorway, and assign him to stand in the

line of men which corresponds to his height. The height is rounded
off to the nearest 10 cm. Thus all the men whose heights fall
between 165 and 175 cm will line up behind the 170-cm mark. After
everyone has been measured and is standing a fixed distance behind
the man in front of him in his assigned line, it will be noted that
a histogram has been generated. The formation as viewed from above
is a histogram of the number of men in each height interval versus
the mean value of the height interval. This is a digital representation of height distribution of the men.

The man at the doorway making the measurement and assigning
the height interval represents the analog-to-digital converter in
the multichannel pulse height analyzer. The x-y graph represents
the analyzer memory, and the viewing of the graph from above is
the same as viewing the analyzer's display of its memory content.
Each height interval on the x axis represents a channel in the
analyzer memory. We can number the channels from 0 to 20, etc., by
noting that a height of 200 cm corresponds to the 20-th interval.

The multichannel analyzer measures and sorts the pulses arriving at its input in the same fashion to generate a digital histogram representation of the x-ray energy spectrum as pictured in
Fig. 4.27(b). Each channel in memory can be associated with a
particular energy interval, and the display can be presented with
the x axis calibrated in terms of the mean energy of each interval.
The y axis of the display gives the number of photons counted in
each energy interval during the entire data accumulation period.
The y axis is usually labeled "counts" which is an abbreviation for
"the number of photons counted per energy interval."

Since deadtimes in this type of spectrometer are quite long
($\sim$60 μs), the system must normally operate with deadtime losses in
the 10 to 60% range. Consequently, most multichannel analyzers are
equipped with an electronic means of deadtime correction, such that
the observed spectrum represents the true number of photons  arriving at the detector during the period of data accumulation. In
addition to the ability to display the spectrum on a cathode-ray

tube or television monitor, the analyzer can usually drive an X-Y
plotter to produce a permanent copy. Alternatively, the contents
of the analyzer memory can be printed on a teletype as the number
of counts in each channel, listed by channel number. Most quantitative fluorescence spectrometers include a small digital computer
with approximately 16,000 words of memory plus some form of mass
storage such as a floppy disk. In such a system the computer may
control specimen presentation, the excitation conditions, and data
accumulation in the multichannel analyzer. At the end of data
acquisition for each specimen the computer analyzes the spectrum in
the multichannel analyzer, computes the raw element intensities,
corrects for interelement effects, and computes the concentration
of each element.

With the long pulse durations used for the Si(Li) detector,
there is a high probability that two or more x-ray photons will
arrive at the detector within the duration of one slow amplifier
output pulse. As a result the amplifier output pulse is distorted.
The resulting pulse shape is the linear sum of the pulse shapes contributed separately by each photon. This distortion process is
called *pulse pileup* since the individual pulses appear to pileup
when they are separated by less than one normal pulse width. To
prevent distorted pulses from being recorded in the analyzed energy
spectrum most energy-dispersive spectrometers incorporate a pileup
rejector. By using a fast amplifier with a much shorter pulse
width, it is possible to detect when two or more pulses have occurred within the normal pulse duration of the slow amplifier's
output. When such a pileup event is detected, the pileup rejector
signals the multichannel analyzer to reject the piled up pulses.
Because the pileup rejector contains circuitry of benefit for operation of the baseline restorer in the slow amplifier, the fast
amplifier, the pileup rejector, and the slow amplifier are usually
contained within the same electronic module.

The remaining block in Fig. 4.25 is the detector high-voltage
bias supply, which is simply used to apply a reverse bias voltage

across the Si(Li) detector diode.  The dc voltage is typically in
the range of -700 to -1500 V.

Since the detector and electronics is relied upon to accurately analyze the x-ray spectrum in the Si(Li) spectrometer, the
performance requirements on the associated electronics are far more
demanding than for the NaI(Tl) detector or the proportional counter.
This is particularly true since the energy resolution of the Si(Li)
detector is much better than the resolution of the other two detectors.  Furthermore, it becomes more important for the spectroscopist to understand the operating limitations peculiar to the
Si(Li) x-ray spectrometer, so that he can avoid faulty data.  The
purpose of Secs. 4.3.1 to 4.3.10 is to delineate the operating
parameters of the system so that the spectroscopist becomes sensitive to where the boundaries between reliable operation and faulty
answers *may* lie for his particular instrument.

### 4.3.1  The Si(Li) Detector Diode

Figure 4.29 is a schematic representation of the sensitive volume
of the Si(Li) detector diode.  The detector diode is fabricated from
a cylindrical section of a single crystal of p-type silicon.  The
sensitive volume ranges from 4 to 16 mm in diameter, and 3 to 5 mm
in thickness, depending on the intended application.  The smaller
diameter detectors yield better energy resolution at low energies,
and the thicker detectors have a higher detection efficiency at
energies above about 20 keV.  To increase the electrical resistivity
the crystal is compensated with lithium.  The drifting of lithium
into the crystal from the rear contact is carried out under high
temperature via a voltage applied between the front and rear detector contacts.  A high concentration of lithium near the rear contact creates an n-type region, over which a layer of gold is
deposited by evaporation to fabricate a nonrectifying contact to
the n-type region.  To the front of the detector a Schottky barrier
contact is applied to produce a p-i-n-type diode.  The front contact
is a window through which the x-ray photons must enter.  It typically consists of a 0.02-μm gold layer followed by a 0.1-μm-thick

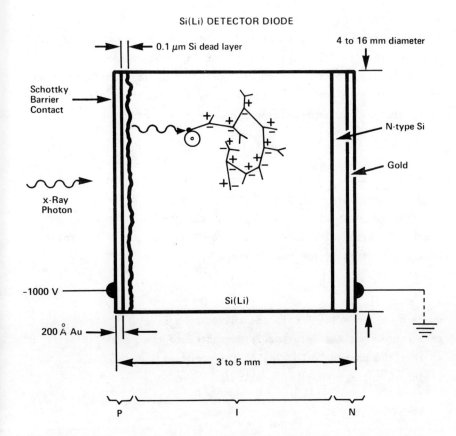

Figure 4.29   A schematic representation of the sensitive volume of the Si(Li) detector diode and the x-ray photon interaction. Reprinted by courtesy of EG&G ORTEC.

silicon deadlayer.  The deadlayer is the region where the electrical resistivity is too low to allow full collection of electron-hole pairs which have been released by the absorption of an x-ray photon.  Thus photons interacting with the detector in the deadlayer will be measured with an abnormally low pulse height, or not measured at all.

This problem is most significant for very low energy x-rays, which have a high probability of being absorbed in the deadlayer. With the normal operating reverse bias of approximately 1000 V the

diode is depleted of the remaining free charge carriers, and it
becomes a solid-state ionization chamber.  X-ray photons enter the
detector through the front contact and interact primarily by the
photoelectric process to produce a cloud of ionization in the form
of electron-hole pairs.  On the average the number of electron-hole
pairs produced n is proportional to the photon energy E:

$$n = \frac{E}{\varepsilon} \qquad (4.25)$$

where $\varepsilon$ = 3.8 eV is the average energy required to produce one
electron-hole pair.  With a negative voltage at the front contact
and the rear contact at ground potential, the holes drift to the
front, while the electrons drift to the rear contact.  The total
charge collected at the rear contact is

$$Q = \frac{E}{\varepsilon} q_e \qquad (4.26)$$

where $q_e$ = $1.6 \times 10^{-19}$ C is the charge on a single electron.  The
rear contact of the detector is connected to the preamplifier input
so that the charge in Eq. (4.26) is stored on a capacitance $C_f$ to
produce a voltage pulse of amplitude

$$V_o = \frac{Q}{C_f} = \frac{E}{\varepsilon} \frac{q_e}{C_f} \qquad (4.27)$$

The charge collection time is quite short; typically lying in the
range of 25 to 100 ns, depending on detector bias voltage, detector
thickness, and the position of the photon interaction within the
detector.  The detector is operated at 77 K to lower the lithium
mobility in the crystal, and to reduce the noise which would be
caused by excessive diode reverse leakage current at higher temperatures.

The actual shape of the Si(Li) diode varies somewhat from the
representation in Fig. 4.29.  Figure 4.30 shows two of the commonly
used diode structures.  The active volume of the diode is formed by
the lithium compensated region.  Because the boundary between the
intrinsic and p-type regions is poorly controlled at the outer

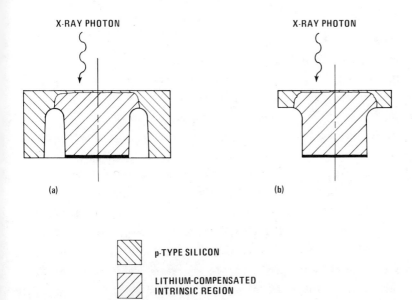

Figure 4.30  Two of the commonly used mechanical structures for the Si(Li) detector diode:   (a)   the annular groove construction; (b)   the top-hat structure.  Both structures are cylindrically symmetric about the center line.  The drawings represent sectional views taken through the diameter of the diode.  Reprinted by courtesy of EG&G ORTEC.

edges near the front of the detector, a collimator is usually added to mask off these regions as shown in Fig. 4.26.  As Goulding and co-workers have shown [15,16], distortion of the electric field lines at the sides of the detector tends to divert electrons toward the side surface as they drift to the rear contact.  Electrons which start out close to the side surface can be trapped at the surface, and cause an abnormally low pulse height to be measured. This effect is a major cause of the incomplete charge collection background shelf observed for higher photon energies (see Sec. 11.3). The conventional solution to this problem, where low background is important, is to reduce the diameter of the detector collimator [17].  This reduced background from incomplete charge collection

is achieved by discarding up to 50% of the detector active volume
at its circumference.

Bulk trapping is another source of incomplete charge collection.  In this case the trapping can occur throughout the active
volume of the detector diode.  As the charge drifts toward the
electrodes, some of it is caught by trapping centers.  As a result,
a lower pulse height is produced.  This effect is most noticeable
as a low-energy tail causing a deviation from the ideal gaussian
peak shape in the energy spectrum.  The contribution from bulk
trapping can be minimized by raising the detector bias voltage well
above the value which is adequate to achieve total depletion of the
diode.  An upper limit to the bias voltage is set by the condition
where an increase in bias voltage causes a noticeable increase in
detector leakage current, resulting in increased noise and poorer
energy resolution.  The bias voltage recommended by the manufacturer usually represents the optimum compromise between the con-
flicting requirements on the bias voltage.

## 4.3.2  Types of Preamplifiers

Two general classes of preamplifiers are in common use with Si(Li)
detectors in x-ray fluorescence spectrometers:  (a) the continuous
feedback preamplifier and (b) the pulsed beedback preamplifier.
Resistive feedback, and the dynamic charge restoration feedback
method [18-21] both fall under the continuous feedback category.
The pulsed feedback system ordinarily encountered in x-ray spectrometry is the pulsed optical feedback [22-25].

Figure 4.31 is a schematic representation of the resistive
feedback preamplifier.  A dc bias voltage of about -1000 V is applied to the p-type contact of the Si(Li) detector diode through a
high-voltage filter network.  The high-voltage filter removes any
ripple or extraneous noise present in the bias voltage from the
high-voltage power supply.  The preamplifier consists of a field-
effect transistor (FET) input stage, further stages of high open-
loop gain A, and the feedback elements $R_f$ and $C_f$.  The components

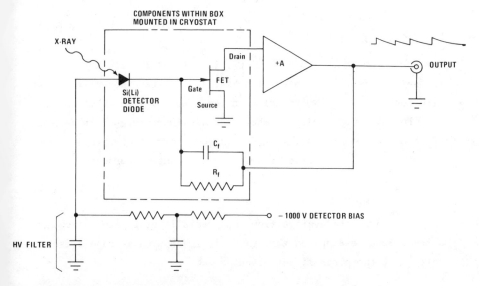

Figure 4.31   The resistive feedback preamplifier.   Reprinted by
courtesy of EG&G ORTEC.

within the dashed box in Fig. 4.31 are mounted in the vacuum cryostat and operated near the boiling temperature of liquid nitrogen.
This lowers the thermal contributions to preamplifier noise and
reduces the lithium mobility in the Si(Li) diode.   All other components are mounted external to the cryostat.   The gate of the FET
is held near ground potential by the feedback components as a result of the large open loop gain.   Consequently, all of the charge
swept from the detector is forced to collect on the capacitor $C_f$
to form a voltage pulse at the output with an amplitude given by
Eq. (4.27).   If no means were provided for subsequently removing
the charge from the feedback capacitor the output voltage would rise
further with each detected x-ray photon, until the preamplifier
power supply voltage was reached, and the preamplifier would
cease to operate.   The function of the feedback resistor is to
slowly remove the charge from $C_f$ after each pulse.   If the finite
detector charge collection time is ignored, a photon of energy E
arriving at time t = 0 results in a preamplifier output pulse shape
given by

$$V = \frac{E}{\varepsilon} \frac{q_e}{C_f} \exp\left(\frac{-t}{R_f C_f}\right) \tag{4.28}$$

for $t \geq 0$, and $q_e = 1.6 \times 10^{-19}$ C. This shape is illustrated by
the series of pulses sketched in Fig. 4.31. The exponential decay
time constant $R_f C_f$ is typically in the range of 1 to 10 ms.

For every preamplifier there is a maximum energy rate beyond
which it cannot pass pulses. Photons of energy E arriving at the
detector with a mean rate of I per second cause a current

$$\bar{i} = \frac{E}{\varepsilon} q_e I \tag{4.29}$$

to be drawn on the average from the preamplifier input. Since this
current must be supplied through $R_f$ it induces an average offset
voltage at the preamplifier output given by

$$\bar{V} = (EI) \frac{q_e R_f}{\varepsilon} \tag{4.30}$$

If the energy rate product EI is high enough, the offset voltage
will reach the preamplifier power supply voltage, and no output
pulses will be produced. A maximum energy rate of $5 \times 10^6$ keV/s is
typical for a resistive feedback preamplifier. For an x-ray spectrum containing more than one energy, the energy-counting rate
product must be summed for all energies [21]. The maximum energy
rate limitation is important where the spectrometer must process
high-energy lines at high counting rates. Improved energy rate
performance can be achieved by reducing the size of the feedback
resistor, at the expense of increased preamplifier noise at long
shaping time constants.

The *dynamic charge restoration* feedback differs from the resistive feedback in that a more complicated, active feedback network
is substituted for the simple resistor feedback. This results in
a lower preamplifier noise at long shaping time constants, but
maintains an energy rate capability at short shaping time constants
similar to the resistive feedback preamplifier. Pulse shapes at

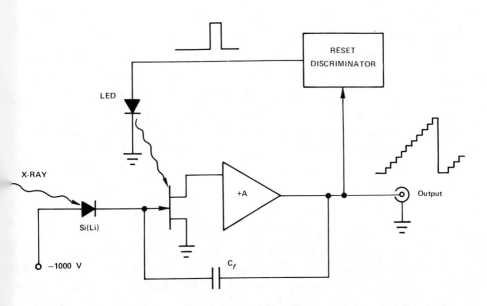

Figure 4.32  The pulsed optical-feedback preamplifier.  Reprinted by courtesy of EG&G ORTEC.

the output of the dynamic charge restoration preamplifier are similar to the pulses from the resistive feedback preamplifier.  In both types of continuous feedback there is no deadtime or deadtime loss associated with the preamplifier.

Figure 4.32 shows the pulsed optical feedback preamplifier of the Landis and Goulding design [25].  It is identical to the resistive feedback preamplifier except that the feedback resistor has been removed.  This change permits lower preamplifier noise at long shaping time constants.  In place of the feedback resistor an optical reset circuit has been incorporated.

The preamplifier functions as follows.  The charge from each detected x-ray photon is allowed to accumulate on the feedback capacitor to produce a staircase of pulses at the preamplifier output.  As before, the amplitude of each step, or pulse, is proportional to the detected photon energy.  When the staircase begins to approach the maximum permissible output voltage a reset

discriminator is triggered.  The discriminator turns on a light-
emitting diode (LED) which is mounted near the FET in the cryostat.
The light from the LED shining on the gate to drain junction of the
FET causes a substantial leakage current to flow from the FET
drain into the gate.  This current discharges the capacitor $C_f$.
When the preamplifier output returns to the initial voltage of the
staircase the LED is turned off, and the preamplifier is free to
process another series of pulses.

Although the resetting of the preamplifier can be complete
within a period of 10 to 100 μs, the spectrometer remains dead for
a much longer period after each reset.  The reset excursion is
typically equivalent in amplitude to a pulse from a 3000-keV photon,
but has the opposite polarity.  Thus the pulse-shaping amplifier is
driven heavily into reverse polarity overload during the reset.
The recovery from overload to a state where further pulses can be
processed without error may require 0.5 to 2 ms in the pulse-shaping
amplifier.  Pulse-height analysis must be prevented during this
recovery period.  Consequently the recovery period constitutes an
additional source of deadtime in the system.  The percent deadtime
caused by the resetting can be computed for photons of energy E
arriving at the detector at a rate of I counts per second as

$$\text{Percent reset deadtime} = \frac{EI\,T_r}{E_r} \times 100\% \qquad (4.31)$$

$E_r$ is the equivalent energy of the reset, and $T_r$ is the reset and
recovery time associated with each reset.  For a situation where
E = 30 keV, I = $10^4$ counts/s, $E_r$ = 3000 keV, and $T_r$ = $10^{-3}$ s, the
percent reset deadtime is 10%.  In the pulsed optical feedback
preamplifier the maximum energy rate which can be handled is determined by the resetting deadtime.  The limiting energy rate is
reached when the percent reset deadtime becomes excessive, or ultimately, when the deadtime becomes 100%.

The Kandiah pulsed optical feedback design [22-24] differs from
the Landis and Goulding design in that the preamplifier is reset

after each x-ray photon is processed.  In this case a short deadtime is added to each processed pulse.

### 4.3.3  The Slow Pulse-Shaping Amplifier

The slow pulse-shaping amplifier normally used with Si(Li) x-ray
spectrometers is the semigaussian shaping amplifier already described in Sec. 4.2.6 and illustrated in Fig. 4.15.  For a basic
description of the semigaussian shaping amplifier Sec. 4.2.6 should
be read, with the following additional information in mind.

The Si(Li) detector differs from the proportional counter in
two respects.  The Si(Li) detector has a much better energy resolution, and the preamplifier noise contribution is much more significant.  As a consequence, much more emphasis is placed on designing
the shaping amplifier to minimize the noise contribution, and on
preserving the fidelity of the analyzed spectrum.  This requires a
substantially more sophisticated amplifier design.  As will be discussed in Sec. 4.3.7, the emphasis on noise reduction results in
long shaping time constants, usually in the range of 4 to 16 μs.
As illustrated in Fig. 4.27, the typical amplifier output pulse
shape is quite long.  In order to achieve a reasonable rate of data
accumulation, the amplifier must be operated at a rather high duty
cycle.  Crudely speaking, the amplifier duty cycle is the fraction of the time pulses are present at the amplifier output above
the noise level near the baseline.  Duty cycles in the range of 40
to 60% are commonly employed.  Under these conditions the demands
on the amplifier performance are severe.  The baseline restorer
must be of higher quality [9,26] and the pole-zero cancellation must
be precisely adjusted.  Note that with the pulsed optical feedback
preamplifier the pole-zero cancellation network is eliminated or
set for an infinite preamplifier decay time constant.  Since the
amplifier output pulse duration is the dominant source of deadtime
in the energy dispersive spectrometer, the high duty cycles cause
high deadtime losses.  Therefore it becomes important to have a
very accurate deadtime correction circuit incorporated in the
spectrometer.

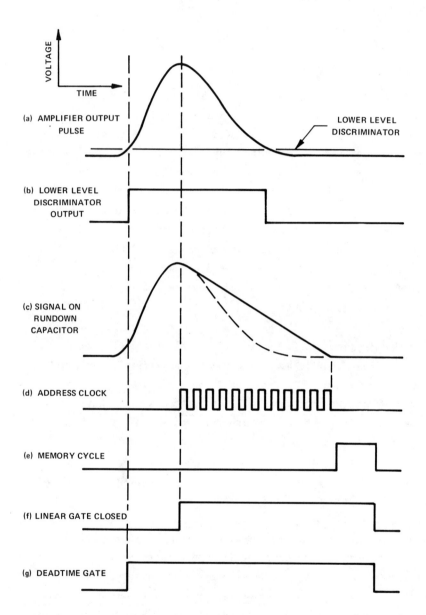

Figure 4.33 Signals occurring during the pulse-height measurement process in the multichannel analyzer [20].

Several references are available for more information on the
shaping amplifier [2,20-24,27-30] and items critical to its performance.

### 4.3.4   The Multichannel Pulse Height Analyzer

The multichannel analyzer (MCA) consists of three main sections:
the analog-to-digital converter (ADC), the memory, and the display.
The analog-to-digital converter measures the amplifier pulse peak
amplitude and converts it to a digital number.  This number represents the address of a memory location in the analyzer memory.  The
memory is similar to a digital computer memory, in that each memory
location can hold a number ranging from 0 up to some maximum value,
say $10^6$ - 1.  The number currently held in a memory location represents the number of times a pulse has been measured with an amplitude corresponding to the address of that memory location.  When
the next pulse for that address is measured the number in the memory
location is increased by one.  This proceeds on a pulse by pulse
basis until the histogram representing the x-ray energy spectrum is
built up in memory.

The analog-to-digital converter is generally of the Wilkinson
type [31-33].  Its function is defined in Figs. 4.33 and 4.34.
Figure 4.33(a) shows the amplifier pulse arriving at the analog-to-
digital converter input.  Some method of recognizing pulses and
rejecting the low-amplitude noise between pulses is required, so
that the analog-to-digital converter does not waste a large fraction of its  available time analyzing noise.  The lower level discriminator [Fig. 4.33(a) and (b)] serves this purpose, and is normally set just above the noise level.  If the linear gate is open
when the amplifier pulse arrives [Fig. 4.33(f)], the rundown capacitor is connected to the input [Fig. 4.34(a)].  The input circuits
force the rundown capacitor to charge up so that its voltage
follows the amplifier signal up to the peak amplitude of the amplifier pulse [Fig. 4.33(c)].  A peak detector circuit senses that the
amplifier pulse has reached its peak amplitude and closes the linear

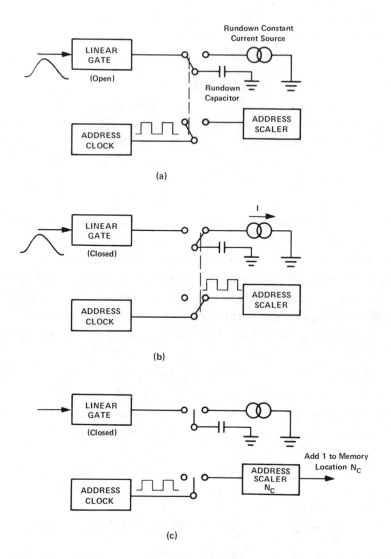

Figure 4.34  Simplified analog-to-digital converter circuits and
modes during the pulse-height measurement process [20]:   (a) capacitor changing, (b) capacitor rundown, and (c) memory cycle.

gate [Fig. 4.33(f)].  At this point the rundown capacitor is disconnected from the input and the voltage on the capacitor is equal
to the peak amplitude of the amplifier pulse [Fig. 4.33(c)].  The

closing of the linear gate prevents further pulses at the input
from interfering with the subsequent measurement process.

Next, the analog-to-digital converter transforms the voltage
on the rundown capacitor into a time interval.  The length of the
time interval is proportional to the voltage on the rundown capacitor, and hence represents the original pulse amplitude.  This interval is measured with a digital clock to produce a number which is
proportional to the time interval, thus yielding a number which is
proportional to the original amplifier pulse height.  When the
linear gate closes at the peak amplitude of the pulse, two events
take place simultaneously [Fig. 4.34(b)].  The rundown capacitor is
connected to a constant current discharging circuit, and an address
clock is connected to the address scaler.  The rundown constant
current source causes the capacitor to discharge linearly to zero
volts [Fig. 4.33(c)].  Therefore, the time taken for the rundown to
zero volts is proportional to the original pulse amplitude.  When
the voltage on the rundown capacitor reaches zero the constant
current source is disconnected from the capacitor, and the address
clock is disconnected from the address scaler [Fig. 4.34(c)].

The address clock is simply an oscillator generating square
pulses at a frequency $f_c$ [Fig. 4.33(d)].  Consequently, the number
of clock pulses counted $N_c$ during the rundown interval is proportional to both the rundown time and the original amplifier pulse
amplitude.  This completes the analog-to-digital conversion.

Next the multichannel analyzer must add one count to memory
location $N_c$.  This is carried out during the memory cycle time
[Fig. 4.33(e)].  The memory location whose address or channel number is $N_c$ is identified and its present contents are read.  If
memory location $N_c$ presently contains m counts, the number m + 1
is written back in.  At the end of the memory cycle the multichannel analyzer is free to process another amplifier pulse.  If a
pulse is already present above the lower level discriminator level
at the linear gate input the linear gate remains closed until the
pulse drops below the discriminator threshold.  At this point the

linear gate opens and the multichannel analyzer is ready to process
the next amplifier pulse which exceeds the lower level discriminator threshold.  The measurement process is repeated on a pulse by
pulse basis to build up the energy spectrum histogram in memory.

Both the amplifier pulse and the multichannel analyzer contribute to the system deadtime.  The deadtime caused by the multichannel analyzer for each analyzed pulse is the sum of the rundown time
and the memory cycle time, i.e.,

$$T_M = \frac{N_c}{f_c} + T_{MC} \qquad\qquad\qquad (4.32)$$

where $T_{MC}$ is the constant memory cycle time, $N_c$ is the channel number into which the pulse is recorded, and $f_c$ is the address clock
rundown frequency.  This deadtime interval starts when the analog-
to-digital converter has sensed the peak amplitude of the amplifier
pulse and closes the linear gate [Fig. 4.33(f)].  Note that the
deadtime is a function of pulse height through $N_c$, and is a nonparalyzable deadtime.

The conversion gain of the analog-to-digital converter is
usually denoted by the number of channels corresponding to a full-
scale input pulse amplitude.  For example, if the maximum input
pulse height which can be analyzed is 10 V, and this is measured
into channel 1024, while channel zero corresponds to zero input
pulse amplitude, then the conversion gain is 1024 channels full
scale.  The analyzer memory may contain a number of channels which
is (a) equal to the conversion gain of the analog-to-digital converter, (b) some binary multiple of the conversion gain, or
(c) some binary fraction of the conversion gain.  Extra memory allows acquisition of more than one spectrum in separate sections of
memory.  These spectra can be overlapped in the display mode to
allow inspection for subtle differences.  If the memory size is
less than the conversion gain the full spectrum must be acquired
and examined in successive energy segments.  A 1024 channel conversion gain with a 1024-channel memory covering a 0- to 40-keV energy

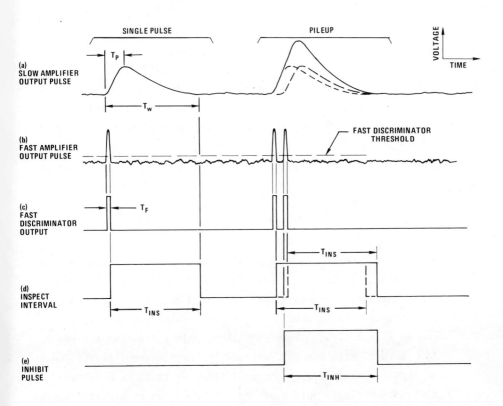

Figure 4.35  Basic waveforms in the pileup rejector.  Reprinted by courtesy of EG&G ORTEC.

range is typical for energy dispersive x-ray fluorescence spectrometers.  This provides a calibration of 40 eV per channel.  That is, each channel covers an energy interval of 40 eV.  In this case the center of channel $N_c$ corresponds to an energy of

$$E_{N_c} = N_c \times 0.04 \text{ keV}.$$

### 4.3.5  The Pileup Rejector

At very high counting rates, the probability of two x-ray photons arriving at the detector within the duration of one amplifier output pulse width is quite high.  When such an overlap occurs the amplifier pulses from the two photons sum, or *pileup*, to produce a distorted pulse of higher amplitude.  Figure 4.35(a) depicts the

pileup pulse which arises from two overlapping pulses of equal
amplitude.  The dashed lines show the response which would have
resulted for each pulse had it occurred by itself.  The solid lines
correspond to the composite response.  If the multichannel analyzer
is allowed to process such events it will record the distorted
pulse amplitude.  The result would be a severe distortion of the
analyzed energy spectrum at high counting rates.

The purpose of the pileup rejector is to prevent the recording
of these distorted pileup pulses.  It functions in the following
way, as illustrated in Fig. 4.35.  The preamplifier signal is fed
to a second amplifier which has a very short shaping time constant,
typically 0.1 μs.  This amplifier is normally referred to as a
"fast amplifier" because it is able to respond quickly and resolve
closely spaced photons.  In contrast, the regular amplifier used
for pulse height analysis is called a *slow amplifier,* because of
its much longer pulse width.  The output of the fast amplifier is
shown in Fig. 4.35(b).  Note that it is able to resolve the two
pileup events which are not resolved by the slow amplifier.  The
fast amplifier pulses are separated from the baseline noise and
converted into logic pulses of the same duration by a fast discriminator whose threshold is set just above the noise level.  Figure
4.35(c) defines the fast discriminator response.  The width of the
fast pulse above the discriminator threshold is $T_F$.  In general
it is a function of energy, with larger amplitude pulses causing
larger values of $T_F$.

To test for pileup the trailing edge of each fast discriminator
pulse triggers an inspection interval $T_{INS}$, long enough to protect
the entire slow amplifier pulse [Fig. 4.35(d)].  If a second fast
discriminator pulse occurs during the inspection interval two
things happen.  First, the inspection interval is lengthened by an
amount $T_{INS}$ starting from the trailing edge of the second discriminator pulse.  Secondly, an inhibit pulse is generated [Fig.
4.35(e)].  The inhibit pulse starts at the leading edge of the
second fast discriminator pulse and ends at the end of the updated
inspection interval.  If the second pulse is followed by more pulses

during the inspection interval, each successive pulse regenerates both the inspection interval and the inhibit interval in the same way as the second pulse did.  The inhibit pulse is used to signal the multichannel analyzer to reject the composite pulse because of pileup distortion.

Figure 4.35 demonstrates the case where the second photon is detected before the slow amplifier pulse from the first photon reaches peak amplitude.  Thus the amplitudes of both pulses are distorted, and both events must be rejected.  If the second photon arrives after the multichannel analyzer has sensed the peak amplitude of the first pulse and closed its linear gate, then the second pulse will not distort the measured amplitude of the first pulse. In this case, only the second pulse must be rejected.  Consequently, the multichannel analyzer only rejects an event which it otherwise would have analyzed if the inhibit pulse occurs before peak amplitude is determined.  Peak amplitude determination is marked by the closing of the linear gate in the analog-to-digital converter.

Figure 4.36 demonstrates the distortion caused by pileup at extremely high counting rates.  A pure iron specimen has been excited by broadband excitation in an x-ray fluorescence spectrometer. Consequently, the major content of the spectrum should be the K lines of iron at 6.4 and 7.1 keV.  Figure 4.36(a) is a multiple trace recording of the slow amplifier output on an oscilloscope. The oscilloscope starts each trace at the beginning of an amplifier output pulse.  As evidenced by the intensity of the traces, the highest counting rate is in the iron K lines with pulse amplitudes between 1.6 and 1.8 V.  These pulse heights correspond to the iron K lines at their proper positions in the spectrum of Fig. 4.36(b).

As can be seen in Fig. 4.36(a) a second pulse can partially overlap the first pulse and rise to an amplitude between 1.6 and 3.6 V.  These events cause the first-order pileup spectrum in Fig. 4.36(b) between 9 and 14.1 keV.  Second-order pileup is barely noticeable in the pulse amplitude range from 3.6 to 5.4 V, but is clearly visible in the spectral content from 14.1 to 21 keV in

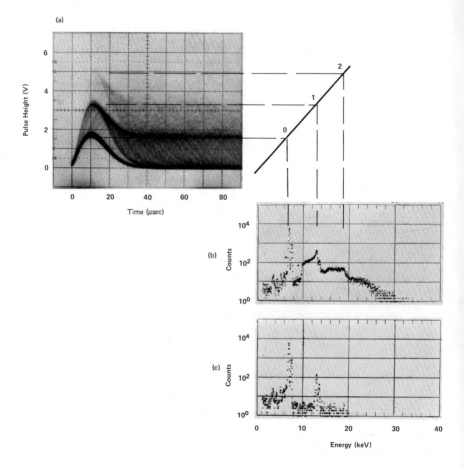

Figure 4.36  Pileup distortion at high counting rates.  A pure iron
specimen has been fluoresced to yield a counting rate of 44,000
counts/sec at the detector.   (a)  A multiple trace picture of the
slow amplifier output on an oscilloscope; first- and second-order
pileup is visible.   (b)  The energy spectrum recorded without a
pileup rejector.   (c)  The spectrum recorded using a pileup rejector.  Reprinted by courtesy of EG&G ORTEC.

Fig. 4.36(b).  Third-order pileup, which consists of three pulses

riding on top of the first pulse, is recorded between 21 and 28 keV

in Fig. 4.36(b), but is not intense enough to be observed in

Fig. 4.36(a).

    Figure 4.36(c) shows the reduction in spectral distortion when

the pileup rejector is employed.  Almost all the pileup spectrum

above 9 keV has been suppressed.  What remains is the full-energy

sum peaks.  Since the fast amplifier is unable to resolve photons
which are separated by a time interval less than $T_F$, the pileup
rejector is unable to suppress the most closely spaced events.
Thus, two photons with energies $E_1$ and $E_2$ detected within the time
interval $T_F$, will produce an analyzed event at the energy $E_1 + E_2$
in spite of the pileup rejector.  In Fig. 4.36(c) the major counting
rate is in the iron K lines at 6.40 and 7.06 keV.  Therefore the
residual, full-energy sum peaks are expected at (6.40 + 6.40) keV,
(6.40 + 7.06) keV, and (7.06 + 7.06) keV in the first-order pileup
spectrum.  Note that a second-order pileup peak is barely detectable
at (6.40 + 6.40 + 6.40) keV.  The full-energy sum peaks will have
the same resolution that a proper x-ray line of the same energy
would have.

The effectiveness of the pileup rejector is measured by its
ability to suppress the pileup spectrum.  For a spectrum containing
only one x-ray energy $E_1$, the ratio of the number of counts observed in the entire pileup spectrum to the number of counts observed at the proper energy is given by

$$\frac{N_s[E > E_1]}{N[E_1]} = e^{I_t T_p} - 1$$

$$\approx I_t T_p \qquad \text{for } I_t T_p \ll 1 \qquad (4.33)$$

where $N_s[E > E_1]$ is the number of counts observed in the pileup
spectrum without a pileup rejector, including all orders of pileup.
$N[E_1]$ is the number of counts observed at the correct peak energy,
$I_t$ is the true counting rate of energy $E_1$ at the detector, and $T_p$
is time at which the analog-to-digital converter senses peak
amplitude and closes the linear gate (Fig. 4.35).  With the addition
of a pileup rejector the ratio is reduced to[*]

$$\frac{n_s[E > E_1]}{N[E_1]} = e^{I_t T_F} - 1$$

$$\approx I_t T_F \qquad \text{for } I_t T_F \ll 1 \qquad (4.34)$$

---

[*] Note that the first two editions of Ref. 21 are in error on this
equation.

where $n_s[E > E_1]$ is the residual number of counts observed in the
pileup spectrum, including all orders of pileup. The factor of improvement in using the pileup rejector is given by the ratio of
Eq. (4.33) to Eq. (4.34), which is approximately $T_p/T_F$. Consequently, very short pulse widths $T_F$ are desired to suppress pileup.
This requires very short shaping time constants in the fast amplifier. As will be discussed in Sec. 4.3.7, shorter shaping time
constants lead to a higher noise level, and the fast discriminator
threshold must be set at an energy which is above some of the
light-element lines in the spectrum. A compromise must be struck
between having a high degree of pileup suppression on high energy
lines and having a threshold low enough to achieve pileup rejection on the lowest energies to be processed [34]. Usually a fast
amplifier shaping time constant of approximately 0.1 μs is chosen,
and $T_F$ lies in the range of 500 to 750 ns. The energy threshold
for pileup rejection commonly falls in the range of 1 to 1.6 keV
depending on the preamplifier noise contribution. The rather high
energy threshold and the finite pulse pair resolving time $T_F$ of
the pileup rejector can have a degrading influence on the accuracy
of the deadtime correction at high counting rates.

Frequently one must be able to identify the residual full-
energy sum peaks in a spectrum and subtract their contribution
from analyte lines with which they interfere. The following equations can be useful for this task. The equations are valid for
the case where the peak amplitude detection time $T_p$ is constant and
independent of pulse height. Depending on multichannel analyzer
design, this condition may or may not be met at very low pulse
heights. The case of $T_p$ being a function of pulse amplitude is
discussed in Sec. 4.6.

A general spectrum will be assumed from which any two lines at
energies $E_1$ and $E_2$ are of interest $I_t(E)$ is the true counting rate
of the photon of energy E at the detector. $T_F(E)$ is the pulse pair
resolving time of the pileup rejector for a second photon of arbitrary energy following the photon of energy E. $N[E]$ is the number

of counts observed at the correct energy E after accumulating data for the livetime $t_\ell$, and $n_s[E]$ is the residual number of counts observed in the pileup sum peak at energy E after counting for a livetime $t_\ell$.

The ratio of the counts observed in the sum peak at $E = (n + 1) E_1$ to the counts observed at $E_1$, as a result of n-th order pileup of $E_1$ on itself, is given by

$$\frac{n_s[E = (n + 1) E_1]}{N[E_1]} = \frac{[I_t(E_1) T_F(E_1)]^n}{n!} \tag{4.35}$$

Since $I_t(E_1)T_F(E_1)$ is normally small compared to unity, second- and higher order pileup are usually negligible. The first-order pileup equations are

$$
\begin{aligned}
n_s[E_1 + E_1] &= N[E_1] \, I_t(E_1) \, T_F(E_1) \\
&= (N[E_1])^2 \frac{T_F(E_1)}{t_\ell}
\end{aligned}
\tag{4.36}
$$

$$
\begin{aligned}
n_s[E_2 + E_2] &= N[E_2] \, I_t(E_2) \, T_F(E_2) \\
&= \{N[E_2]\}^2 \frac{T_F(E_2)}{t_\ell}
\end{aligned}
\tag{4.37}
$$

$$
\begin{aligned}
n_s[E_1 + E_2] &= N[E_1] \, I_t(E_2) \, [T_F(E_1) + T_F(E_2)] \\
&= N[E_2] \, I_t(E_1) \, [T_F(E_1) + T_F(E_2)] \\
&= N[E_1] \, N[E_2] \frac{[T_F(E_1) + T_F(E_2)]}{t_\ell}
\end{aligned}
\tag{4.38}
$$

where the livetime relationship $I_t(E) = N[E]/t_\ell$ has been used (see Sec. 4.7.2). Equations (4.36) through (4.38) can be used to measure $T_F(E_1)$ and $T_F(E_2)$. Once these parameters are known, the same equations can be used to predict the counts in the pileup peaks for identification purposes or for interference removal.

### 4.3.6   Detection Efficiency

Figure 4.37 gives the calculated detection efficiency for the Si(Li) detector. Only the effects of the detector diode thickness (3 and 5 mm), and the beryllium window thickness have been

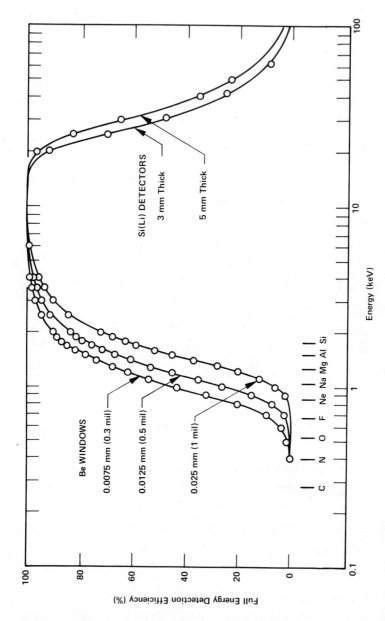

Figure 4.37 The calculated detection efficiency for a Si(Li) detector. Only the effects of detector thickness and beryllium window thickness have been incorporated. Reprinted by courtesy of EG&G ORTEC.

incorporated.  The diagram illustrates the importance of thick detectors for efficiency at high energies.  For light-element analysis a thin beryllium entrance window is essential.  The more rugged
12.5-μm-thick beryllium window is sometimes used where optimum
light-element sensitivity is not crucial.  If an accurate absolute
efficiency calibration is required, other effects such as incomplete
charge collection, escape peak losses, edge losses, silicon deadlayer absorption and fluorescence, and gold contact absorption and
fluorescence must be taken into account [35].

### 4.3.7  Energy Resolution

Equation (4.25) gives the average number of electron-hole pairs
produced in the detector by a photon of energy E as

$$n = \frac{E}{\varepsilon} \qquad\qquad\qquad (4.25)$$

The expected variance $\sigma_n^2$ in this average number can be expressed as

$$\frac{\sigma_n^2}{n} = F \qquad\qquad\qquad (4.39)$$

where F is the Fano factor.  If all of the incident photon's energy
were always converted into electron-hole pairs the Fano factor would
be zero.  However, there are competing processes during the conversion of the photon energy to ionization, which consume energy without producing electron-hole pairs [11].  If the ionizing electrons
found the probability of an ionizing interaction was very small
compared to these competing options for energy loss, the process
would be described by Poisson statistics, and the Fano factor would
be unity.  In practice the Fano factor lies between 0.115 and 0.05
for the silicon detector [29].  Hence the ionization process is
more nearly described by deterministic laws than by Poisson statistics.

Since the relative standard deviation in the analyzed pulse
height $\sigma_E/E$ is equal to $\sigma_n/n$, the contribution to energy resolution
from ionization statistics can be computed as

$$\sigma_E = E \frac{\sigma_E}{E} = E \frac{\sigma_n}{n}$$

$$= E \frac{\sqrt{Fn}}{n}$$

$$= \sqrt{\epsilon F E} \qquad\qquad (4.40)$$

Resolution is normally measured as the full width of the peak at half its maximum height (abbreviated: fwhm or FWHM). Since n is large the peak is approximately gaussian and the contribution of ionization statistics to the resolution is

$$\Gamma_{ionization} = 2.35 \ \sigma_E = 2.35 \ \sqrt{\epsilon F E} \qquad\qquad (4.41)$$

The total FWHM resolution on a peak at energy E is a combination of the preamplifier noise contribution $\Gamma_{noise}$, the ionization statistics $\Gamma_{ionization}$, and other line broadening contributions $\Gamma_{other}$, such as incomplete charge collection. The total FWHM resolution is given by

$$\Gamma = \sqrt{(\Gamma_{noise})^2 + (2.35 \ \sqrt{\epsilon F E})^2 + (\Gamma_{other})^2} \qquad\qquad (4.42)$$

It is common practice to ignore the third term and lump its effect into the second term. This results in an inflated Fano factor lying in the range 0.115 to 0.13. In fact, an inflated Fano factor of 0.125 can be used as a typical value for most calculational purposes. The convention of dropping the third term will be followed in this book.

All of the energy dependence is contained in the second term, which is independent of the slow amplifier shaping time constant. While the first term is not a function of energy, it is a strong function of the slow amplifier shaping time constant. Figure 4.38 illustrates the relationship for a small detector. Clearly, the noise contribution is at a minimum for a shaping time constant of approximately 14 μs. This explains why the amplifier pulses are so long with the Si(Li) detector. Generally, a time constant of

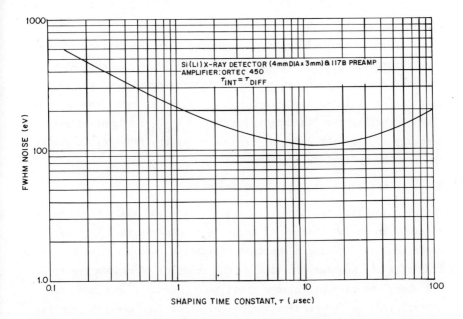

Figure 4.38  The preamplifier noise contribution $\Gamma_{noise}$ as a function of the amplifier shaping time constant for a small detector [20].

about 6 μs is chosen to sacrifice a small amount on the energy resolution in favor of shorter pulse widths for high counting rates.  Figure 4.38 also demonstrates the drastic increase in pre-amplifier noise contribution at the short shaping time constants used in the fast amplifier for the pileup rejector, i.e., 0.1 μs.

The preamplifier noise depends on several parameters besides the shaping time constant.  Among these are the quality of the cooled FET, other sources of noise on the preamplifier input, and capacity to ground on the preamplifier input.  Since the detector capacity is proportional to detector area, and inversely proportional to detector thickness, large-area, thin detectors cause a higher preamplifier noise contribution.  Crudely speaking, the effect of higher detector capacity is to move the curve in Fig. 4.38

upward and to the right.

For the detectors typically employed with x-ray fluorescence spectrometers, $\Gamma_{noise}$ falls in the range of 90 to 120 eV. At high energies the term from ionization statistics dominates, and resolution is fairly insensitive to $\Gamma_{noise}$. At low energies the resolution approaches the limiting value set by $\Gamma_{noise}$. Manufacturers normally specify the detector resolution as the FWHM resolution on the 5.9-keV Mn K$\alpha$ line.

## 4.3.8  Limitations at High Counting Rates

The need for high counting rates arises from the analytical error inherent in the statistics of counting random events. In fluorescence analyzers having a constant excitation intensity the detection of x-ray photons constitutes the counting of random events, and obeys the laws of Poisson statistics. Crudely speaking, the contribution to the percent analytical error from counting statistics is inversely proportional to the square root of the number of x-ray photons counted. In other words, the percent error from counting statistics is roughly proportional to $1/\sqrt{I_m t}$, where $I_m$ is the observed counting rate for the analyte line, and t is the real time during which the x-ray photons are counted. Generally, the available counting time t is limited. Therefore minimization of the error from counting statistics requires maximizing the observed counting rate.

It is important to point out that this is but one aspect of the problem of achieving analytical accuracy. In fact, the analyst can be grossly misled by this simplistic view, unless he maintains a broader perspective on the problem. First, it must be remembered that the error from counting statistics may be negligible compared to the other sources of analytical error, even at moderate counting rates. The dominant error may lie in sampling errors, specimen preparation technique, errors in the calibration standards, uncompensated interelement interferences, or other repeatability problems in the instrument. Second, at high counting rates systematic distortions of the quality of the data become significant, as a result of imperfections in the performance of the electronic instrumentation.

The purpose of this section is to outline the factors which limit the accuracy of the analysis at high counting rates with the electronics in energy-dispersive x-ray spectrometers. The limitations can be grouped into three categories: spectral distortions, loss of throughput efficiency, and systematic errors in the deadtime correction.

Several types of spectral distortion can occur at high counting rates. Perhaps the most obvious is pileup distortion, which has already been discussed in Sec. 4.3.5. In addition, the energy resolution will degrade at high counting rates, and the peak positions will shift toward lower energy. If all pulse heights do not experience the same percent deadtime losses, the detection efficiency as a function of energy will appear to be counting rate dependent.

Figure 4.39 is an example of the resolution broadening and peak shifting experienced at high counting rates [26]. While the shorter shaping time constants yield poorer energy resolution at low counting rates, they maintain a constant resolution and peak position to much higher counting rates. The spectroscopist should be aware that the performance with respect to spectral distortion at high counting rates can vary substantially from one spectrometer to another. The amount of distortion will also depend upon the spectral content and the energy of the line being examined. The analyst should be prepared to establish the safe upper limit on counting rate for his particular instrument and analytical problem.

A number of items are important in achieving constant energy resolution and peak position up to very high counting rates. The preamplifier gain must be independent of the detected energy rate, and there must be no noise sources whose contributions increase with increasing energy rate. The pole-zero cancellation in the amplifier must be adjusted correctly and precisely, and the

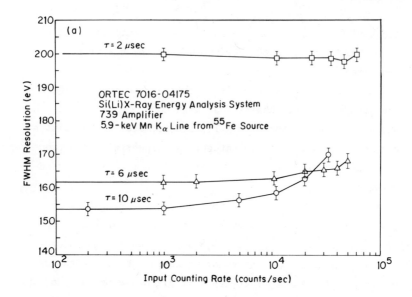

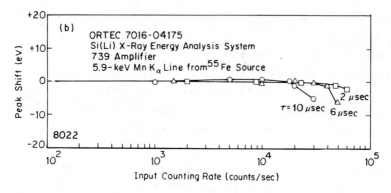

Figure 4.39  The dependence of resolution (a), and peak position
(b), on counting rate for three shaping amplifier time constants
($\tau$ = 2, 6, and 10 μsec).  The measurements were made on the Mn Kα
line from an [55]Fe radioisotope source.  Reprinted from Ref. 26 by
courtesy of EG&G ORTEC.

multichannel analyzer calibration must be insensitive to counting
rate.  Even if these conditions are met, the counting rate at which
the resolution broadening and peak shifting becomes significant
is ultimately determined by the ability of the baseline restorer in
the shaping amplifier to cope with the problem of a vanishing
baseline at high counting rates.

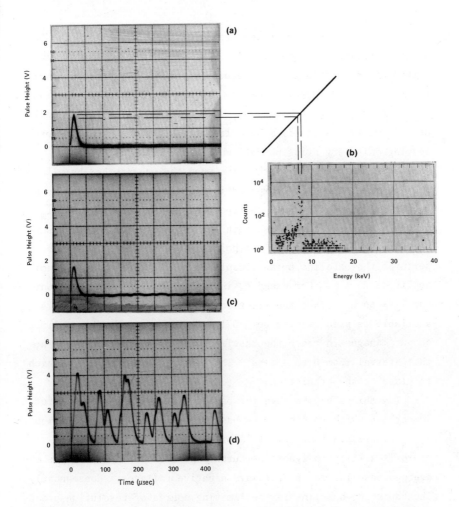

Figure 4.40  Pileup distortion and the problem of a vanishing baseline at high counting rates.  The iron K lines from a pure iron
specimen are being monitored.  (a) A multiple trace oscilloscope
picture of the shaping amplifier output at 440 counts/sec.
(b)  The energy spectrum corresponding to (a).  (c)  A single trace
picture of the amplifier output at 440 counts (sec$^{-1}$).  (d)  A
single trace picture of the amplifier output at a counting rate of
44000 counts/sec at the detector.  Reprinted by courtesy of EG&G
ORTEC.

Figure 4.40 illustrates the phenomena of pileup and a vanishing baseline at high counting rates.  The task assigned to the
baseline restorer is to inspect the average baseline between pulses

and hold it at ground potential using electronic negative feedback. At low counting rates, such as in Fig. 4.40(c), the mean spacing between pulses is large. Consequently, there is plenty of baseline information to sample in order to achieve accurate baseline restoration. At very high counting rates [Fig. 4.40(d)] the mean spacing between pulses is small. The pulses tend to pile up on each other, leaving only very brief moments when the true baseline can be sampled. The fraction of the time when the unperturbed baseline is present for restoration purposes is given by $\exp(-R\,\overline{T_W})$, where R is the total counting rate in the spectrum at the detector, and $\overline{T_W}$ is the average amplifier pulse width in the spectrum. $T_W$ is measured near the baseline, between the points where the amplifier pulse perturbs the baseline by an acceptably small fraction of the noise amplitude $\Gamma_{noise}$. For example, in Fig. 4.40(d) R = 44,000 counts/s and $\overline{T_W}$ = 60 μs. Thus the fraction of the time when baseline is available is predicted to be 0.071. In Fig. 4.40(d) the amplifier output manages to reach the baseline only from 385 to 415 μs in the interval from 0 to 415 μs. This is a fraction of 0.072 which is close to the prediction.

Two things happen when the baseline restorer has only brief moments in which to sample the baseline [9]. The noise contribution $\Gamma_{noise}$ increases, and slight imperfections in the baseline restorer's ability to ignore perturbations from the pulses cause the average baseline to shift toward negative voltages. Consequently, the vanishing baseline is the limiting source of resolution broadening and peak shifting at high counting rates.

Actually Fig. 4.40(d) represents a rather extreme example. The amplifier duty cycle, which is computed as 100% × [1 - $\exp(-R\,\overline{T_W})$], is running at an excessively high value of 92.9%. For throughput reasons discussed below, the normal duty cycle range is from 0 to about 56%. The duty cycle measures the average fraction of the time the amplifier output is perturbed from the baseline as a result of an output pulse being present.

In addition to controlling spectral distortions, the shaping amplifier pulse width determines the throughput efficiency at high

counting rates. With the long shaping time constant usually employed, the spectrometer deadtime is typically set by the amplifier
pulse shape. This deadtime is paralyzable. Consequently, the
measured counting rate in a peak of energy E is given by

$$I_m(E) = I_t(E) \ e^{-R[T_p(E) + \overline{T_W}]} \qquad (4.43)$$

where $I_t(E)$ is the true counting rate at energy E as seen by the
detector. R is the total, true counting rate in the spectrum at
the detector. $T_p(E)$ is the time from the start of the slow amplifier output pulse of energy E to the point at which the analog-to-
digital converter senses the peak amplitude and closes the linear
gate (see Fig. 4.35). $\overline{T_W}$ is the average pulse width for the entire
spectrum, computed from

$$\overline{T_W} = \frac{1}{R} \int_{E=0}^{\infty} T_W(E) I_t(E) \ dE \qquad (4.44)$$

where $T_W(E)$ is the width of the pulse whose energy is E. In this
case $T_W(E)$ is measured as the minimum duration of the pulse of
energy E beyond which a subsequent pulse can occur and be measured
without significant distortion. Frequently, $T_W(E)$ is established
by the pileup rejector as shown in Fig. 4.35. Since $I_m(E)$ is the
measured counting rate of undistorted events in the peak at energy
E in the multichannel analyzer memory, the ratio $I_m(E)/I_t(E)$ defines
the throughput efficiency.

Figure 4.41 illustrates Eq. (4.43) as measured on a typical
energy-dispersive spectrometer with three different shaping time
constants ($\tau$ = 2, 6, and 10 $\mu$s). The graph applies to a case
where the entire spectrum consists of only the Mn K lines. The
"input rate" of the Mn K lines denotes $I_t(E)$, while $I_m(E)$ is labeled
as the "output rate" of the Mn K lines. For a particular time constant, the output rate increases as the input rate is increased,
until a maximum output rate is achieved. When the input rate is
increased beyond this maximum point the output rate decreases.
Thus, for any shaping time constant there is a maximum obtainable

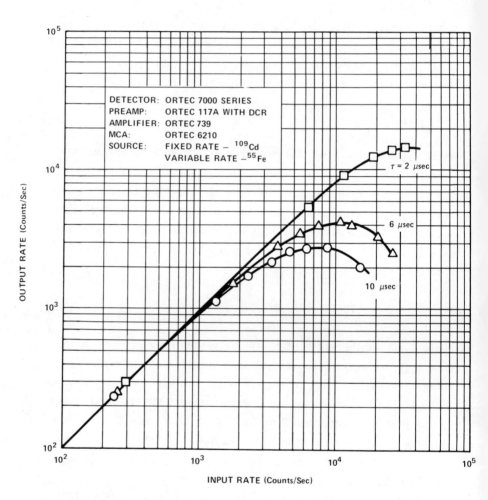

Figure 4.41 Output versus input rate for a typical energy-dispersive spectrometer with pileup rejection. The input rate is the true counting rate at the detector. The output rate is the measured counting rate in the multichannel analyzer memory for events not distorted by pileup. Shaping time constants of 2, 6, and 10 μsec have been used. Reprinted by courtesy of EG&G ORTEC.

output counting rate. Shorter shaping time constants yield higher maximum output rates, because of the shorter amplifier pulse widths.

These same relationships can be noted in Eq. (4.43) for a particular spectrum if it is assumed that $I_t(E)$ is a constant fraction of R, independent of counting rate. Assume, for example, that

$I_t(E) = kR$, where $k \leq 1$ and $k$ is a constant.  This corresponds to
a situation where the x-ray tube voltage is held constant while the
current is varied.   Ideally, the shape of the energy spectrum
emitted by the specimen remains constant, while the true counting
rate R increases in direct proportion to the x-ray tube current.
It can be shown that the maximum value of the measured counting
rate is

$$[I_m(E)]_{max} = \frac{k}{[T_p(E) + \overline{T_W}] \, e}$$

$$= \frac{\left[ \dfrac{I_t(E)}{R} \right]}{[T_p(E) + \overline{T_W}] \, e} \qquad (4.45)$$

and is obtained when

$$R = \frac{1}{[T_p(E) + \overline{T_W}]} \qquad (4.46)$$

Since $T_p(E) + \overline{T_W}$ is approximately proportional to the slow amplifier
shaping time constant, shorter time constants will result in higher
maximum measured counting rates $[I_m(E)]_{max}$ in Eq. (4.45).  It is
interesting to note that the maximum observable counting rate
corresponds to the condition where the deadtime losses are 63.2%.
Consequently, there is no point in operating the spectrometer at
counting rates which cause deadtime losses greater than 63.2%.  The
percent deadtime losses can be computed as

$$\text{Percent deadtime losses} = \frac{I_t(E) - I_m(E)}{I_t(E)} \times 100\%$$

$$= 100\left(1 - e^{-R[T_p(E) + \overline{T_W}]}\right) \qquad (4.47)$$

More specific numbers can be generated from the general equations if a shaping time constant of $\tau = 6$ μs is assumed with
$T_p(E) = 14$ μs and $\overline{T_W} = 60$ μs.  In this typical case the maximum
measured counting rate in the total spectrum would be 4971 counts/s,

and would occur at a true input rate of 13,514 counts/s.  Under
these conditions the deadtime losses [Eq. (4.47)] would be 63.2%,
and the amplifier duty cycle would be approximately 56%.  This set
of numbers corresponds fairly closely to Fig. 4.41.  It is evident
from the figure that the spectrometer could be operated at somewhat
lower true counting rates without significantly reducing the
measured counting rate.  In fact, operating the spectrometer at a
45% deadtime loss yields a measured counting rate which is 89% of
the maximum which can be obtained.  This will correspond to an
amplifier duty cycle of about 38%.  The 45% deadtime loss condition
is a reasonable target for setting up instrument operation, providing all other performance requirements are fulfilled.

The spectral distortions which can arise from nonuniform deadtime losses are apparent in Eqs. (4.43) and (4.47).  If $T_p(E)$ is
the same for all pulse heights then the throughput efficiency and
deadtime losses are the same for all pulse heights.  Under such an
ideal situation there will be no spectral distortion which is directly caused by the deadtime.  Unfortunately, $T_p(E)$ is a function
of pulse height to some degree in all spectrometers.  Usually $T_p(E)$
is fairly constant over the top 90% of the pulse amplitude range
(1 to 10 V), but deviates more and more strongly from this value
as zero pulse height is approached in the bottom 10% of the pulse
amplitude range (under 1 V).  This means that both the throughput
efficiency and the deadtime losses will be a function of pulse
height.  Consequently the shape of the measured spectrum will be
distorted when the deadtime losses are high.  The degree of distortion depends upon the instrument design, since the pulse height
dependence of $T_p(E)$ arises from systematic errors in the pileup
rejector timing and the time of peak amplitude detection in the
analog-to-digital converter.  The analyst should be aware of this
potential error and operate in such a manner as to minimize its
importance.  Under worst case conditions the appropriate solution
may be to operate at lower counting rates.

Most energy-dispersive spectrometers include a means for automatically correcting for deadtime losses. The residual error in
the deadtime loss correction is generally an increasing function of
counting rate. Thus, systematic errors in the deadtime correction
scheme can limit the counting rates which can be employed. This
topic will be examined in Sec. 4.8.

### 4.3.9 Escape Peaks and Other Artifacts

The escape peaks for the Si(Li) detector are generated by the same
mechanism as in the gas flow proportional counter and the NaI(Tl)
scintillation detector. Section 4.2.4 should be consulted for a
more detailed description of this mechanism. There are two important differences with the Si(Li) detector. Since the detector is
composed primarily of silicon it is the escape of the silicon K
x-rays which causes the escape peaks in the spectrum. Secondly,
the pulse height spectrum is used for elemental analysis and usually
contains a large number of lines. Each of these parent lines will
have an escape peak associated with it. The escape peak from an
element with high concentration can interfere with the analysis of
trace amounts of an element of lower atomic number.

As illustrated in Fig. 4.29 a photon is detected in the Si(Li)
diode primarily by ionizing the K shell of a silicon atom in a
photoelectric interaction. The photoelectron carries off the energy
of the original photon less the binding energy of the silicon
K-shell electron and proceeds to lose this energy by causing further
ionization. The original silicon atom is left in an excited state
with a vacancy in its K shell. One mode of decay is for the
vacancy to be filled by an electron from the L or M shells, with
the subsequent emission of a K x-ray. If this silicon K x-ray is
absorbed within the detector active volume, the resulting amplifier pulse will have an amplitude which is proportional to the
original photon energy. Consequently the event will be recorded at
the proper energy in the spectrum. If, on the other hand, the
silicon K x-ray escapes the active volume of the detector, the

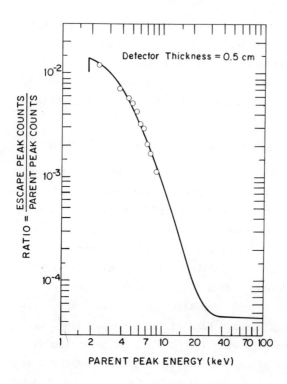

Figure 4.42   The dependence of the silicon escape-peak counts to parent-peak counts ratio on the parent-peak energy.  Adapted from Refs. 36 and 37.  Experimental data are from Ref. 36.  Reprinted by courtesy of EG&G ORTEC.

event will be recorded in the spectrum at an energy which is too low by an amount equal to the energy of the escaping K x-ray.  Thus, a peak at energy E in the spectrum will have a silicon escape peak associated with it at an energy E - 1.74 keV.  The escape peak will have a resolution which is essentially the same as a proper peak at the same energy.  Several workers [36,37] have computed the ratio of the escape peak counting rate to the counting rate of the parent peak at energy E as a function of the parent peak energy. The results are summarized in Fig. 4.42.  The ratio decreases rapidly with increasing energy.  Consequently the effect can be ignored for high-energy lines.  The reason for the rapid decrease in the ratio can be found in the low energy of the silicon K x-ray.

For the silicon K x-ray to escape the detector volume the original photoelectric interaction must take place fairly close to the detector surface. Higher energy x-rays penetrate more deeply into the detector on the average, leading to a reduced probability for the silicon K x-rays to escape. At very high energies where the detection efficiency is small compared to 100% the escape peak to parent peak counting rate ratio is independent of energy [37].

Several other characteristic peaks may be viewed in the spectrum depending on detector construction and the energy spectrum viewed by the detector. These peaks are caused by the incoming x-rays fluorescing the materials in the vicinity of the detector. Normally the peaks are quite small, and are only important when undertaking trace analysis. Weak gold L lines are sometimes observed from the thin gold layers on the detector front and back electrodes. Occasionally palladium replaces the gold layers. Indium-tin alloys are commonly used to improve electrical contacts to the detector, and their K lines may be visible. In addition, lead-tin alloy solders are used for permanent electrical connections in the detector mount, and their characteristic x-rays can be observed under favorable circumstances. Aluminum is a common construction material in the vicinity of the detector, and the aluminum K line may be observed when there is an intense silicon or phosphorous line in the spectrum. Silicon in the inactive regions of the detector diode can also be excited to cause a silicon K line [36]. This is only observed when there is an intense phosphorous or sulfur K line in the spectrum. Other characteristic lines from less commonly used construction materials may be present, depending on the detector mount and cryostat design.

## 4.3.10  Ge(Li) and High-Purity Ge Detectors

Over the energy range of 1 to 40 keV the Si(Li) detector diode is the preferred device for energy dispersive x-ray spectroscopy. The low atomic number of silicon minimizes the absorption of low-energy x-rays in the detector deadlayer. In addition the silicon

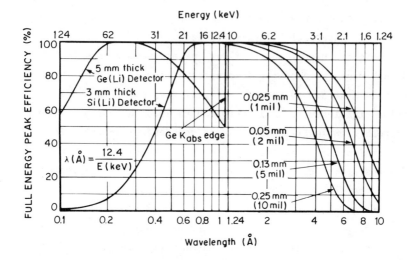

Figure 4.43  A comparison of full energy detection efficiencies for
Si(Li) and Ge(Li) detectors.  The transmissions of various
beryllium window thicknesses are plotted for low-energy x-rays.
Reprinted by courtesy of EG&G ORTEC.

escape peak probability is quite low so that escape peaks do not
cause a substantial complication of the observed spectrum.  As illustrated in Fig. 4.43, the detection efficiency of the Si(Li)
detector decreases rather abruptly for energies above 20 keV.

Where the K lines of heavy elements are to be measured, it is
worthwhile to take advantage of the better detection efficiency of
the germanium detector at high energies.  The K lines of the elements from cesium (Z = 55) to fermium (Z = 100) fall in the energy
range from 30 to 140 keV.  In this energy range the germanium detector has a much better detection efficiency than provided by the
silicon detector (see Fig. 4.43).

The germanium detector is not a very attractive alternative
for energies below 40 keV for two reasons.  First, its dominant
escape peaks cause much more complication of the energy spectrum,
and strongly reduce the full energy detection efficiency from
11.103 to 30 keV.  Second, the higher atomic number makes it more
difficult for low-energy x-rays to pass through the front contact
deadlayer.

The lithium-drifted germanium detector, denoted Ge(Li), is similar in construction to the Si(Li) detector. However, a striking difference is found in the very high lithium mobility in the germanium detector at room temperature. A Ge(Li) detector will be destroyed if it is allowed to warm up from the normal 77 K operating temperature. The Si(Li) detector and preamplifier on the other hand, usually can withstand a warm-up, providing the detector bias voltage is turned off.

By 1975 the Ge(Li) detector had been supplanted by a newer, more rugged device: the high-purity germanium detector. This germanium detector construction is produced without the lithium-drift compensation. Consequently it is as impervious to warm up as the Si(Li) detector.

Resolutions for the germanium detectors can be calculated using a Fano factor F = 0.125, and $\varepsilon$ = 2.95 eV in Eq. (4.42).

## 4.4  ENERGY DISCRIMINATION WITH BALANCED FILTERS

On portable fluorescence spectrometers balanced filters [38] are often used in conjunction with a proportional counter or scintillation detector in order to provide energy discrimination [39]. The principle is illustrated in Fig. 4.44. Two filters are made from adjacent elements in the periodic table, such that their K absorption edges span the energy interval to be analyzed. In Fig. 4.44 the zinc K$\alpha$ x-rays at 8.6 keV are to be counted; so nickel and copper have been selected as the two filters. The thicknesses of the two filters are adjusted so that the transmissions are virtually identical for energies below the nickel K absorption edge and above the copper K absorption edge. The nickel K absorption edge lies at 8.331 keV and the copper K absorption edge is at 8.980 keV.

The specimen is normally excited by a radioisotope. For example, the silver K lines from a $^{109}$Cd source may be used to excite the zinc K$\alpha$ line in an ore sample. First the radiation from the specimen is counted for a time t with the copper filter in

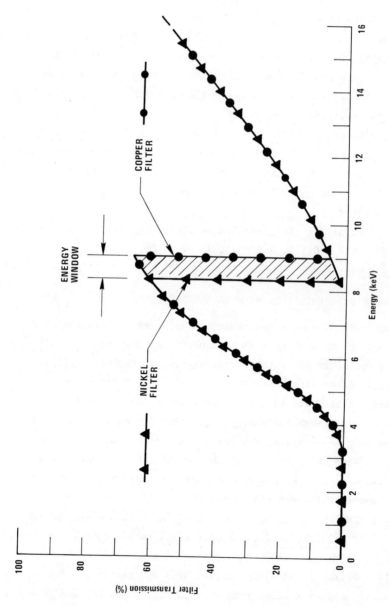

Figure 4.44 The principle of balanced filters. The difference in transmission between the copper and nickel filters is in the cross-hatched energy window between the K absorption edges of nickel and copper. Reprinted by courtesy of EG&G ORTEC.

front of the detector.  Then the copper filter is removed, and the
radiation is counted again for a time t with the nickel filter in
front of the detector.  The difference in these two counts is a
result of the difference in transmission between the two filters,
which is represented by the crosshatched area in Fig. 4.44.  Consequently, the difference between the two counts will represent the
zinc Kα intensity, to the exclusion of all other energies.

Accurate balancing of the filters is not always easily
achieved.  The reader should consult Ref. 39 for further guidance
if this method is to be employed.


## 4.5  COUNTING STATISTICS WITH ZERO DEADTIME

The time distribution of the x-ray photons incident on the detector
in an x-ray fluorescence spectrometer is described by Poisson
statistics, providing the half-life of the radioisotope excitation
source is long compared to the data accumulation period, or providing the x-ray tube output is truly constant.  This means that both
the x-ray tube current and voltage must be constant, with negligible drift or line frequency ripple.  The arrival of photons at the
detector is random in time as illustrated in Fig. 4.45.

In this section the random statistical error will be outlined
for systems with zero deadtime.  These results can be applied to
systems where the deadtime losses are less than 10% with confidence that ignoring the deadtime will contribute less than 5% to
the inaccuracy in estimating the *random* error.  Section 4.6 defines
the systematic error arising from deadtime losses, and Sec. 4.7
deals with the modification of the random error which must be considered at high deadtime losses.  Deadtime loss correction schemes
are discussed in Sec. 4.8.

### 4.5.1  The Poisson Probability Density Function

The random arrival of x-ray photons at the detector is described
by the Poisson probability density function.  More specifically,

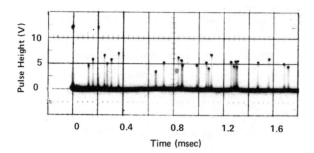

- NaI(Tl) Detector
- Delay-Line Amplifier Output
- Counting Rate: 15,000 Counts/sec
- Single Trace

Figure 4.45.  The random arrival of Ag K x-rays from a radioactive $^{109}$Cd source.  Reprinted by courtesy of EG&G ORTEC.

the probability of observing n photons arriving at the detector in the arbitrary time interval t is

$$P_n(t) = \frac{(\rho t)^n}{n!}\, e^{-\rho t} \tag{4.48}$$

where the constant $\rho$ is the expected mean arrival rate.  $P_n(t)$ is the Poisson probability density function.  This law can be derived from the following, simple postulates [11,40-42] which reveal the fundamental principles involved in x-ray photon counting.

1.  The arrival times of the photons are distributed at random in time.  The probability that a given number of photons arrive in a given time interval depends only on the length of that time interval, and is independent of any information concerning the arrival of photons in any other time intervals.

2.  If $P_{n>1}(\Delta t)$ is the probability of more than 1 photon arriving during the time interval $\Delta t$, then

$$\frac{P_{n>1}(\Delta t)}{\Delta t} \to 0 \qquad \text{as } \Delta t \to 0$$

3.  If $P_1(\Delta t)$ is the probability of one photon arriving during the time interval $\Delta t$, then

$$\frac{P_1(\Delta t)}{\Delta t} \to \rho \qquad \text{where } \rho \text{ is a constant as } \Delta t \to 0$$

Conditions 1 and 3 require that the underlying physical phenomenon responsible for generating the mean photon rate stays constant
over the period of time Eq. (4.48) is tested or employed.  This
means that the half-life of the radioisotope must be long compared
to the measurement time in the case of radioisotope excitation, and
the x-ray tube spectral content and intensity must be constant over
the measurement time in the case of x-ray tube excitation.  X-ray
photons from radioisotopes with very short half-lives, from pulsed
x-ray tubes, or from x-ray tubes exhibiting significant residual
line frequency ripple, will not be distributed according to Eq.
(4.48).

The probability density function $P_n(t)$ is normalized, since
$\sum_{n=0}^{\infty} P_n(t) = 1$.  This fact is  used in computing the mean value of
$P_n(t)$ which is given by

$$\mu = \sum_{n=0}^{\infty} n \, P_n(t) = \rho t \qquad (4.49)$$

The population mean $\mu$ is sometimes called the expectation value of
n and may be denoted by $E(n)$ or $<n>$.  The constant $\mu$ may be regarded as a fundamental parameter of the physical phenomenon causing
the observed events.  Thus Eq. (4.48) can also be written

$$P_n(t) = \frac{\mu^n}{n!} e^{-\mu} \qquad (4.50)$$

The dispersion of $P_n(t)$ about the mean $\mu$ is described by the population variance

$$\sigma^2 = \sum_{n=0}^{\infty} (n - \mu)^2 P_n(t) = \rho t = \mu \qquad (4.51)$$

which is also a fundamental parameter of the physical phenomenon
being sampled by the measurement process.  The population standard
deviation is defined as

$$\sigma = \sqrt{\sigma^2} = \sqrt{\rho t} = \sqrt{\mu} \qquad (4.52)$$

If $\mu$ is known absolutely, then Eq. (4.50) predicts the probability that a particular value of n will be observed in a single measurement. However, in practice, the value of $\mu$ is not known. The spectroscopist seeks to estimate the true value of $\mu$ from a limited number of measurements. It can be shown [42] that for a single measurement yielding an observed number of counts n in the time interval t the most probable value of $\mu$ is estimated by

$$\hat{\mu} = n \qquad\qquad\qquad (4.53)$$

That $\hat{\mu} = n$ is an unbiased estimator of $\mu$ follows from the fact that the probability of obtaining a value $\hat{\mu}$ in a single measurement is equal to the probability of observing the value n. Therefore the expectation value, or mean value, of $\hat{\mu}$ for an infinite number of measurements is

$$\langle\hat{\mu}\rangle = \sum_{n=0}^{\infty} \hat{\mu}P_n(t)$$
$$= \sum_{n=0}^{\infty} nP_n(t)$$
$$= \mu \qquad\qquad\qquad (4.54)$$

It can also be shown that the variance in the estimate $\hat{\mu}$ can be estimated as

$$\hat{\sigma}_{\hat{\mu}}^2 = \hat{\mu} = n \qquad\qquad\qquad (4.55)$$

Just as for the estimate of the mean, it can be shown that the expectation value or mean value of $\hat{\sigma}_{\hat{\mu}}^2$ for an infinite number of samples is the population variance $\sigma^2 = \mu$. Thus the estimate of the standard deviation in $\hat{\mu}$ is

$$\hat{\sigma}_{\hat{\mu}} = \sqrt{n} \qquad\qquad\qquad (4.56)$$

Frequently $\hat{\sigma}_{\hat{\mu}}$ is written as $\sigma_n$.

What the above discussion means is that *when a single measurement of an x-ray intensity is made, and a number of counts n is measured in the time interval t, the most probable value of the desired parameter $\mu = \rho t$ is just the measured value n. The stand-*

*ard deviation associated with this measured value is to be reported as $\sqrt{n}$.*

More often it is the mean rate parameter $\rho$ which is sought and not $\mu$. Since t is assumed to be a precisely determined time interval, not subject to significant statistical variation, the most probable value of $\rho$ is estimated as

$$\hat{\rho} = \frac{\hat{\mu}}{t} = \frac{n}{t} \qquad (4.57)$$

with an estimated standard deviation

$$\hat{\sigma}_{\hat{\rho}} = \frac{\sqrt{n}}{t} \qquad (4.58)$$

It should be noted that n/t is the calculated "true" counting rate at the detector (before deadtime losses) which is defined as $I_t$. Consequently,

$$I_t = \frac{n}{t} \qquad (4.59)$$

is the most probable estimate of the mean counting rate at the detector, with an estimated standard deviation

$$\sigma_{I_t} = \frac{\sqrt{n}}{t} = \sqrt{\frac{I_t}{t}} \qquad (4.60)$$

From Eqs. (4.53), (4.56), (4.59), and (4.60) it can be seen that the relative statistical precision of the measurement is given by

$$\frac{\sigma_n}{n} = \frac{1}{\sqrt{n}} = \frac{1}{\sqrt{I_t t}} = \frac{\sigma_{I_t}}{I_t} \qquad (4.61)$$

Therefore the statistical precision can be improved by increasing the counting rate and/or the counting time. Increasing either of these parameters by a factor of 4 will improve the relative statistical precision by a factor of 2.

Equations (4.59) and (4.60) together with the italicized statement following Eq. (4.56) define the fundamental procedures

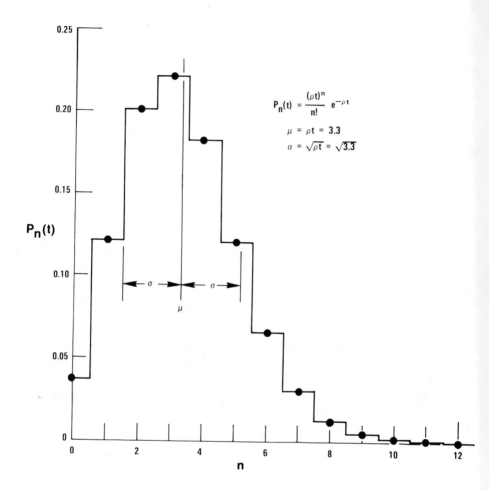

$$P_n(t) = \frac{(\rho t)^n}{n!} \, e^{-\rho t}$$

$$\mu = \rho t = \mathbf{3.3}$$

$$\sigma = \sqrt{\rho t} = \sqrt{\mathbf{3.3}}$$

Figure 4.46   The Poisson probability density function for $\mu$ = 3.3.
Reprinted by courtesy of EG&G ORTEC.

for reporting the counts or counting rate when the x-ray photons
are counted for a preset real time in a system with zero or negligible deadtime.   For deadtime losses up to 10% the random error
expressed in Eq. (4.60) will be distorted by less than 5%, but the
measured counting rate must be corrected for the systematic error
caused by the deadtime losses.

     Figure 4.46 shows the Poisson probability density function
$P_n(t)$ for the case $\mu$ = 3.3.   Although the envelope of the function

has been drawn as a histogram, dots have been added to emphasize
that the function exists only at positive integer values of n.
Note that the Poisson probability density function is quite
assymetric about its mean for low values of μ. For large values of
μ the distribution becomes more symmetric as will be shown in
Fig. 4.48.

4.5.2 The Interval Probability Density Function

Knowledge of the single interval probability density function is
required to predict the statistics with a livetimer, or when operating in the preset count mode. The interval probability density
function defines the probability of observing a time interval t
between two successive x-ray photons arriving at the detector. The
probability of observing a time interval of length t between a
photon at time zero and the succeeding photon can be computed from
the probability of zero photons in the time interval 0 to t followed by one photon in the infinitessimal time interval from t to
t + dt. From Eq. (4.48) the probability of zero photons in the
time interval 0 to t is

$$P_o(t) = e^{-\rho t} \tag{4.62}$$

The probability of one photon in the interval t to t + dt is

$$P_1(dt) = \rho \, dt \tag{4.63}$$

The combined probability is

$$P_t(n = 1)dt = P_t(1)dt = P_o(t)P_1(dt)$$
$$= \rho e^{-\rho t} \, dt \tag{4.64}$$

The probability density function for a single interval (n = 1) is
the probability per unit time, i.e.,

$$P_t(1) = \rho e^{-\rho t} \tag{4.65}$$

Equation (4.65) is the interval probability density function, which
is sometimes loosely labeled "the interval distribution." Figure

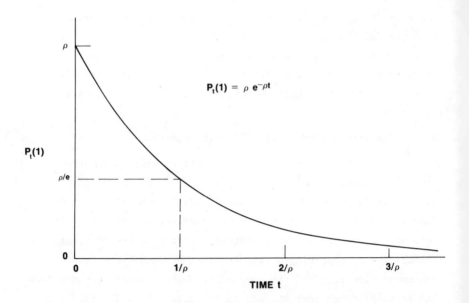

Figure 4.47  The interval probability density function.  Reprinted by courtesy of EG&G ORTEC.

4.47 demonstrates the shape of the function.  Note that short intervals are the most probable.  This is why the x-ray photons seem to arrive in bunches when observed at very low counting rates. Note also that $P_t(1)$ is a *continuous* probability density function.

Equation (4.65) is already normalized since

$$\int_{t=0}^{\infty} P_t(1) \, dt = 1 \tag{4.66}$$

and the population mean for t can be calculated as

$$<t> = \int_{t=0}^{\infty} t P_t(1) \, dt = \frac{1}{\rho} \tag{4.67}$$

This result indicates that the expected mean spacing between photons is just the inverse of the expected mean arrival rate.  The population variance is computed as

$$\sigma_t^2 = \int_{t=0}^{\infty} (t - \langle t \rangle)^2 \, P_t(1) \, dt = \frac{1}{\rho^2} \tag{4.68}$$

### 4.5.3   The n-Fold Interval Probability Density Function

Instead of measuring the number of photons which arrive during a
fixed time interval (preset time mode), one can record the time
required to count n photons.  This second method is called the
*preset count mode*.  It is used when a predetermined statistical
precision is desired in the counting rate estimate.  The n-fold
interval probability density function $P_t(n)$ defines the probability
that a particular value of the random variable t will be measured
as the time required to count n photons in a single measurement.
It is derived as follows.  In general, the time interval will start
somewhere between two photon arrival times.  However, since zero
deadtime is assumed, condition 1 in Sec. 4.5.1 guarantees that the
probability of seeing a time interval of length $t_1$ before the first
photon is counted is given by Eq. (4.65) as

$$P_{t_1}(1) = \rho e^{-\rho t_1} \tag{4.69}$$

independent of when the counting interval starts.  The probability
of seeing a time interval of length $t_2$ between the first and second
photons is given by

$$P_{t_2}(1) = \rho e^{-\rho t_2} \tag{4.70}$$

Consequently the probability of observing a time $t = t_1 + t_2$ to
record the second photon is given by the sum of the probabilities
of all combination of $t_1$ and $t_2$ which add up to the value t.  Since
the probability density functions are continuous the sum becomes an
integral.  Thus the probability of a two-fold interval is

$$P_t(2) = P_{t=t_1+t_2} \quad (n = 2)$$

$$= \int_{t_1=0}^{t} P_{t_1}(1) P_{t_2=t-t_1}(1) \, dt_1$$

$$\equiv P_t(1) * P_t(1) \tag{4.71}$$

The integral in Eq. (4.71) is simply the convolution of $P_t(1)$ with itself, which is the meaning of the notation in the last line of Eq. (4.71). The symbol $*$ stands for the convolution operation performed in the second line of Eq. (4.71). Following this method three-fold interval distribution can be computed as

$$P_t(3) = P_{t_3}(1) * P_{t_1+t_2}(2)$$

$$= P_t(1) * P_t(1) * P_t(1) \tag{4.72}$$

This is just a third-order self-convolution of the single-interval probability density function. Extending this sequence to the n-th order convolution, and performing the integrations, the n-fold interval probability density function becomes

$$P_t(n) = \underbrace{P_t(1) * P_t(1) * \cdots * P_t(1) * P_t(1)}_{\text{n-th order convolution}}$$

$$= \frac{\rho^n t^{(n-1)}}{(n-1)!} e^{-\rho t} \tag{4.73}$$

This is the probability of observing a time interval of length t to record n photons. The continuous random variable is t, while $\rho$ and n are constants.

It can be verified that $P_t(n)$ is a normalized function, and the expectation value of t is given by

$$\langle t \rangle = \int_{t=0}^{\infty} t P_t(n) \, dt = \frac{n}{\rho} \tag{4.74}$$

which is simply n times the mean of the single-interval probability density function, as might have been expected. The population variance is

$$\sigma_t^2 = \int_{t=0}^{\infty} (t - <t>)^2 \, P_t(n) \, dt = \frac{n}{\rho^2} \qquad (4.75)$$

Here again it must be pointed out that Eq. (4.73) predicts the probability of measuring a particular value of t *if the population parameter $\rho$ is known*. Unfortunately, the spectroscopist does not know the expected mean rate $\rho$. In fact he is hoping to estimate $\rho$ from his measurement of the time interval t. It can be shown that if t is the measured time required to count the preselected number of photons n, then

$$\hat{\rho} = \frac{n}{t} \qquad (4.76)$$

is the most probable estimate of the population mean rate $\rho$ for a single measurement of t. This is an unbiased estimator since the average value of $\hat{\rho}$ for k repetitive measurements of t approaches the population mean rate $\rho$ as k tends to infinity. Similarly,

$$\hat{\sigma}_t^2 = \frac{n}{\hat{\rho}^2} = \frac{t^2}{n} \qquad (4.77)$$

and $\quad \hat{\sigma}_t = \frac{\sqrt{n}}{\hat{\rho}} = \frac{t}{\sqrt{n}} \qquad (4.78)$

are unbiased estimators of the variance in t and the standard deviation in t, respectively. Note that n/t is what has been defined as the calculated true counting rate at the detector $I_t$. Consequently the estimated relative statistical precisions in $\hat{\rho}$, t, and $I_t$ can be calculated as

$$\frac{\sigma_{I_t}}{I_t} = \frac{\sigma_{\hat{\rho}}}{\hat{\rho}} = \frac{\sigma_t}{t}$$

$$= \frac{t/\sqrt{n}}{t}$$

$$= \frac{1}{\sqrt{n}} \qquad (4.79)$$

The result is rather fortunate. It shows that the precision of the

measurement can be predetermined by selecting the appropriate value
of the preset count n.

It will be obvious that the single-interval probability density function and its population parameters are given by the n-fold
probability density function by setting n = 1.  At low values of n,
$P_t(n)$ is strongly asymmetric.  At high values of n the function becomes symmetric and is accurately approximated by a guassian probability density function having the same mean and variance.  This
latter result could also be predicted from the *central limit
theorem* [42].  Evans [11] has plotted the n-fold probability density function for a range of values of n, and provides further discussion on the application of this function to photon counting
systems.  One important aspect described by Müller [43] is the fact
that the Poisson probability density function can be derived from
the n-fold interval probability density function.  This result is
important to the understanding of the statistics of a system with
a livetime clock.

### 4.5.4  The Normal Approximation to the Poisson Probability Density Function

The *central limit theorem* [42] shows that the normal or gaussian
probability density function is a good approximation to the Poisson
probability density function for large values of $\mu$.  That is,

$$P_n(t) = \frac{\mu^n}{n!} e^{-\mu} \rightarrow \frac{1}{\sqrt{2\pi\mu}} e^{-(n-\mu)^2/2\mu} \qquad (4.80)$$

as $\mu \rightarrow \infty$.  Here, the gaussian function

$$f(n) = \frac{1}{\sqrt{2\pi\sigma^2}} e^{-(n-\mu)^2/2\sigma^2} \qquad (4.81)$$

has been restricted, by setting $\sigma^2 = \mu$.  Figure 4.48 demonstrates
that the approximation is reasonably good even for values of $\mu$ as
low as 30.  Note that the Poisson function is defined only for
positive integer values of n, whereas the gaussian function is continuous.  This approximation is useful for computing statistical
errors in complicated functions of the measured counting rate, and

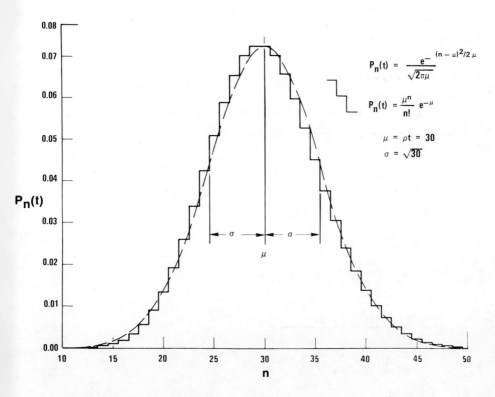

Figure 4.48  The Poisson (solid line) and the gaussian (dashed line) probability density function for $\mu = 30$ and $\sigma = \sqrt{\mu}$. Reprinted by courtesy of EG&G ORTEC.

for computing detection limits (Chap. 11).  See Chap. 5 for further information on the use of the normal distribution.

### 4.5.5  Superposition and the Sum of Two Counts

Consider two independent, random series of photons occuring at population mean rates $\rho_1$ and $\rho_2$.  Suppose series one is counted for a time interval $t_1$, resulting in $n_1$ counts being recorded, and series two is counted for a time interval $t_2$ resulting in $n_2$ counts being recorded.  The probabilities of measuring counts $n_1$ and $n_2$ will be given by

$$P_{n_1}(t_1) = \frac{(\rho_1 t_1)^{n_1}}{n_1!} e^{-\rho_1 t_1} \tag{4.82}$$

and $P_{n_2}(t_2) = \dfrac{(\rho_2 t_2)^{n_2}}{n_2!} e^{-\rho_2 t_2}$                                     (4.83)

respectively, where $\mu_1 = \rho_1 t_1$ and $\mu_2 = \rho_2 t_2$ are the respective
population means.

Consider the number N formed from values $n_1$ and $n_2$ obtained
from single measurements of $n_1$ and $n_2$

$$N = n_1 + n_2$$                                                        (4.84)

Since $n_1$ and $n_2$ are random variables, N is also a random variable.
The probability of obtaining a particular value of N is given by

$$
\begin{aligned}
P_N &= P_{N=n_1+n_2} \\
&= \sum_{n_1=0}^{N} P_{n_1}(t_1) P_{n_2=N-n_1}(t_2) \\
&= \frac{(\rho_1 t_1 + \rho_2 t_2)^{n_1+n_2}}{(n_1 + n_2)!} e^{-(\rho_1 t_1 + \rho_2 t_2)} \\
&= \frac{(\mu_1 + \mu_2)^{N}}{N!} e^{-(\mu_1 + \mu_2)}
\end{aligned}
$$                                                        (4.85)

Equation (4.85) shows that N is also distributed with a Poisson
probability density function whose mean is given by

$$\mu_N = \mu_1 + \mu_2$$

and with a variance

$$\sigma_N^2 = \sigma_1^2 + \sigma_2^2 = \mu_1 + \mu_2$$                                     (4.86)

In other words, the mean for N is the sum of the means for $n_1$ and
$n_2$, and the variance for N is the sum of the variances for $n_1$
and $n_2$.

This result can be extended to k values $n_i$ such that

$$N = \sum_{i=1}^{k} n_i$$                                                (4.87)

The resulting probability density function for N is Poissonian

$$P_N = \frac{(\mu_N)^N}{N!} e^{-\mu_N} \tag{4.88}$$

with population mean

$$\mu_N = \sum_{i=1}^{k} \mu_i \tag{4.89}$$

and population variance

$$\sigma_N^2 = \sum_{i=1}^{k} \sigma_i^2$$

$$= \sum_{i=1}^{k} \mu_i \tag{4.90}$$

Equations (4.87) to (4.90) show how to sum the results of two or more counts and predict the statistical error in the sum. The most probable value of $\mu_N$ is estimated by

$$\hat{\mu}_N = N = \sum_{i=1}^{k} n_i \tag{4.91}$$

and the unbiased estimate of the variance in $\hat{\mu}_N$ is

$$\hat{\sigma}_N^2 = N = \sum_{i=1}^{k} n_i \tag{4.92}$$

from which the standard deviation $\hat{\sigma}_N$ may be calculated.

A special case of the above result, which is important in energy dispersive x-ray spectrometry, is the *superposition* of Poisson distributed events. Consider all the counting intervals to be the same, i.e., $t_i = t$. Then Eq. (4.89) becomes

$$\mu_N = \rho_N t = \sum_{i=1}^{k} \rho_i t = t \sum_{i=1}^{k} \rho_i \tag{4.93}$$

or     $\rho_N = \sum_{i=1}^{k} \rho_i \tag{4.94}$

and Eq. (4.90) becomes

$$\sigma_N^2 = \sum_{i=1}^{k} \sigma_i^2 = \sum_{i=1}^{k} \rho_i t = t \sum_{i=1}^{k} \rho_i = \rho_N t \qquad (4.95)$$

Since the k different series of photons were assumed to be independent random variables, Eqs. (4.93) to (4.95) show that the $\rho_i$ can
be considered as the population mean rates for k different energy
components of a spectrum, all being counted simultaneously by the
same detector. Each component has its own population mean rate
$\rho_i$ and variance $\sigma_i^2$ which can be estimated in spite of the presence
of the other components. In fact, with zero deadtime, or with an
ideal livetime clock, the population parameters $\rho_i$ and $\sigma_i^2$ for each
component are independent of whether or not the other components
are present. The population parameters of each component can be
estimated from the measurement of the $n_i$ counts recorded in the
time t. The estimates of the population mean count and variance
for the i-th component are

$$\hat{\mu}_i = n_i \qquad (4.96)$$

$$\hat{\sigma}_{\hat{\mu}_i}^2 = n_i \qquad (4.97)$$

and the estimate of the population mean rate and its standard
deviations are

$$\hat{\rho}_i = \frac{n_i}{t} \equiv I_t(E_i) \qquad (4.98)$$

$$\hat{\sigma}_{\hat{\rho}_i} = \frac{\sqrt{n_i}}{t} \equiv \sigma_{I_t(E_i)} \qquad (4.99)$$

The addition of the $E_i$ in parenthesis signifies the i-th component
has an x-ray energy $E_i$.

The discussions above show that *each component of the energy
spectrum has a Poissonian distribution, and the sum of any number
of components also has a Poissonian distribution.* This also means
that the i-th component can be considered to be defined by a
channel at energy $E_i$ in the multichannel analyzer memory. If the

$n_i$ counts in each of s adjacent channels are summed to arrive at the total counts within a peak associated with a particular characteristic x-ray line, then the number of counts in the peak can be reported as

$$M = \sum_{i=j}^{j+s-1} n_i = \hat{\mu}_M \qquad (4.100)$$

with an estimated standard deviation

$$\hat{\sigma}_M = \sqrt{M} \qquad (4.101)$$

where j and j + s - 1 are the first and last of the s channels being summed. In the day by day work where it is not so important to emphasize that M and $\hat{\sigma}_M$ are estimates of the true population parameters the $\hat{\mu}_M$ notation is usually omitted and $\hat{\sigma}_M$ is reported as $\sigma_M = \sqrt{M}$.

## 4.5.6 The Difference of Two Counts

Consider the same two Poisson distributions as defined by Eqs. (4.82) and (4.83) in Sec. 4.5.5. In this case single samples of the two distributions yield the values $n_1$ and $n_2$, and the distribution of the difference

$$N = n_1 - n_2 \qquad (4.102)$$

is sought.

The probability density function of N is given by

$$P_N = P_{N=n_1-n_2} = \sum_{n_2=0}^{\infty} P_{n_1=N+n_2}(t_1) P_{n_2}(t_2) \qquad \text{(for } N \geq 0)$$

$$= \sum_{n_1=0}^{\infty} P_{n_1}(t_1) P_{n_2=n_1-N}(t_2) \qquad \text{(for } N \leq 0) \qquad (4.103)$$

The function $P_{N=n_1-n_2}$ is not a Poisson probability density function, nor is it particularly simply in analytical form. Consequently, another method must be used to compute the most probable value of the population mean $\mu_N$ and its variance $\sigma_N^2$.

General theorems exist in the mathematics of statistics to predict the means and variances for probability density functions of
arbitrary form.  These theorems may be applied where the probability
density function is not known or is inconvenient to calculate.  It
can be shown [42] for the difference expressed by Eq. (4.102) that
the expectation values and variances are related by

$$\langle N \rangle = \langle n_1 \rangle - \langle n_2 \rangle = \mu_1 - \mu_2 \tag{4.104}$$

$$\sigma_N^2 = \sigma_{n_1}^2 + \sigma_{n_2}^2 = \mu_1 + \mu_2 \tag{4.105}$$

Hence the difference N will be reported as

$$N = n_1 - n_2 \tag{4.102}$$

with a standard deviation

$$\sigma_N = \sqrt{n_1 + n_2} \tag{4.106}$$

Note that the magnitude of N is smaller than either $n_1$ or $n_2$ but
the standard deviation of N is larger than the standard deviation
for either $n_1$ or $n_2$.

Equations (4.102) and (4.106) are applicable to background
subtraction, where $n_1$ is the peak plus background counts and $n_2$ is
the separate background count.

More meaningful results can be obtained by assuming both $n_1$
and $n_2$ are large compared to 30, and the gaussian approximation to
the Poisson probability density function is valid.  In this case
Eq. (4.103) becomes

$$
\begin{aligned}
P_{N=n_1-n_2} &= \int_{n_2=-\infty}^{+\infty} P_{n_1=N+n_2} P_{n_2} \, dn_2 \\
&= \frac{1}{\sqrt{2\pi(\mu_1 + \mu_2)}} e^{-(N-\mu_1-\mu_2)^2/2(\mu_1+\mu_2)} \\
&= \frac{1}{\sqrt{2\pi\sigma_N^2}} e^{-(N-\mu_N)^2/2\sigma_N^2}
\end{aligned}
\tag{4.107}
$$

The difference N in this case also has a gaussian probability density function with a population mean

$$\mu_N = \mu_1 - \mu_2 \tag{4.108}$$

and a population variance

$$\sigma_N^2 = \mu_1 + \mu_2 = \sigma_{n_1}^2 + \sigma_{n_2}^2 \tag{4.109}$$

This result is identical to Eqs. (4.104) and (4.105), but with one important addition. Equation (4.107) provides a convenient probability density function which allows computation of the confidence that a small positive difference $N = n_1 - n_2$ signifies the detection of a peak above background. This result will be used to calculate detection limits in Chap. 11.

4.5.7 The Ratio of Two Counts

In x-ray spectrometry it is often necessary to form the ratio of two counts $n_1$ and $n_2$, i.e.,

$$z = \frac{n_1}{n_2} \tag{4.110}$$

For example, $n_2$ may be the number of counts obtained on a standard, while $n_1$ is the number of counts from the unknown. For similar compositions z is the ratio of the concentration of the element in the unknown to the known concentration in the standard. Using conventional, first-order, error analysis formulas [41,42] the relative standard deviation of the ratio z can be reported as

$$\frac{\sigma_z}{z} = \left[ \left( \frac{\sigma_{n_1}}{n_1} \right)^2 + \left( \frac{\sigma_{n_2}}{n_2} \right)^2 \right]^{\frac{1}{2}}$$

$$= \sqrt{\frac{1}{n_1} + \frac{1}{n_2}} \tag{4.111}$$

providing $n_1$ and $n_2$ are independent random variables, which is the normal case in the example quoted.

4.5.8  Variance of a Function of Random Variables

Occasionally it is necessary to compute a result which is a complicated function of one or more random variables.  For example,

$$y = f(x_1, x_2, x_3, \ldots, x_n) \qquad (4.112)$$

is a function of the random variables $x_i$ for i = 1 to n.  The requirement is to compute the variance in y, i.e., $\sigma_y^2$, where estimates
of the variances $\sigma_{x_i}^2$ in each $x_i$ are already known.  Using a first-
order Taylor expansion the approximate variance in y is given by
[41,42]

$$\sigma_y^2 = \sum_{i=1}^n \left(\frac{\partial y}{\partial x_i}\right)^2 \sigma_{x_i}^2 + \sum_{\substack{i=1 \\ i \neq j}}^n \sum_{j=1}^n \left(\frac{\partial y}{\partial x_i}\right)\left(\frac{\partial y}{\partial x_j}\right) \sigma_{x_i x_j}^2 \qquad (4.113)$$

where $\sigma_{x_i x_j}^2$ is the covariance between the variables $x_i$ and $x_j$
defined by

$$\sigma_{x_i x_j}^2 = \langle (x_i - \langle x_i \rangle)(x_j - \langle x_j \rangle) \rangle \qquad (4.114)$$

If all the $x_i$ are independent variables then the covariances are
all zero and the second term in Eq. (4.113) disappears.  It is
usually possible to structure $y = f(x_1, x_2, \ldots, x_n)$ when dealing
with counting statistics such that the covariances are all zero.

It should be appreciated that Eq. (4.113) is a first order
approximation to the variance in the function.  It is reasonably
accurate if the partial derivatives $\partial y/\partial x_i$ are constant over the
interval $\pm\sigma_{x_i}$ about the value of $x_i$, and if $\sigma_{x_i}$ is a small fraction
of $x_i$.

4.5.9  Warning Concerning Nonindependent Random Variables and
       Computer-Enhanced Spectra

The equations in Secs. 4.5.1 to 4.5.7 apply strictly to cases
where the random variables are independent.  In energy-dispersive
x-ray spectrometry computers are sometimes used to modify the raw

data in the acquired spectrum. In many cases the modified spectrum is no longer simply described by Poisson statistics, and the counts in the various channels cannot be considered independent variables. Smoothing is a particular example, where the variance of the counts in each channel is reduced at the expense of resolution broadening and an introduction of a correlation in the data between neighboring channels.

Another example which is often misinterpreted is the computer algorithm for resolution enhancement [44,45]. It is tempting to conclude that application of this algorithm results in an improved peak resolution while preserving the same statistical precision in the peak intensity [45]. Actually, the improved resolution is gained at the expense of the statistical precision in the peak intensity. Consequently, the statistical precision available for separating two overlapping peaks is not improved over the intrinsic precision in the raw, unprocessed spectrum.

Finally, it should be noted that deadtime losses alter the variance in the observed counts. If the deadtime losses are higher than 10%, the effect on the statistical variance must be accounted for. Section 4.7 discusses this latter effect.

## 4.6 THE SYSTEMATIC ERROR CAUSED BY DEADTIME

For all photon counting systems a finite length of time is required to measure and record each photon. During this time the system is unable to recognize the arrival of subsequent photons. Consequently the time taken to process one photon is called the deadtime of the system, and is denoted by $t_d$. The most noticeable effect caused by deadtime is the reduction of the number of observable events at the output of the deadtime generating element in the system compared to the number of events occurring before the deadtime generating element. This is the systematic error caused by deadtime. The systematic error is largest at high counting rates where the deadtime losses are greatest. To predict the systematic error

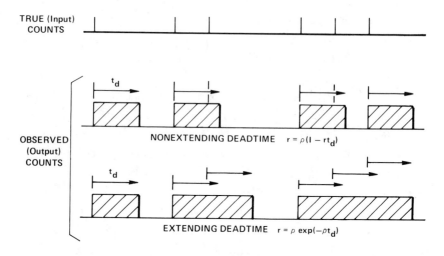

Figure 4.49  The two basic types of deadtime.  Reprinted by courtesy of EG&G ORTEC.

it is necessary to relate the expected mean counting rate after deadtime losses to the expected mean counting rate before deadtime losses.

### 4.6.1  Basic Types of Deadtimes

There are two basic types of deadtime:   (1) nonextending or non-paralyzable and (2) extending or paralyzable deadtime.  All photon counting systems can be described by one of these two types, or by a combination of these basic types.  Figure 4.49 demonstrates the difference between extending and nonextending deadtime.  The top line in Fig. 4.49 represents the photons arriving at the detector (zero deadtime).  The crosshatched rectangles on the second and third lines describe the response of the deadtime generating element to the photons.  The length of the rectangle is the result-ing deadtime interval.  One event is counted for each deadtime interval.  The arrows have a length $t_d$, denoting the deadtime interval associated with a single isolated event.

Both types respond in the same way to a single isolated photon, as depicted by the first event in the figure.  The difference in

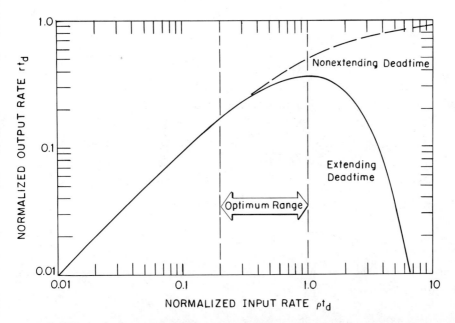

Figure 4.50  Throughput curves for extending and nonextending deadtimes.  Reprinted by courtesy of EG&G ORTEC.

response arises when successive photons arrive during the deadtime
interval caused by a previous photon.  With nonextending deadtime,
a photon which arrives during a previously generated deadtime interval is simply ignored and does not alter the deadtime interval.
This effect is depicted in the arrival of the third and fifth
photons in Fig. 4.49.  With extending deadtime a photon which arrives during a previously generated deadtime interval is also not
counted.  However, the deadtime interval is *extended* by a length $t_d$
from the arrival time of the uncounted photon.  This effect is depicted by the third, fifth, and sixth photons in Fig. 4.49 on the
bottom line.

A number of methods are available for calculating the expected
mean rate r at which the deadtime intervals are generated and
counted, given that the arrival of photons at the detector is described by Eq. (4.48) with an expected mean rate $\rho$ [11,43].  Only
the results will be quoted here.

For a nonextending deadtime the expected mean rate to be measured after deadtime losses is given by

$$r = \frac{\rho}{1 + \rho \, t_d} = \rho(1 - r t_d) \tag{4.115}$$

This relationship is plotted as a dashed line in Fig. 4.50. The equation has been cast in a dimensionless form by multiplying both sides of Eq. (4.115) by $t_d$. Thus, $\rho t_d$ is the normalized input rate, and $r t_d$ is the normalized output rate for the deadtime generating element. Note that the maximum obtainable mean rate after deadtime losses is

$$r_{max} = \frac{1}{t_d} \tag{4.116}$$

which is achieved when $\rho = \infty$. The percent deadtime losses are predicted to be

$$\frac{\rho - r}{\rho} \times 100\% = \frac{\rho \, t_d}{1 + \rho \, t_d} \times 100\%$$

$$= r \, t_d \times 100\% \tag{4.117}$$

as illustrated in Fig. 4.51.

For an extending deadtime the expected mean rate to be measured after deadtime losses is given by

$$r = \rho \, e^{-\rho t_d} \tag{4.118}$$

This equation is plotted as the solid line in Fig. 4.50. Note that the maximum mean output rate is

$$r_{max} = \frac{1}{e \, t_d} \tag{4.119}$$

and occurs at a finite value for the mean input rate $\rho = 1/t_d$. Up to the input rate $\rho = 1/t_d$ increasing the input rate causes an increase in the output rate. Beyond $\rho = 1/t_d$ an increase in $\rho$ causes a *decrease* in the output rate. This is the reason the extending

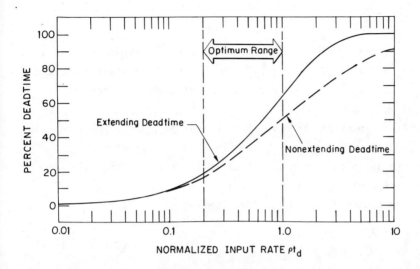

Figure 4.51  Predicted percent deadtime losses for simple extending and nonextending deadtime systems.   Reprinted by courtesy of EG&G ORTEC.

deadtime is often called a paralyzable deadtime.   Operation at input rates above $\rho = 1/t_d$ should be avoided.   The percent deadtime losses for a paralyzable deadtime are given by

$$\frac{\rho - r}{\rho} \times 100\% = \left(1 - e^{-\rho t_d}\right) \times 100\% \qquad (4.120)$$

as illustrated in Fig. 4.51.

Note that the deadtime losses are 63.2% for the maximum out-put rate condition at $\rho = 1/t_d$.

For $\rho t_d$ less than 0.2 both the extending and nonextending equations give approximately the same results since the exponential in Eq. (4.118) can be expanded as an infinite series to give

$$r = \frac{\rho}{1 + \rho t_d + (1/2!)(\rho t_d)^2 + (1/3!)(\rho t_d)^3 + \cdots} \qquad (4.121)$$

For $\rho t_d \ll 1$ the expansion can be truncated to give

$$r \approx \frac{\rho}{1 + \rho t_d} \tag{4.122}$$

for an extending deadtime. The error in this approximation is 1.8% for $\rho t_d = 0.2$ and 0.47% for $\rho t_d = 0.1$. This approximation is useful when deadtime corrections must be calculated and applied manually.

It is important to note that the values $r$ and $\rho$ are theoretical population means, and are not known by the spectroscopist. He seeks to estimate $\rho$ as a result of estimating $r$. Typically, a number of photons $n_m$ are counted in a *real* time interval $t$. The measured counting rate after deadtime losses

$$I_m = \frac{n_m}{t} \tag{4.123}$$

is considered to be the estimate of $r$. Subsequently, $I_t$, which is the calculated estimate of $\rho$, can be obtained from Eqs. (4.115), (4.118), or (4.122) by replacing $r$ with $I_m$ and $\rho$ with $I_t$.

$$
\begin{aligned}
I_m &= \frac{I_t}{1 + I_t\, t_d} \\
&= I_t(1 - I_m\, t_d)
\end{aligned}
\qquad
\begin{aligned}
&\text{nonextending deadtime} \\
&\tag{4.124}
\end{aligned}
$$

$$
\begin{aligned}
I_m &= I_t \exp(-I_t\, t_d) \\
&\approx \frac{I_t}{(1 + I_t\, t_d)} \quad \text{for } I_t\, t_d < 0.2
\end{aligned}
\qquad
\begin{aligned}
&\text{extending deadtime} \\
&\tag{4.125}
\end{aligned}
$$

The appropriate equation can be solved for $I_t$ provided $t_d$ and the type of deadtime are known from previous measurements. This calculated value of $I_t$ is the estimate of the true x-ray counting rate. Equations (4.124) and (4.125) represent the basic means of manually correcting the measured counting rate $I_m$ for the systematic error caused by deadtime losses.

### 4.6.2  Cascaded Deadtimes in Practical Systems

The single-photon counting electronics used in x-ray fluorescence spectrometry involves several deadtime generating elements in

cascade.   Generally only two deadtime elements are significant,
and the system can be described by two deadtimes in series.   In the
single-channel analyzer, or pulse height selector system the two
deadtimes are determined by (1) the shaping-amplifier pulse duration and (2) the output pulse width from the pulse-height selector.
In the multichannel analyzer system with energy-dispersive spectrometers the two deadtime elements are (1) the slow amplifier pulse
width with its associated pileup rejector protection interval and
(2) the multichannel analyzer deadtime.   A number of authors have
treated the problem of cascaded deadtimes.   Two recent reviews are
recommended [10,43].   It should be pointed out that the expressions
for several of the cases treated by Beaman et al. [10] are in disagreement with the more rigorous derivations by Müller [43].   It
appears likely that the derivations of Beaman et al. for their
cases "nsp → nsp," "sp → sp," and "nsp → sp" are in error for the
classification "$\lambda > \tau$ or $\lambda > \tau'$."

  The actual deadtime losses in a practical system strongly depend on the design of the signal processing electronics.   Approximate expressions for the deadtimes are presented in this section so
that the spectroscopist can become aware of situations where the
ritualistic application of a standard deadtime correction can lead
to the wrong answer.

Single-Channel Pulse Height Selectors

Figure 4.52 defines the deadtime intervals in a single-channel
analyzer system with the pulse-height selector operated in the
integral mode (see Sec. 4.2.7).   This operation is typical of the
wavelength dispersive spectrometer.   The integral discriminator
threshold is set at a pulse height $V_L$, and the duration of the
amplifier pulse above this threshold is a type of paralyzable deadtime.   The pulse-height selector produces an output pulse of duration $T_\delta$ when the trailing edge of the amplifier pulse crosses the
threshold $V_L$.   The pulse-height-analyzer output deadtime $T_\delta$ will be
assumed to be nonparalyzable, although it can also be paralyzable,

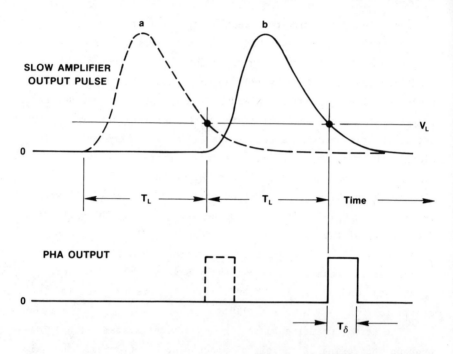

Figure 4.52 Definition of deadtimes in the single-channel analyzer system with the pulse-height selector operating in the integral mode. Reprinted by courtesy of EG&G ORTEC.

depending on design. The time from the start of the amplifier pulse (the arrival time of the x-ray photon) to the time when the amplifier pulse decays back through $V_L$ will be denoted $T_L$. If the solid amplifier pulse (b) arrives after the dashed pulse (a) has decayed below $V_L$, both pulses will be counted. If the solid pulse occurs before the dashed pulse has decayed below $V_L$, only one of the two pulses will be counted. Consequently $T_L$ is the extendable deadtime interval which can be associated with the dashed pulse. Using the formula for cascaded deadtimes the expected mean rate of counted events is given by

$$r = \frac{\rho}{e^{\rho T_L} + \rho(T_\delta - T_L)U(T_\delta - T_L)} \tag{4.126}$$

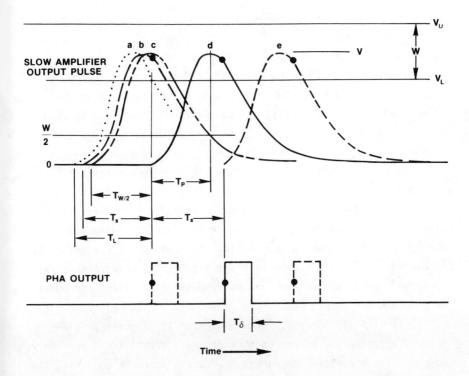

Figure 4.53 Definition of deadtimes in the single-channel analyzer system with the pulse-height selector operating in the window mode. Reprinted by courtesy of EG&G ORTEC.

where $U(T_\delta - T_L)$ is Heaviside's step function $U(x)$ defined by

$$U(x) = 1 \quad \text{for} \quad x \geq 0$$
$$\qquad\ = 0 \quad \text{for} \quad x < 0 \tag{4.127}$$

and $\rho$ is the expected mean rate at the detector. Frequently it is possible to ensure that $T_\delta < T_L$ and the relationship simplifies to

$$r = \rho e^{-\rho T_L} \qquad (\text{for } T_\delta < T_L) \tag{4.128}$$

This is just the simple paralyzable deadtime formula given by Eq. (4.118) with $t_d = T_L$. For a case where the amplifier output

contains a spectrum of pulse heights with amplitudes all greater
than $V_L$, $T_L$ is replaced with $\overline{T_L}$ in Eq. (4.128). $\overline{T_L}$ is the average
value of $T_L$ over the entire range of pulse heights.

Figure 4.53 defines the deadtimes for the single-channel
analyzer system with the pulse-height selector operating in the
window mode. This is typical for the wavelength-dispersive spectrometer where the window ($V_L$ to $V_U$) is set symmetrically on a peak
in the pulse-height spectrum (see Sec. 4.2.7). The pulse height V
represents the centroid of the peak. The time from the start of
the amplifier output pulse (the arrival time of the x-ray photon)
to the point of peak amplitude is defined as $T_p$. The time from the
start of the amplifier pulse to the point where the pulse decays
through the lower level discriminator threshold $V_L$ is designated $T_L$,
as before. The pulse-height selector strobes the condition of the
upper and lower level discriminator memories at a time $T_S$ in order
to generate the output pulse of duration $T_\delta$. As discussed in Sec.
4.2.7, $T_S$ will lie somewhere between $T_p$ and $T_L$, depending on design. To define the appropriate deadtime intervals it is necessary
to determine how close the dashed pulses can approach the solid line
pulse (d) and still permit an undistorted analysis and counting of
the solid pulse. For the pulse preceeding the solid pulse there
are three possible criteria for the minimum separation time. First
of all, if the trailing edge of the preceeding dashed pulse has an
amplitude greater than W/2 at the peak amplitude time of the pulse
(d), the composite pileup pulse will rise above the upper level discriminator $V_U$ and will not be counted. This defines the deadtime
$T_{W/2}$, which is the minimum time the dashed pulse (c) must preceed
the solid pulse to have a trailing-edge amplitude less than W/2 at
the peak-amplitude point of the solid pulse. Here, W is the window
width $W = V_U - V_L$, and the window is set symmetrically about the
pulse height V. In most pulse-height selectors a second output
cannot be generated unless the input signal falls below $V_L$ before
the peak of the second amplifier pulse occurs. This results in the
condition that the preceeding pulse must occur at least a time

$T_L$ before the solid pulse in order to count the solid pulse.  This
criterion is illustrated with the dotted pulse (a) in Fig. 4.53.
In a few pulse-height selectors this latter condition is relaxed
somewhat, and it is only necessary that the previous output strobe
has taken place before the solid pulse starts.  This condition results in a deadtime $T_S$ associated with the preceeding pulse (b).
For simplicity the deadtime associated with the preceeding pulse
will be designated T, with the understanding that the longest of
the applicable values $T_{W/2}$, $T_S$, or $T_L$ must be substituted for T.
Note that T is a paralyzable deadtime.

   Pileup distortion from a subsequent pulse must also be considered.  If the subsequent dashed pulse (e) occurs before the
strobing time $T_S$ on the solid pulse, the strobed information may be
distorted by pileup.  Combining all the deadtimes, the expected
mean rate measured at the pulse-height-selector output for the
window mode becomes

$$r = \frac{\rho}{e^{\rho(T + T_s)} + \rho(T_\delta - T)U(T_\delta - T)} \tag{4.129}$$

where $\rho$ is the expected mean rate at the detector, and $U(T_\delta - T)$ is
the step function $U(x)$ defined in Eq. (4.127).  Frequently it is
possible to ensure that $T_\delta < T$ so that Eq. (4.129) is simplified to

$$r = \rho e^{-\rho(T+T_s)} \qquad \text{(for } T_\delta < T) \tag{4.130}$$

This is just the simple extendable deadtime Eq. (4.118) with
$t_d = T + T_s$.

   An important case occurs when the amplifier output contains a
spectrum of pulse heights.  Suppose that the pulse-height-selector
window is set to count only the photons in the peak centered at
energy E corresponding to a pulse height V, but the spectrum contains an appreciable counting rate of photons with pulse heights
lying outside the window.  In this situation the expected mean rate
for photons of energy E measured at the output of the pulse

height selector is

$$r(E) = \rho(E) \; e^{-R[\overline{T} + T_\delta(E)]}$$  (4.131)

where the simplifying assumption $T_\delta < T$ has been employed. The expected mean rate of photons of energy E at the detector is $\rho(E)$,
while R is the expected mean rate of photons of all energies at the
detector (total spectrum rate). $T_s(E)$ is the time interval $T_s$ for
the pulse amplitude V corresponding to the energy E, and $\overline{T}$ is the
average value of the interval T over the entire spectrum of pulse
heights. Note that the deadtime losses are determined by the total
spectrum counting rate R, not just the counting rate at energy E.
The important point to be learned from Eq. (4.131) is that *the
measured counting rate from the pulse-height selector operating in
the window mode contains insufficient information to calculate the
true counting rate at the detector if the counting rate outside the
window is not negligible.* The measured counting rate from the
window contains no information about the total spectrum counting
rate, whereas it is the total spectrum counting rate which determines the deadtime losses. This effect can be important when the
crystal spectrometer and pulse-height-selector window are set to
measure an element at trace concentrations using first order
diffraction. If an element of high concentration or an intense
scattered x-ray tube line happens to have approximately half the
wavelength of the trace element, then second-order diffraction of
these lines will produce a significant counting rate outside the
pulse-height-selector window. Consequently, deadtime corrections
based on the observed trace-element counting rate will estimate too
low a concentration for the trace element.

From the above discussion it should be obvious that the details of the deadtime effects for pulse-height selector systems on
wavelength dispersive spectrometers can be rather complicated.
Operationally, it is not necessary to remember the details in
making deadtime loss corrections, providing the following general

points are used as guidelines in applying the deadtime correction
formula.

1.   The effective deadtime $t_d$ is a function of the following
     parameters:
     a.   Amplifier pulse length
     b.   Integral versus window pulse-height-selector modes
     c.   Upper and lower level discriminator settings on
          the pulse-height selector.
     d.   Amplifier output pulse height
     If any of these parameters are changed, it will change
     the effective deadtime.
2.   The deadtime is determined by the total pulse-height-
     spectrum content at the amplifier output, not just by the
     component analyzed by the pulse-height selector.
3.   Although the system consists of cascaded deadtimes, the
     deadtime losses can usually be calculated with acceptable
     accuracy by assuming a single *paralyzable* deadtime.

*Multichannel Analyzers*

Figure 4.54 depicts the sources of deadtime in a multichannel pulse-
height analyzer as used on energy-dispersive spectrometers.  Pulse
(d) is the event to be counted.  The deadtime intervals are defined
by determining the closest spacing of pulse (d) and preceeding or
succeeding pulses which just allows an undistorted analysis of
pulse (d).  The time from the start of the amplifier pulse to the
point at which the multichannel analyzer has measured the pulse
height and closed its linear gate is denoted $T_p$.  The dotted ramp
from the peak amplitude of the pulse back to the baseline represents the conversion time of the analog-to-digital converter.  The
dotted rectangle at the end of the ramp accounts for the memory
cycle time.  From Sec. 4.3.4, Eq. (4.32) the deadtime interval
caused by the multichannel analyzer is

$$T_M = \frac{N_c}{f_c} + T_{MC} \qquad\qquad (4.32)$$

and is a nonparalyzable deadtime.  $N_c$ is the channel number into
which the pulse is analyzed, $f_c$ is the address clock frequency, and
$T_{MC}$ is the memory cycle time.  The multichannel-analyzer lower
level discriminator threshold $V_L$ is usually set just above the noise

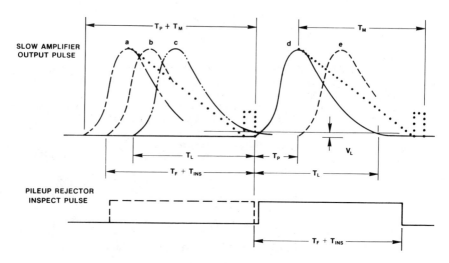

Figure 4.54 Definition of deadtimes in the multichannel analyzer system on energy-dispersive spectrometers. Reprinted by courtesy of EG&G ORTEC.

level. The time from the start of the amplifier pulse to the point at which it falls back through $V_L$ is called $T_L$. The period $T_L$ is a paralyzable deadtime. Several conditions apply to the closest spacing of the preceeding pulse in order that pulse (d) be analyzed. If the previous pulse was analyzed the memory cycle must be complete before pulse (d) can be accepted. This results in a deadtime interval $T_p + T_M$ as shown by pulse (a). If the previous pulse was not analyzed, it must have decayed below $V_L$ before pulse (d) can be accepted. This results in the paralyzable deadtime $T_L$ illustrated by pulse (c). The pileup rejector inspection interval must also be finished before pulse (d) starts (see Sec. 4.3.5, Fig. 4.35). This establishes a paralyzable deadtime interval $T_F + T_{INS}$ as depicted by pulse (b) and the dashed pileup rejector inspect pulse. The longer of the two intervals $T_L$ and $T_F + T_{INS}$ will dominate. Consequently the deadtime from this effective amplifier pulse width will be denoted $T_W$, with the understanding that the longer of the two intervals $T_L$ and $T_F + T_{INS}$ will be used for $T_W$. Pulse (e) must not arrive before the height of pulse (d) has been measured, or else the measurement of pulse (d) will be distorted and lost because of pileup.  Thus the paralyzable deadtime $T_p$ accounts for leading-edge pileup losses.

Applying the general methods for calculating deadtime losses with cascaded deadtimes, the expected mean rate of counting undistorted pulses in the multichannel analyzer memory is

$$r = \frac{\rho}{e^{\rho(T_W + T_p)} + \rho(T_M + T_p - T_W)U[T_M - (T_W - T_p)]} \qquad (4.132)$$

where $\rho$ is the expected mean rate at the detector and $U[T_M - (T_W - T_p)]$ is the step function $U(x)$ defined in Eq. (4.127). For the long shaping time constants usually employed, and for reasonably high address clock frequencies $T_M$ is commonly less than $T_W - T_p$, and Eq. (4.132) simplifies to

$$r = \rho e^{-\rho(T_W + T_p)} \qquad \text{for } T_M < T_W - T_p \qquad (4.133)$$

which is the simple paralyzable deadtime equation with $t_d = T_W + T_p$.

For a spectrum containing more than one pulse height, Eq. (4.133) can be generalized to

$$r(E) = \rho(E)\, e^{-R[\overline{T_W} + T_p(E)]} \qquad (4.134)$$

where $\rho(E)$ is the expected mean counting rate at energy E before deadtime losses, $r(E)$ is the expected mean undistorted counting rate at energy E in the multichannel analyzer memory, and R is the expected mean counting rate in the *total* spectrum before deadtime losses.  $T_p(E)$ is the value of $T_p$ for the pulse at energy E, and $\overline{T_W}$ is the average effective amplifier pulse width over the entire spectrum.

$$\overline{T_W} = \frac{1}{R} \int_{E=0}^{\infty} T_W(E)\rho(E)\, dE \qquad (4.135)$$

and   $$R = \int_{E=0}^{\infty} \rho(E)\, dE \qquad (4.136)$$

Since almost all energy-dispersive spectrometers utilize automatic electronic compensation for deadtime losses, the only important points to remember about the deadtime loss equations are

1.  The maximum analyzed counting rate is obtained at a finite
    value of the total spectrum counting rate at the detector,
    i.e., at $R = 1/[\overline{T_W} + T_p(E)]$.
2.  The deadtime losses are the same for all pulse heights
    only if $T_p(E)$ is the same for all pulse heights. This
    condition is usually violated for the very low energy
    pulse heights.

4.6.3  Methods for Measuring the Effective Deadtime

In order to make manual corrections for deadtime losses, or to
check the accuracy of an automatic deadtime correction circuit, it
is necessary to measure the effective deadtime of the spectrometer.
A number of methods are available, of which the more useful ones
are outlined below.

*Wavelength-Dispersive Spectrometers*

The most easily applied technique for measuring the effective deadtime for a wavelength dispersive x-ray fluorescence spectrometer
assumes that the expected mean counting rate of x-ray photons at
the detector is proportional to the x-ray tube current i.  A stable
specimen which will generate the characteristic lines and the pulse
height spectrum of interest is placed in the fluorescence spectrometer.  All conditions of the experiment and all the settings of the
adjustable or selectable spectrometer controls are recorded for
later cross-reference, including the pulse-height-selector settings.
The x-ray tube voltage and excitation mode are selected and held
constant throughout the experiment.  A series of measurements is
made of the observed counting rate in the line of interest as a
function of x-ray tube current.  As shown in the previous two sections it can be assumed that the measured counting rate is given by

$$I_m = \rho e^{-\rho t_d} \tag{4.137}$$

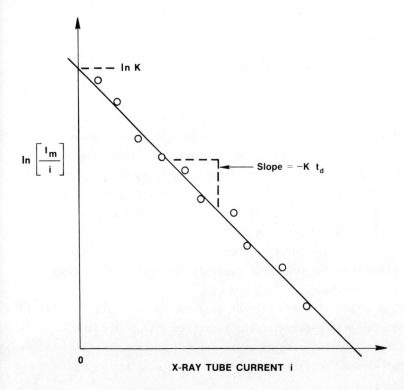

Figure 4.55   The graphical method for measuring deadtime.   Reprinted by courtesy of EG&G ORTEC.

providing there is negligible counting rate at pulse heights other than those falling within the pulse height selector window.   If the expected mean rate at the detector $\rho$ is truly proportional to the x-ray tube current, then $\rho$ can be replaced with $Ki$ in Eq. (4.137), where $K$ is a constant, and $i$ is the x-ray tube current.

$$I_m = Kie^{-Kit_d} \qquad (4.138)$$

Dividing both sides by $i$ and taking the logarithm, Eq. (4.138) can be expressed as

$$\ln \left(\frac{I_m}{i}\right) = -Kt_d i + \ln K \tag{4.139}$$

Equation (4.139) is a linear equation of the form

$$y = mx + b \tag{4.140}$$

where $y = \ln (I_m/i)$ and $x = i$. The slope is $m = -Kt_d$, and the y intercept is $b = \ln K$ at $x = 0$. Consequently the measured counting rate $I_m$ and the tube current can be plotted as shown in Fig. 4.55. The value of K is derived from the antilogarithm of the y intercept, $K = e^b$. The effective deadtime $t_d$ is calculated by dividing the measured slope of the straight line by -K. As pointed out in Sec. 4.6.2, this measured deadtime is correct only for the conditions under which it was measured. Thus a separate deadtime measurement is required when the equipment controls are changed to measure a different x-ray line.

This method of measuring the deadtime is not strictly valid if the pulse height spectrum at the amplifier output contains a significant counting rate of pulse heights outside the pulse-height-selector window. For this case the general equation is obtained from Eq. (4.131):

$$I_m(E) = \rho(E) \, e^{-Rt_d} \tag{4.141}$$

where $I_m(E)$ is the measured counting rate within the window at the pulse-height-selector output. The expected mean rate at the detector for those events which fall within the window is $\rho(E)$, while R is the expected mean rate at the detector for all events in the pulse-height spectrum. When Eq. (4.141) is rearranged and plotted as in Fig. 4.55, it takes the following form

$$\ln \left(\frac{I_m}{i}\right) = -K \frac{R}{\rho(E)} t_d i + \ln K \tag{4.142}$$

When the measured slope of this straight line is divided by the antilogarithm of the y intercept an erroneously high deadtime $t_d'$ is calculated:

$$t'_d = \frac{R}{\rho(E)} t_d \qquad\qquad\qquad (4.143)$$

That is, the computed deadtime is in error by the ratio of the total counting rate in the pulse height spectrum to the counting rate within the window.

Fortunately, the spectroscopist can recover from this error when making the deadtime loss correction *if the pulse-height spectrum to receive the correction is identical to the one on which the deadtime was measured*. The compensation for the error in the measured deadtime is achieved by assuming no counting rate exists outside the pulse-height-selector window when making the deadtime correction. This means that the analyst will assume the deadtime losses are described by

$$I_m(E) = \rho(E)\, e^{-\rho(E) t'_d} \qquad\qquad\qquad (4.144)$$

By substitution from Eq. (4.143), Eq. (4.144) will reduce to the correct relationship as expressed in Eq. (4.141).

$$I_m(E) = \rho(E)\, e^{-\rho(E) t'_d}$$
$$= \rho(E)\, e^{-\rho(E) \frac{R}{\rho(E)} t_d}$$
$$= \rho(E)\, e^{-R t_d} \qquad\qquad\qquad (4.145)$$

This is one case where ignorance of the true situation can still lead to the correct answer. However, it must be emphasized that *compensation for the error in measuring the deadtime is only achieved when the pulse-height spectra are identical for measuring the deadtime and applying the deadtime loss correction*. This condition cannot be guaranteed in many applications.

In most cases where deadtime loss corrections are to be applied manually the deadtime losses are low enough that the approximation in Eq. (4.122) can be used. A further simplification arises from noting that Eq. (4.122) is identical to the nonparalyzable deadtime Eq. (4.115). Consequently, the deadtime can be measured by

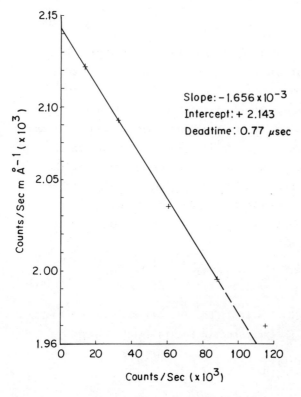

Figure 4.56 Deadtime measurement by the graphical method for low deadtime losses.

the graphical method by starting with the approximate relation from Eq. (4.115):

$$I_m \approx \rho(1 - I_m t_d) = Ki(1 - I_m t_d) \tag{4.146}$$

Dividing by the tube current leads to the linear equation

$$\frac{I_m}{i} = -K\, t_d I_m + K \tag{4.147}$$

which is of the form

$$y = mx + b \tag{4.148}$$

with $y = I_m/i$, and $x = I_m$. The slope of the line is $m = -Kt_d$, and

the y intercept is b = K.  Thus plotting of the measured data as
in Fig. 4.56 allows computation of the deadtime.  The effective
deadtime is calculated from the graph by dividing the slope of the
straight line by the y intercept and multiplying by -1.  In the
case illustrated a deadtime of 0.77 µs was obtained.

The major difficulty encountered in using the x-ray tube current with the graphical method of measuring the deadtime is the
failure to achieve proportionality between the x-ray tube current
and the true counting rate at the detector.  This problem may arise
because the instrument does not permit accurate reading or setting
of the tube current.  More fundamentally, one cannot expect the
proportionality to hold over a wide range of tube current.  Even if
the x-ray tube voltage can be held constant, large changes in the
current significantly modify the focal spot position and size on
the x-ray tube anode.  These effects lead to geometry and spectral
alterations which perturb the proportional relationship between
tube current and counting rate.  This objection can be overcome by
using filters with known transmission factors to adjust the counting rate at the detector.  The x-ray tube current is adjusted to
give the highest counting rate desired with no filters inserted.
The lower counting rates are obtained by inserting filters of increasing attenuation between the specimen and the detector.  In
this case the true counting rate at the detector is proportional to
the total transmission of the filters inserted.  Thus the known
transmissions of the filters can be substituted for the tube current in Eqs. (4.139) and (4.147), and the analysis performed as
usual.  Where possible it is best to insert the filters between the
specimen and the crystal spectrometer to avoid errors caused by
fluorescence of the filter.  The drawback in using the filter
method is the enhancement of second- and higher order diffraction
with increasing filter thickness.

Shifts in the apparent pulse height at the amplifier output can
cause additional losses if the pulse-height-selector window is not
set properly.  Figure 4.57 illustrates the effect.  At high

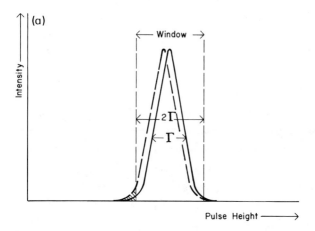

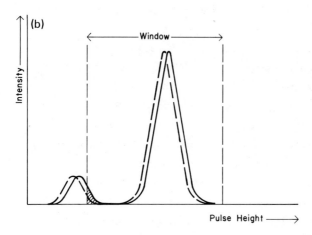

Figure 4.57  Loss of counting rate as a result of a shift in the
pulse-height spectrum:  (a) too narrow a window and (b) too wide a
window.

counting rates the pulse height spectrum at the amplifier output
tends to shift toward zero pulse height as a result of the detector
and baseline restorer effects previously discussed in this chapter.
If the pulse-height-selector window is set too narrow as depicted
in Fig. 4.57(a) some of the events which should have been counted
will shift out of the window.  If the escape peak is well separated

from the parent peak as shown in Fig. 4.57(b), it is also possible
to set the window too wide and encounter a similar problem.  Proper
setting of the pulse-height-selector window is clearly important
for stability.

*Energy Dispersive Spectrometers*

The graphical method described above using the x-ray tube current
can also be employed with the multichannel analyzer systems in
energy-dispersive spectrometers.  The filter technique cannot be
used, however, since it distorts the measured spectrum.  The general equation (4.142) must be used.  Fortunately, the value of the
ratio $R/\rho(E)$ can be measured directly from the observed spectrum
on the multichannel analyzer at low counting rates.  This allows
the true deadtime to be calculated from Eq. (4.143).

A second method is often used to check the accuracy of the
livetimer in the energy dispersive spectrometer.  A low-activity
(1 to 10 µCi) $^{109}$Cd radioactive source is rigidly mounted to the
detector, so that the counting rate in the silver K lines from this
source remains constant throughout the measurement.  Provision is
made for measuring the net counts above background in the silver Kα
peak in a precisely repeatable fashion for a preset time interval.
The total counting rate of the spectrum is varied with the manganese
K lines.  This can be achieved by fluorescing a pure Mn specimen
with a x-ray tube voltage less than 25.5 kV or by using a high-
activity $^{55}$Fe radioisotope source.  The Mn K line counting rate is
varied by adjusting the x-ray tube current or by controlling the
distance between the $^{55}$Fe source and the detector.  In each case
the net number of counts in the Ag Kα peak above background is recorded for the same preset time interval.  The decrease in the number of Ag Kα counts recorded as the spectrum counting rate increases
represents the deadtime losses.  From this data the effective deadtime can be calculated.  This is the method by which Figs. 4.41 and
4.62 were generated.  Note that the calculated deadtime applies only
to the specific case measured.

To check other energies a spectrum having a suite of well-
defined characteristic lines must be run at a variety of counting
rates to determine whether the relative intensities of the various
lines change with counting rate.  This is a difficult measurement
to make.  First of all it is hard to guarantee that the spectrum
incident on the detector is not distorting with counting rate because of nonideal excitation conditions.  Secondly, particular care
must be exercised in separating the pileup spectrum from the true
peaks being monitored.


## 4.7  COUNTING STATISTICS WITH A FINITE DEADTIME

The finite deadtime in x-ray spectrometers has an effect on the
random error in the measured counting rate because deadtime tends
to eliminate the shortest intervals between pulses.  Consequently
it should be expected that the probability density functions describing the counting of pulses after deadtime losses are no longer
the Poisson function [Eq. (4.48)] or the interval probability
density functions given by Eqs. (4.65) and (4.73).

To treat the effects of deadtime on the random counting error,
it is necessary to distinguish between two types of clocks for recording elapsed time.  The *real-time clock*, is the one which is used
in everyday life.  It is the same as the clock on the laboratory
wall which measures time as the analyst normally experiences it.
The *livetime clock* is quite different.  It measures elapsed time
only when the x-ray spectrometer is not dead.  That is, it measures
elapsed time from the end of one deadtime interval to the start of
the next deadtime interval.  During each deadtime interval it is
turned off.  At the end of each deadtime interval it is turned on
again.  Consequently, the livetime clock measures the total time the
spectrometer is "live" and able to respond to the next pulse that
arrives.  Livetime clocks form an important basis for automatic
deadtime correction in energy dispersive spectrometers.

### 4.7.1 Real-time Clocks

With a real-time clock the average number of counts recorded in a
real time t is defined by Eqs. (4.115) and (4.118), depending on
whether the deadtime is nonparalyzable or paralyzable. The expected number of measured counts $n_m$ is related to the expected number of photons at the detector $n_t$ by

Extending deadtime:

$$n_m = rt = \rho t e^{-\rho t_d} = n_t e^{-\rho t_d} \qquad (4.149)$$

Nonextending deadtime:

$$n_m = rt = \frac{\rho t}{1 + \rho t_d} = \frac{n_t}{1 + \rho t_d} \qquad (4.150)$$

The probability density function for the measured number of counts
in a realtime t after deadtime losses is no longer the Poisson function given by Eq. (4.48). Consequently, the expected variance in
the observed number of counts is no longer equal to the observed
number of counts. The variance in the measured counts has been
derived by several authors [43,46-48]. Only the assymptotic approximations are quoted here. These equations are valid for
$t \gg t_d$ and $t \gg \rho^{-1}$, which is the usual case for x-ray fluorescence
spectrometry. The variance in the measured counts $n_m$ is given by

Extending deadtime:

$$\sigma_{n_m}^2 = n_m(1 - 2\rho t_d e^{-\rho t_d}) \qquad (4.151)$$

Nonextending deadtime:

$$\sigma_{n_m}^2 = n_m(1 + \rho t_d)^{-2} \qquad (4.152)$$

Note that the variance in the measured counts differs from $n_m$ by a
factor depending on the product of the expected mean input rate at
the detector, $\rho$, and the deadtime $t_d$. This product is referred to
as the normalized input rate. The relative variance $\sigma_{n_m}^2/n_m$ from
Eqs. (4.151) and (4.152) is plotted as a function of the normalized

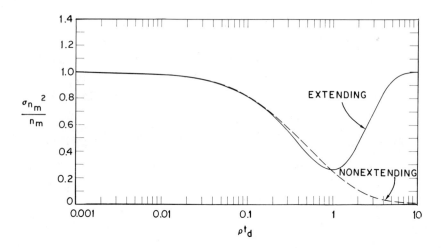

Figure 4.58  The relative variance in the measured counts as a
function of the normalized input rate.  Reprinted by courtesy of
EG&G ORTEC.

input rate $\rho\, t_d$ in Fig. 4.58.  In general, it can be seen that
$\sigma_{n_m} \leq \sqrt{n_m}$.  For deadtime losses less than 20% it is reasonable to
use the approximation $\sigma_{n_m} \approx \sqrt{n_m}$, since this estimate of the standard deviation will be in error by an amount less than 22%.  For
normalized input rates above $\rho t_d = 0.2$ it is important to use the
more rigorous expression for $\sigma_{n_m}$.

   With the simple real-time clock the measured counts must always be corrected for deadtime losses.  The true number of counts
$n_t$ is calculated from the measured counts using Eq. (4.149) or
(4.150) and the known value of the deadtime $t_d$.  Using Eq. (4.113),
the variance in the calculated counts $n_t$ caused by $\sigma_{n_m}^2$ can be
computed to be

   Extending deadtime:

$$\left[\sigma_{n_t}^2\right]_{calc} = n_t \left[\frac{e^{\rho t_d} - 2\rho t_d}{(1 - \rho t_d)^2}\right] \qquad (4.153)$$

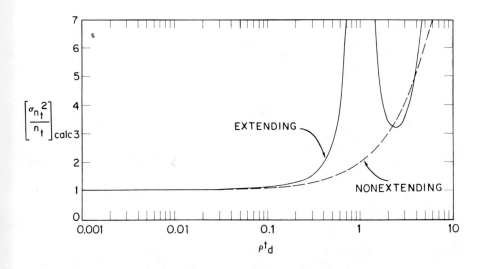

Figure 4.59  The relative variance in the calculated true counts as a function of the normalized input rate.  Reprinted by courtesy of EG&G ORTEC.

Nonextending deadtime:

$$\left[\sigma_{n_t}^2\right]_{\substack{n_t \\ \text{calc}}} = n_t(1 + \rho t_d) \tag{4.154}$$

The relative variance in the calculated true counts expressed by Eqs. (4.153) and (4.154) is plotted in Fig. (4.59) as a function of the normalized input rate.  Clearly $\sigma_{n_t} \geq \sqrt{n_t}$, and the existence of deadtime increases the random error in estimating the true input counts.  For $\rho t_d < 0.2$, or deadtime losses less than 20%, the approximation $\sigma_{n_t} \approx \sqrt{n_t}$ can be used.  This approximation under-estimates the actual random error by less than 14% for deadtime losses less than 20%.  Equation (4.153) approaches infinity at $\rho t_d = 1$ as a result of the zero-slope condition in Eq. (4.149).

The effect of deadtime on the random error leads to the question of whether there is a counting rate which provides the best precision in estimating the true counting rate at the detector.

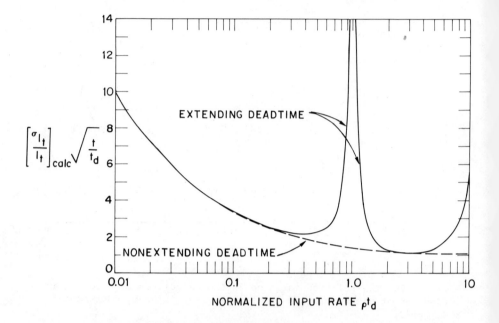

Figure 4.60  The normalized relative precision in the calculated true counting rate as a function of the normalized input rate. Reprinted by courtesy of EG&G ORTEC.

The relative precision in the calculated true counting rate arising from the random error is given by Eqs. (4.155) and (4.156) and is plotted in Fig. (4.60).

Extending deadtime:

$$\left[\frac{\sigma_{I_t}}{I_t}\right]_{calc} = \left(\frac{t_d}{t}\right)^{\frac{1}{2}} \left(\frac{e^{\rho t_d} - 2\rho t_d}{\rho t_d (1 - \rho t_d)^2}\right)^{\frac{1}{2}} \qquad (4.155)$$

Nonextending deadtime:

$$\left[\frac{\sigma_{I_t}}{I_t}\right]_{calc} = \left(\frac{t_d}{t}\right)^{\frac{1}{2}} \left(\frac{1 + \rho t_d}{\rho t_d}\right)^{\frac{1}{2}} \qquad (4.156)$$

Normally the deadtime is of the extending type, and amplifier performance does not permit normalized input rates above unity. Within this restricted range the optimum precision in estimating

the true input rate is obtained when $\rho t_d$ = 0.39.  For other practical reasons it is desirable to hold the counting rate below
$\rho t_d$ = 0.2.  At $\rho t_d$ = 0.2 the precision is only 16% worse than the
optimum.  Consequently, restricting the deadtime losses to less
than 18%, which corresponds to $\rho t_d$ = 0.2, is a reasonable goal.
It provides a precision which is within 16% of the optimum, it
allows the use of the approximation $\sigma_{n_t} \approx \sqrt{n_t}$, and the approximate
expression for deadtime losses given by Eq. (4.122) can be employed
to compute the true input rate.

Equations (4.149) through (4.156) and Figs. 4.58 through 4.60
have been previously summarized [49].  However the extendable deadtime Eqs. (4) and (6) and Fig. 2 of Ref. 49 are in error.  The corrected versions have been presented here.

## 4.7.2  The Ideal Livetime Clock

The ideal livetime clock measures elapsed time only between deadtime intervals.  To distinguish between realtime and livetime the
symbol t will be used for real time, and $t_\ell$ will designate livetime.  The probability density functions describing the counting of
events in livetime can be derived in a simple fashion.

The probability of observing a single livetime interval of
length $t_\ell$ between two deadtime intervals can be computed as follows.
As a result of condition (1) in Sec. 4.5.1 and Eq. (4.48), the
probability that no photons will arrive at the detector in the time
interval of length $t_\ell$, starting with $t_\ell$ = 0 at the end of a previous
deadtime interval, is

$$P_0(t_\ell) = e^{-\rho t_\ell} \qquad\qquad (4.157)$$

The probability of observing one, and only one photon arriving at
the detector in the infinitessimal time interval from $t_\ell$ to
$t_\ell + dt_\ell$ is

$$P_1(dt_\ell) = \rho\, dt_\ell \qquad\qquad (4.158)$$

Therefore the probability density function for the single livetime interval t is

$$P_{t_\ell}(1) = P_o(t_\ell) \frac{P_1(dt_\ell)}{dt_\ell} = \rho e^{-\rho t_\ell} \tag{4.159}$$

Note that this result is the same as Eq. (4.65) except that real time has been replaced by livetime. Consequently, the expected mean livetime interval is $\langle t_\ell \rangle = \rho^{-1}$.

The n-fold livetime interval probability density function can be deduced by the method outlined in Sec. 4.5.3. Thus the probability of observing a total livetime $t_\ell$ to record n deadtime intervals (n measured events) is

$$P_{t_\ell}(n) = \underbrace{P_{t_\ell}(1) * P_{t_\ell}(1) * \cdots * P_{t_\ell}(1) * P_{t_\ell}(1)}_{\text{n-th order convolution}}$$

$$= \frac{\rho^n t_\ell^{(n-1)}}{(n-1)!} e^{-\rho t_\ell} \tag{4.160}$$

This result is identical to Eq. (4.73) except that livetime replaces realtime. The expection value of $t_\ell$, or the expected mean livetime to record n events is

$$\langle t_\ell \rangle = \frac{n}{\rho} \tag{4.161}$$

and the variance in $t_\ell$ is

$$\sigma_{t_\ell}^2 = \frac{n}{\rho^2} \tag{4.162}$$

Following the arguments in Sec. 4.5.3 leads to the following conclusion for an ideal livetime clock. *If a livetime $t_\ell$ is measured as the time required to count n events then the most probable estimate of the mean rate $\rho$ is*

$$\hat{\rho} = \frac{n}{t_\ell} \equiv I_t \tag{4.163}$$

*and the unbiased estimate of the relative standard deviation in the*
*calculated true input rate* $I_t$ *is*

$$\frac{\sigma_{I_t}}{I_t} = \frac{\hat{\sigma}_{\hat{\rho}}}{\hat{\rho}} = \frac{\hat{\sigma}_{t_\ell}}{t_\ell} = \frac{\sigma_{t_\ell}}{<t_\ell>} = \frac{1}{\sqrt{n}} \qquad (4.164)$$

Using the method outlined by Müller [43] the probability of
observing a number of recorded events n in the livetime $t_\ell$ can be
derived from Eq. (4.160). The result is

$$P_n(t_\ell) = \frac{(\rho t_\ell)^n}{n!} e^{-\rho t_\ell} \qquad (4.165)$$

This expression is identical to the case with zero deadtime as
described by Eq. (4.48), except that real time is replaced with
livetime. The expected mean number of observed counts is

$$<n> = \mu = \rho \, t_\ell \qquad (4.166)$$

and the variance is

$$\sigma_\mu^2 = \rho t_\ell = \mu \qquad (4.167)$$

Following the arguments in Sec. 4.5.1 several conclusions may be
drawn. *If a number of events n is recorded in the livetime* $t_\ell$ *the*
*unbiased estimate of the standard deviation in n is*

$$\sigma_n = \sqrt{n} \qquad (4.168)$$

*The most probable estimate of the expected mean photon rate at the*
*detector is*

$$\hat{\rho} = \frac{n}{t_\ell} \simeq I_t \qquad (4.169)$$

*The unbiased estimate of the relative standard deviation in the*
*calculated true rate* $I_t$ *is*

$$\frac{\sigma_{I_t}}{I_t} = \frac{\hat{\sigma}_{\hat{\rho}}}{\hat{\rho}} = \frac{\sigma_\mu}{\mu} = \frac{\sigma_n}{n} = \frac{1}{\sqrt{n}} \qquad (4.170)$$

The principle of the ideal livetime clock contains several
surprises.  First of all, it provides an automatic correction for
deadtime losses.  Dividing the measured counts by the elapsed livetime directly provides an unbiased estimate of the true counting
rate at the detector.  Second, the standard deviation in the counts
measured during a livetime $t_\ell$ is simply the square root of the
number of counts.  Third, a livetime clock makes a system with deadtime behave like a system with zero deadtime.  Consequently all of
the results in Secs. 4.5.4 to 4.5.9 apply equally to the system with
an ideal livetime clock.  The only prerequisites for an ideal livetime clock are

1.  The photons arriving at the detector must be distributed
    according to the Poisson probability density function.
2.  The livetime clock must be turned off for the entire time
    that the spectrometer is unable to record a photon arriv-
    ing at the detector.
3.  One event must be recorded for each deadtime interval.

Interestingly, the general results are quite independent of the type
and amount of deadtime.  The livetime clock principle is a very
powerful tool in dealing with complicated sources of deadtime.

A livetime clock compensates for the deadtime losses by extending the elapsed realtime.  In other words, if a preset livetime
of 100 s is chosen to count the x-ray photons, the elapsed real
time will exceed 100 s.  The relationship between elapsed real time
and livetime is

Elapsed real time = elapsed livetime + total deadtime   (4.171)

Therefore knowledge of the true nature of the deadtime is required
to predict the relationship between the elapsed livetime and the
elapsed real time.

In any measurement process where a given level of accuracy is
required, throughput is determined by the amount of *real time* required to obtain the desired level of measurement accuracy.  For an
ideal livetime clock the relative precision caused by the random
error in estimating the true counting rate at the detector for a

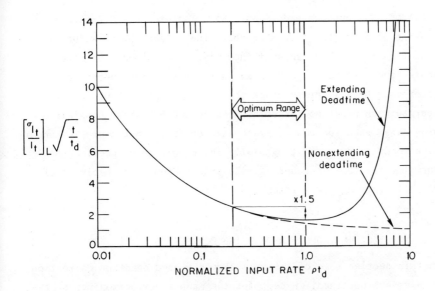

Figure 4.61  The normalized relative standard deviation in the measured input rate with an ideal livetime clock operating for a specified real time t.  Reprinted by courtesy of EG&G ORTEC.

specified amount of real time t has been shown to be [49]

Extending deadtime:

$$\left[\frac{\sigma_{I_t}}{I_t}\right]_L = \left(\frac{t_d}{t}\right)^{\frac{1}{2}} \left(\frac{e^{\rho t_d}}{\rho t_d}\right)^{\frac{1}{2}} \tag{4.172}$$

Nonextending deadtime:

$$\left[\frac{\sigma_{I_t}}{I_t}\right]_L = \left(\frac{t_d}{t}\right)^{\frac{1}{2}} \left(\frac{1 + \rho t_d}{\rho t_d}\right)^{\frac{1}{2}} \tag{4.173}$$

These expressions are plotted in Fig. 4.61 as a function of the normalized input rate.  Since most x-ray spectrometers are characterized by an extending deadtime, the optimum precision is obtained at $\rho = 1/t_d$.  There is no point in operating above this optimum rate since systematic errors from pulse pileup and spectral distortion increase rapidly.  Actually, a rather wide operating range

provides near optimum precision. Rates between $\rho = 0.2/t_d$ and
$\rho = 1/t_d$ degrade the precision above the optimum by a factor of
less than 1.5. This range of input rates corresponds to deadtime
losses from 18% to 63%. With an energy dispersive spectrometer
incorporating a livetime clock the spectroscopist should aim to
operate with deadtime losses between 40 and 50% where possible,
providing other sources of systematic error are not a problem under
such conditions. This target will leave some tolerance for the
normal variations in counting rate experienced with specimens of
slightly varying composition.


## 4.8  PRACTICAL METHODS FOR DEADTIME CORRECTION

In this section some of the more commonly used deadtime correction
schemes are outlined. Except for the manual correction method they
all employ electronics to apply the correction automatically.

### 4.8.1  The Manual Deadtime Correction

This method is commonly used on wavelength-dispersive spectrometers
where no automatic means of correction is available. The system
is operated so that deadtime losses are kept below 20%. This condition permits the use of the approximate Eq. (4.122) in the form

$$I_t \simeq \frac{I_m}{1 - I_m t_d} \qquad (4.174)$$

The true counting rate $I_t$ is calculated from the measured counting
rate $I_m$ by inserting the previously measured deadtime $t_d$ in Eq.
(4.174). To provide accurate answers the deadtime must have been
measured for an identical pulse height spectrum at the pulse-height-
selector input, and identical settings for all of the controls on
the pulse-height-spectrometry electronics. If the deadtime losses
are 20%, a 20% error in the deadtime $t_d$ will cause a 5% error in the
calculated true counting rate. In general, slightly different
deadtimes will have to be used for each element measured. Sections

4.6.2, 4.6.3, and 4.7.1 should be reviewed for more specific information on this method.  With deadtime losses held below 20% the
approximate expression for the relative random error from counting
statistics

$$\frac{\sigma_{I_t}}{I_t} \approx \frac{1}{\sqrt{n_m}} \tag{4.175}$$

can be used, where $n_m$ is the number of counts measured in the real
time t.

### 4.8.2  The Philips Deadtime Corrector

This scheme is used with a wavelength-dispersive spectrometer.  The
deadtime corrector is essentially an electronic box inserted between
the pulse-height-selector output and the scaler input.  The input to
the box is the uncorrected counting rate $I_m$ and the output to the
scaler is the corrected rate $I_t$.  The corrector assumes the approximate relationship in Eq. (4.174) which can be expressed as

$$I_m \approx I_t - I_m I_t \, t_d \tag{4.176}$$

A trigger signal is fed to the corrector from the output of the
first decade of the scaler.  If the corrector is working properly
these trigger pulses will occur at a rate $I_t/10$.  Each trigger pulse
causes a gating period of length $10t_d$ to be generated within the
corrector.  For each pulse-height-selector output pulse which occurs
during the gate period a second pulse is generated and fed to the
scaler input in addition to the regular pulse-height-selector pulse.
These second pulses represent the deadtime correction and will occur
at an average rate

$$I_{CORR} = I_m\left(\frac{I_t}{10}\right)10t_d \tag{4.177}$$

Consequently, the total counting rate presented to the scaler input
will be

$$I_m + I_{CORR} = I_t - I_m I_t t_d + I_m I_t t_d = I_t \tag{4.178}$$

which is the desired true input rate. Compensation for the fact
that the deadtime is actually paralyzable is achieved by making the
gating interval $10t_d$ increase with counting rate. The only shortcoming in this system is the need to pick the correct gating interval to compensate for the actual deadtime. Where spectrometer settings are changed to measure different lines the gating period
should be readjusted to reflect the change in the deadtime. Alternatively, the deadtime losses can be kept low to reduce the sensitivity to errors in the applied deadtime correction. The discussions in Secs. 4.6.2, and 4.6.3 should be reviewed and related
to this method of deadtime correction.

Up to the time of the original writing of this book a definition of the variance to be expected in the recorded counts using
this method had not been published. However, if the raw deadtime
losses are kept below 18% the approximation

$$\frac{\sigma_{I_t}}{I_t} \approx \frac{1}{\sqrt{n_t}} \tag{4.179}$$

should be adequate, where $n_t$ is the corrected number of counts recorded in the real time t.

### 4.8.3 The Simple Livetime Clock

Many of the automatic deadtime correctors in energy dispersive
spectrometers are based on the ideal livetime clock. As shown in
Sec. 4.7.2 the simple livetime clock provides exact compensation
for the loss of pulses which arrive during the time the system is
processing a previous pulse. It does this by turning off the time
clock for the period of time the amplifier pulse is above the
multichannel analyzer's lower level discriminator threshold. The
clock is turned on again when the amplifier output falls below the
discriminator threshold, or when the multichannel analyzer memory
cycle is complete, whichever occurs later. Unfortunately, the
simple livetime clock does not compensate for the loss of the first
pulse when a second pulse arrives before the height of the first

pulse can be measured without pileup distortion.  In other words
the simple livetime clock would provide exact compensation for Eq.
(4.132) or Eq. (4.134) only if $T_p(E) = 0$.  Livetime clocks in
practical systems have to be modified to include compensation for
leading-edge pileup losses, i.e., the case where $T_p(E)$ is finite.
Sections 4.3.5 and 4.6.2 can be reviewed for background on this
problem.  The remaining deadtime correctors described in this section involve various schemes to include compensation for leading
edge pileup losses.

### 4.8.4  Lowes' Livetime Corrector

The Lowes' livetime corrector involves a minor modification of the
simple livetime clock to provide compensation for leading-edge
pileup losses [50,51].  Compensation for the loss of pulses which
occur while a previous pulse is being processed is achieved by a
simple livetime clock as described in Sec. 4.8.3  When a pulse which
would have been counted is lost because a second pulse arrives before the peak amplitude of the first pulse can be measured, the
system carries out an additional pileup loss correction.  The need
for a special pileup loss compensation is signaled by the occur-
rence of an inhibit (reject) pulse from the pileup rejector before
the first pulse's amplitude is measured.  In this case the first
pulse is discarded, and no event is recorded in memory.  To compensate for the loss of the first pulse, another pulse of the same
amplitude must be found and added to memory.  During this correction
process the analyzer is not receiving new information.  It is simply
replacing previous rejected information.  Consequently the spectrometer must be considered to be dead during the correction procedure, and the livetime clock is turned off until the lost event has
been replaced.  The correction procedure assumes that the next
undistorted pulse which arrives is identical to the pulse which was
lost.  Although this appears to be an unjustified assumption, it is
an equivalent representation of the average effect in the spectrum
when a large number of events are processed.  In essence, the

correction procedure consists of holding off the livetime clock until the next undistorted pulse has been processed whenever the pileup rejector discards the first pulse.

Equations (4.132) and (4.134) define the relationship between the true counting rate and the measured counting rate in real time for this system. Lowes' livetime corrector is an ideal livetime clock, since one event is stored for each deadtime interval which is generated. Consequently the true counting rate is estimated by dividing the measured counts by the elapsed livetime. Also the variance in the measured number of counts is equal to the measured number of counts.

At high counting rates, Lowes' livetime corrector shows a deviation from the performance of an ideal livetime clock as a result of the finite deadtime of the fast amplifier in the pileup rejector. This limiting deadtime is typically 0.5 μs. Therefore at high counting rates the performance will approach that of a realtime clock system with a paralyzable deadtime of 0.5 μs. The residual error in the deadtime correction will be typically 1% at about 20,000 counts/s.

Because of the short shaping time constant employed in the pileup rejector the lower energy threshold for pileup rejection is quite high, typically lying in the range of 1 to 2 keV. Pulses in the spectrum below this energy threshold will cause pileup losses but will not receive proper deadtime loss corrections. Thus a significant error in the measured counting rate can occur if the spectrum contains intense sodium, magnesium or aluminum lines.

4.8.5  The Gedcke-Hale Livetime Clock

The Gedcke-Hale livetime clock [21,52] is a modification of the simple livetime clock to include compensation for leading-edge pileup losses. The simple livetime clock accounts for the loss of a second pulse during the deadtime caused by a previous pulse by turning off the clock for the duration of the pulse. This provides exact compensation for the relationship

$$n_m = rt = \rho t e^{-\rho T_w} \tag{4.180}$$

where $T_w$ is the duration of the amplifier pulse above the multichannel-analyzer lower level discriminator (see Sec. 4.6.2). The expected mean rate at the detector is $\rho$, and $n_m$ is the number of counts recorded in the multichannel analyzer memory. The counting rate of the recorded pulses in real time is r. The simple live-time clock works because of the following reasons. Turning off the clock during the deadtime intervals measures the elapsed live time. Furthermore the elapsed live time $t_\ell$ is related to the elapsed real time t by the same factor as in Eq. (4.180), i.e.,

$$t_\ell = t e^{-\rho T_w} \tag{4.181}$$

Consequently, when the recorded number of counts $n_m$ is divided by the livetime $t_\ell$ the result is

$$\frac{n_m}{t_\ell} = \frac{\rho t e^{-\rho T_w}}{t e^{-\rho T_w}} = \rho \tag{4.182}$$

which is the desired estimate of the counting rate at the detector. When pileup losses are experienced, Eq. (4.180) becomes

$$n_m = rt = \rho t e^{-\rho(T_p + T_w)} \tag{4.183}$$

where $T_p$ is the time from the start of the pulse to the point at which the analog to digital converter measures the pulse amplitude and closes the linear gate. To compensate for Eq. (4.183), the livetimer must record a shorter livetime given by

$$t_L = t e^{-\rho(T_p + T_w)} \tag{4.184}$$

Note that the effective deadtime per pulse with pileup losses can be written as

$$t_d = T_p + T_w = 2T_p + (T_w - T_p) \tag{4.185}$$

In other words, the rising portion of the pulse $T_p$ receives double
weighting as a deadtime interval (see Fig. 4.54). Simply turning
off the livetime clock applies a single weight to the measurement
of a deadtime interval. Double weighting of a deadtime interval
can be achieved by making the livetime clock *subtract* time during
the deadtime interval. Thus the required double weighting of the
interval $T_p$ in Eq. (4.184) is achieved by subtracting time from the
clock during $T_p$, and the single weighting of the interval $T_w - T_p$
is obtained by turning off the livetime clock for the interval
$T_w - T_p$. This is the principle of the Gedcke-Hale livetimer in its
simplest form. When a pulse is accepted at the multichannel-analyzer
input, the livetimer begins subtracting time. When the peak amplitude is sensed and the linear gate is closed, the livetime clock is
turned off. The clock is turned on again when the analyzer memory
cycle is complete, or when the amplifier output falls below the
lower level discriminator, whichever occurs later. This causes the
livetime $t_L$ given by Eq. (4.184) to be recorded. Simultaneously the
number of undistorted pulses which is recorded is given by Eq.
(4.183). If the recorded number of pulses is divided by the recorded livetime the result is

$$\frac{n_m}{t_L} = \frac{\rho t e^{-\rho(T_p + T_w)}}{t e^{-\rho(T_p + T_w)}} = \rho \qquad (4.186)$$

which is the desired input rate. Consequently the modified livetimer compensates for the pileup losses.

One of the advantages of the Gedcke-Hale livetimer is the fact
that it does not require a pileup rejector in order to compensate
for pileup losses from the true peak positions in the energy spectrum. For this reason it will work with proportional counters and
NaI(Tl) detectors providing the amplifier pulse shapes are unipolar.
Most of the other automatic methods described here require the
conventional pileup rejector and are not applicable to the NaI(Tl)
detector and the proportional counter.

The accuracy of the Gedcke-Hale livetimer can be improved at very high deadtimes by including information from the pileup rejector when it is available [21,49]. As discussed in Sec. 4.6.2, the intervals $T_w$ and $T_p$ should be measured from the time at which the photon arrives. The analyzer's lower level discriminator is triggered slightly later, which causes a small undercorrection for the true deadtime [21]. Both the multichannel analyzer lower level discriminator and the pileup rejector fast discriminator can be allowed to start the subtraction mode in order to improve the correction. For energies which are above the fast discriminator threshold, an earlier triggering occurs on the fast discriminator, resulting in a more accurate correction [49].

At exceptionally high deadtimes the system tends to overcorrect [21]. This occurs because the analog-to-digital converter peak amplitude detector is used to turn off the subtraction mode. When a pileup occurs the peak-amplitude point is delayed beyond the normal $T_p$ interval, and the subtraction lasts too long. Compensation for this effect can be achieved by stopping the subtraction when a pileup event is detected [49].

Figure 4.62 shows the raw deadtime losses, and the residual deadtime loss error obtained with the Gedcke-Hale livetimer. The data in this figure were obtained using the fixed and variable radioisotope sources method described in Sec. 4.6.3. Normally an error of less than 1% can be maintained up to 80% raw deadtime losses [49]. Larger errors may be experienced in the low-energy range (<2 keV) where timing delays can become significant.

At the date of writing a derivation of the expected variance in the measured counts had not been published. However, Fig. 4.63 represents an experimental test for a perturbation of the standard deviation from the ideal livetimer value $\sigma_{n_m} = \sqrt{n_m}$. The counts in the silver K$\alpha$ line from a low counting rate, fixed position, $^{109}$Cd source were measured while the total spectrum counting rate was set by fluorescing a pure iron specimen. At each counting rate a series of 51 measurements of the Ag K$\alpha$ counts were recorded for

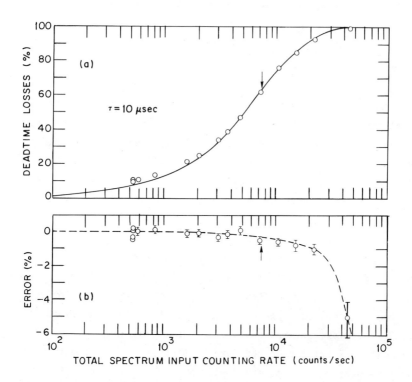

Figure 4.62   (a)   The raw deadtime losses and (b) the residual deadtime error with the Gedcke-Hale livetimer.   The method described in
Sec. 4.6.3 using a $^{109}$Cd and an $^{55}$Fe source was used.   A negative
residual error signifies overcorrection.   The arrows mark the conditions for maximum throughput.   Reprinted by courtesy of EG&G
ORTEC.

10-s preset livetime.   The sample mean $\bar{n}$ and sample standard deviation s were computed for each series.   The ratio $s/\sqrt{\bar{n}}$ obtained for
each counting rate is plotted as a dot in Fig. 4.63.   The expectation value of $s/\sqrt{\bar{n}}$ is unity if the system behaves as an ideal livetimer.   Curve (a) and the confidence levels in Fig. 4.63 relate to
this case.   Curve (b) shows the expectation value of $s/\sqrt{\bar{n}}$ if the
system were an ideal realtime clock.   The experimental data is in
reasonably good agreement with the hypothesis that $\sigma_{n_m} = \sqrt{n_m}$.
Further analysis shows that the average relative variance for all
ten data points is 1.0875.   This falls just below the 5% confidence

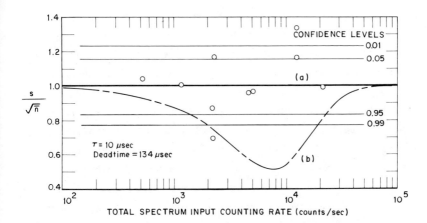

Figure 4.63  Experimental test of the standard deviation in the measured counts with the Gedcke-Hale livetimer.  Reprinted by courtesy of EG&G ORTEC.

level of 1.089 for the hypothesis that $\sigma_{n_m} = \sqrt{n_m}$, indicating reasonable agreement with the hypothesis.  On the other hand, the expectation value of the average relative variance for the 10 data points would be 0.528 if the system behaved as an ideal realtime clock.  This value is in sharp disagreement with the measured data.  Therefore, it is adequate to conclude that the system behaves as an ideal livetime clock and $\sigma_{n_m} = \sqrt{n_m}$ is valid.

This test can be applied to other deadtime correction methods as a means of establishing the relationship between the variance in the measured counts, and the measured number of counts.

### 4.8.6  Harms' Deadtime Correction

The Harms' deadtime corrector attempts to compensate for deadtime losses by counting the events which were actually lost during the deadtime [53,54].  In its usual application to multichannel energy-dispersive x-ray spectrometers it is assumed that the output of the fast discriminator in the pileup rejector represents the true number of photons arriving at the detector.  There are a variety of ways to carry out the correction [53,54].  The following particular

method is described in order to outline the general principle. The
fast discriminator pulses from the pileup rejector are counted in
a "correction" scaler which can be read by the multichannel analyzer. When data acquisition commences the contents of the analyzer
memory and the correction scaler are zero. During data acquisition
the correction scaler counts all the fast discriminator pulses as
they occur. Each time the multichannel analyzer has analyzed a
valid pulse and is ready to add one to a particular memory location,
it interrogates the correction scaler and stores a number in that
memory location which is equal to the current number in the correction scaler. The correction scaler is quickly reset to zero and
counting continues. Thus, each time the analyzer attempts to store
an event, it stores that event plus all the events which have been
lost in the previous deadtime interval. This procedure also compensates for leading-edge pileup losses.

     It may be perturbing to note that the events which were stored
along with the analyzed event did not necessarily have the same
pulse height as the analyzed event, although they were all stored
in the same pulse-height channel. For an ideal system it can be
shown [53] that the method provides rigorously correct compensation
for deadtime losses for the total counts in the spectrum. If the
number of recorded counts is large the system will also provide
correct compensation *on the average* for the deadtime losses in each
channel of the multichannel analyzer. Note that this deadtime loss
correction is accomplished in real time. The system uses a realtime clock. Thus the number of counts recorded in the real time t
is equal to the number of photons which arrived at the detector.
Dividing the recorded number of counts by the *real* time t provides
an unbiased estimate of the true counting rate $\rho$.

     A theoretical derivation of the variance in the measured counts
had not been published by the date this book was written. It is
easy to show that the variance in the number of counts recorded in
the total spectrum is equal to the total number of counts in the
spectrum. This follows since the system counts all the events

without deadtime losses.  On the other hand, it is anticipated that
the variance in the number of counts recorded in any fraction of the
total spectrum is greater than the number of counts recorded in that
fraction of the spectrum.  Consequently, the operator of this type
of system must experimentally determine the relationship defining
the variance in the recorded counts.

The Harms' deadtime corrector deviates from the ideal case in
a practical system for several reasons.  First of all, the fast discriminator exhibits a finite paralyzable deadtime, typically about
0.5 μs.  As a result, the deadtime loss correction will be in error
by 1% at approximately 20,000 counts/s.  Further residual deadtime
can be experienced in the time it takes to read the correction
scaler and reset its content to zero.  A more severe problem occurs
because of the rather large energy threshold for the fast discriminator in the pileup rejector.  Since short shaping time constants
are used in the fast amplifier the fast discriminator threshold
usually lies between 1 and 2 keV.  Absolutely no deadtime correction
is applied for photons below the fast discriminator threshold.  If
a spectrum contains intense sodium, magnesium, or aluminum K lines
the residual systematic deadtime loss error can be quite large.
The error in this case is more severe than with Lowes' livetime
correction or the Gedcke-Hale livetime clock since the latter two
systems provide partial correction for events below the fast discriminator threshold.

### 4.8.7  The Barnhart Deadtime Correction

The Barnhart deadtime correction method [55] uses the same approach
as Harms' deadtime correction for measuring the true counting rate,
but employs a different technique for compensating for the losses.
The output of the pile-up rejector fast discriminator is assumed to
represent the true counting rate at the detector.  The fast discriminator output pulses are counted at the "add" input of an up/
down scaler.  At the start of data acquisition (time zero) the content of this scaler is set to zero.  During data acquisition elapsed

real time is recorded as the fast discriminator pulses are added to
the scaler content.  Simultaneously, one count is subtracted from
the scaler content via the "subtract" input each time an event is
stored in the multichannel analyzer memory.  At any point in time
the scaler content reflects the difference in counts between the
fast discriminator and the total multichannel analyzer memory.  In
other words, the content of the up/down scaler is the number of
pulses lost in the total spectrum as a result of deadtime.  When the
counts in the up/down scaler reach some predetermined maximum level
a deadtime loss correction procedure is initiated.  Both the time
clock and the add input are disabled.  The multichannel analyzer
continues to acquire a spectrum and subtract counts from the scaler
until the content of the up/down scaler reaches zero.  At this point
the time clock is turned on again, and the add input to the up/down
scaler is activated.  During the correction procedure the analyzer
has added a number of counts to memory equal to the number of counts
previously lost, and the elapsed time clock has been held off.  The
system alternates between the normal data acquisition mode and the
deadtime loss correction mode until the time clock has recorded
the desired elapsed time and the up/down scaler has reached zero
counts.  At the end of the data acquisition period the total counts
in the spectrum is equal to the total counts produced by the fast
discriminator during the recorded measurement time.  Although the
relationship between the recorded measurement time and the elapsed
real time is the same as the relationship between livetime and real
time expressed by Eqs. (4.181) and (4.184), the system cannot be
described as a true livetime clock.  The elapsed time clock is not
turned off during the deadtime intervals.

At the date this book was written a theoretical derivation of
the variance in the recorded counts had not been published.  As
with Harms' deadtime correction, it is easy to show that the variance in the number of counts recorded in the total spectrum is
equal to the number of counts in the total spectrum.  Unfortunately,
it is not so obvious what the variance is for the counts recorded

in any fraction of the total spectrum.

The system does compensate for leading-edge pileup losses. However, it has the same problems as Harms' deadtime correction for pulses falling below the fast discriminator threshold. Absolutely no deadtime correction is applied for events below the fast dis- criminator threshold, which typically lies between 1 and 2 keV. In a spectrum containing intense sodium, magnesium, or aluminum K lines a large  systematic deadtime loss error can occur.

At high counting rates the paralyzable deadtime of the fast amplifier and fast discriminator in the pileup rejector becomes a limiting factor.  With a typical deadtime of 0.5 μs in the fast channel the systematic deadtime loss error will be 1% at about 20,000 counts per second.

### 4.8.8  The Constant Reference Source

When all else fails, the deadtime losses in an energy dispersive multichannel spectrometer can be measured by introducing into the spectrum, an artificial line whose intensity does not vary.  The most obvious method is to place a low-activity radioisotope near the detector.  The radioactive source is chosen to have a suitable γ-ray or x-ray line at an energy which can be analyzed without interference in the spectrum.  If an x-ray emitting source is chosen the conditions must be chosen so that radiation from the specimen cannot fluoresce the radioisotope and enhance the intensity of the emitted line.  If the source has a fixed position and a long half- life, any change in the recorded counts should be due to counting statistics or a change in the average deadtime of the spectrometer. Thus the deadtime measured by the fixed source can be applied as a correction to the rest of the spectrum.  The correction can be achieved by multiplying the observed counts in any line by the ratio of the number of counts which should have been counted in the reference line to the number of counts actually recorded in the reference line.  A small, calculated deadtime correction may be necessary when deducing the number of reference counts which should have been recorded.

Basically this is the fixed and variable source method used in Sec. 4.6.3 to measure the deadtime. A number of drawbacks are encountered with this technique in addition to those outlined in Sec. 4.6.3. Only a limited number of counts are permitted in the reference line. The relative statistical precision in the reference line can become the limiting random error in measuring elements with high concentrations. Extreme care must be exercised to avoid the errors in measuring the net counts above background in the reference line caused by changing backgrounds and other types of spectral interferences.

Occasionally a pulser is injected into the slow amplifier input in parallel with the preamplifier signal in order to provide a reference line which has a more selectable energy. Although the pulser makes it easier to avoid spectral interferences it is hard to exactly duplicate the real pulse shape and deadtime with a pulser.

Whether a radioisotope or a pulser is used for a reference line the method does not provide accurate compensation for deadtime losses over the entire spectrum. It shares this fault with all the other deadtime correction schemes, since the deadtime is a function of pulse height; a fact which is most noticeable at low pulse heights.

## 4.9  MEASUREMENT OF A CHARACTERISTIC LINE INTENSITY

In this section a number of conclusions are drawn based on the random counting error. In order for these conclusions to be valid a simplifying assumption must be made. It will be assumed that the variance in the measured number of counts is equal to the measured number of counts. As shown in Secs. 4.7 and 4.8 this assumption is reasonable if the deadtime losses are negligible, or if the system incorporates an ideal livetime clock. Where the simplifying assumption is not applicable the more complicated relationships can be derived from the information in Sec. 4.7. Only the simpler results will be presented in this section.

### 4.9.1 Loose versus Strict Definitions of Intensity

It has become common practice in x-ray fluorescence spectrometry to use the word *intensity* and the symbol I when referring to the photon counting rate at the detector.  Based on the strict classical definition of intensity [56] this is a misuse of the term.  On the other hand, the word *intensity* is so convenient when used to mean counting rate, that the looser use can be expected to persist. Therefore it becomes important for the spectroscopist to understand and avoid the confusion which can arise from the dual use of the term.

The strict classical definition of intensity is the amount of energy passing through a surface of unit area per unit time [56]. For example, suppose a uniform beam of photons of energy E passes through a small segment of a surface in a direction normal to the surface.  If the area of the segment is $\Delta A$, and the number of photons passing through the segment in the short time interval $\Delta t$ is $\Delta N$, then the intensity at the segment of the surface is

$$I = \frac{E \ \Delta N}{\Delta A \ \Delta t} \qquad (4.187)$$

The units of intensity will be some form of energy per unit area per unit time, i.e., ergs/(cm$^2$·s), or J/(m$^2$·s).

A term which is readily related to the intensity is the *flux density* $\phi$.  A flux density is a measure of the number of photons passing through a unit area per unit time.  Thus, the same example yields an expression for the flux density at the segment of the surface given by

$$\phi = \frac{\Delta N}{\Delta A \ \Delta t} = \frac{I}{E} \qquad (4.188)$$

Now assume that the total surface normal to the incident beam direction is the front surface of a detector with sensitive area A, and detection efficiency $\eta(E)$ at the energy E.  If the intensity is uniform over the entire area of the detector then the number of photons impinging on the detector in the time t will be

$$N = \phi At = \frac{I}{E} At \qquad\qquad (4.189)$$

Consequently the detected counting rate from the detector for photons of energy E will be

$$\rho(E) = \eta(E)A\phi = \frac{\eta(E)A}{E} I \qquad\qquad (4.190)$$

For a particular detector and a fixed photon energy the detected counting rate is proportional to both the flux density and the intensity. This is the relationship which leads to the use of the loose definition of intensity. However, the simple proportionality is lost when the beam contains photons of more than one energy. Consider a case where the beam contains an intensity $I_1$ of photons of energy $E_1$ and an intensity $I_2$ of photons of energy $E_2$. The total number of photons arriving at the detector area A in the time interval t is

$$
\begin{aligned}
N &= N_1 + N_2 \\
&= \phi_1 At + \phi_2 At \\
&= (\phi_1 + \phi_2)At \\
&= \left(\frac{I_1}{E_1} + \frac{I_2}{E_2}\right)At \qquad\qquad (4.191)
\end{aligned}
$$

where the subscripts 1 and 2 define the parameters associated with photons of energy $E_1$ and $E_2$, respectively. Note that the total number of photons arriving at the detector is *not* proportional to the sum of the intensities. It is, however, proportional to the sum of the flux densities. When dealing with the counting rate of photons arriving at the detector flux density is the more generally useful description of the beam of photons.

The total *detected* counting rate is given by

$$
\begin{aligned}
R &= \rho(E_1) + \rho(E_2) \\
&= [\phi_1 \eta(E_1) + \phi_2 \eta(E_2)]A \\
&= \left[\frac{I_1}{E_1} \eta(E_1) + \frac{I_2}{E_2} \eta(E_2)\right]A \qquad\qquad (4.192)
\end{aligned}
$$

In this case, even proportionality to flux density is lost unless
the detection efficiency $\eta(E)$ is the same for all energies counted.

For the commonly used single-photon-counting x-ray spectrometers it is the detected photon counting rate which is measured.
Consequently, it is convenient to derive formula in terms of the
detected photon counting rate, or the arrival rate of photons at
the detector. When the term *intensity* is loosely applied for this
situation the real meaning should be defined by noting that the
units of the measurement are photons per second, counts per second
or simply $s^{-1}$.

Some photon sensors provide a response which is actually proportional to the detected intensity. For example, if the average
current produced by a Si(Li) detector is taken as the measured
parameter it will be given by

$$i = \frac{\rho(E_1)E_1 q_e}{\varepsilon} + \frac{\rho(E_2)E_2 q_e}{\varepsilon}$$

$$= [\eta(E_1)I_1 + \eta(E_2)I_2] \frac{q_e}{\varepsilon} \qquad (4.193)$$

where $q_e$ is the charge on a single electron, and $\varepsilon$ is the average
energy required to produce one electron-hole pair in the detector.
If the detection efficiency $\eta(E)$ is constant over the measured
spectrum then the total measured current will be proportional to the
total intensity. Many optical detectors operate on this principle,
and the older, integrating electrometer systems described in Sec.
4.2 had this same basic response. Methodology borrowed from such
instruments may have been a major influence in the widespread use
of the loose application of the term *intensity*.

Note that Secs. 2.2 to 2.6 dealing with scattered intensities
in Chap. 2 are the only places in this book where the strict
definition of intensity is implied. Most diffraction and scattering theories have been developed around the strict definition of
intensity [56]. This must be kept in mind when applying the results of such theories to single-photon-counting detectors.

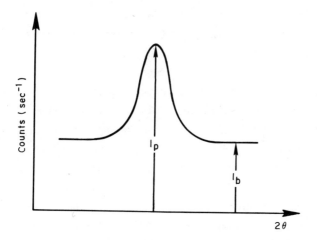

Figure 4.64  Measurement of the net peak intensity with the wavelength dispersive spectrometer.  Reprinted by courtesy of EG&G
ORTEC.

## 4.9.2  Wavelength-Dispersive Spectrometers

To obtain the net peak intensity the background contribution must be
subtracted from the total counting rate measured at the peak position.  Figure 4.64 illustrates the measurement for a sequential
wavelength-dispersive spectrometer.  A time interval $t_p$ is spent
counting the total counts at the correct characteristic line position.  The number of counts recorded in the time $t_p$ is $N_p$ and the
gross peak counting rate is computed as

$$I_p = \frac{N_p}{t_p} \tag{4.194}$$

Similarly a time $t_b$ is used to count $N_b$ photons at a position which
gives a reasonable estimate of the background under the characteristic line.  The background counting rate is computed as

$$I_b = \frac{N_b}{t_b} \tag{4.195}$$

The net peak intensity is

$$I_{p-b} = I_p - I_b \qquad (4.196)$$

which is the value used for quantitative analysis.  The expected
standard deviation in $I_{p-b}$ caused by the random counting error can
be derived via the formula in Sec. 4.5.  The result is

$$\sigma_{I_{p-b}} = \sqrt{\frac{I_p}{t_p} + \frac{I_b}{t_b}} \qquad (4.197)$$

For the case where the peak-to-background ratio $I_p/I_b$ is large
it is evident in Eq. (4.197) that the error can be reduced by
spending more time counting the peak, and less time counting the
background.  It can be shown that $\sigma_{I_{p-b}}$ is minimized, where the
total counting time $t = t_p + t_b$ is fixed, by choosing

$$\frac{t_p}{t_b} = \sqrt{\frac{I_p}{I_b}} \qquad (4.198)$$

The minimized value of $\sigma_{I_{p-b}}$ is

$$\left[ \sigma_{I_{p-b}} \right]_{min} = \frac{\sqrt{I_p} + \sqrt{I_b}}{\sqrt{t_p + t_b}} \qquad (4.199)$$

From which the minimized percent relative precision in the net peak
counting rate can be calculated as

$$(\sigma\%)_{net} = \frac{\left[ \sigma_{I_{p-b}} \right]_{min}}{I_{p-b}} \times 100\%$$

$$= \frac{100\%}{\sqrt{t_p + t_b} \, (\sqrt{I_p} - \sqrt{I_b})} \qquad (4.200)$$

From equation (4.200) it is clear that the percent error in measuring $I_{p-b}$ is minimized if the factor $(\sqrt{I_p} - \sqrt{I_b})$ is maximized.
Therefore $(\sqrt{I_p} - \sqrt{I_b})$ is a useful figure of merit on wavelength
dispersive spectrometers for optimizing the excitation and analysis
conditions when the random error from counting statistics is
significant.

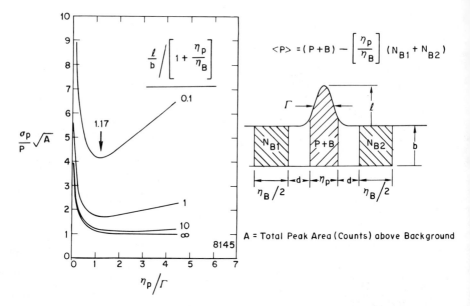

Figure 4.65 The dependence of the relative precision in the net peak counts on the number of channels utilized in the peak and the background measurements. Reprinted by courtesy of EG&G ORTEC.

If the peak-to-background ratio is low accurate background subtraction is important and the guidance of Eqs. (4.198) and (4.200) should be employed. For cases where the peak-to-background ratio is high, it is sometimes possible to omit the background measurement and rely on the quantitative regression analysis to estimate and remove the background contribution.

### 4.9.3  Energy-Dispersive Spectrometers

In energy-dispersive fluorescence spectrometers both the peak and the background counts are acquired simultaneously in the spectrum. Consequently there is no question of a division of time between peak and background counting. However, since the entire spectrum is recorded a question arises concerning how much of the available spectrum should be used to measure the net peak intensity. More specifically, how many channels within the characteristic x-ray peak should be utilized, and how many channels of background should be summed to get the optimum statistical precision?

These questions are addressed in Fig. 4.65.  A number of channels $\eta_P$ is chosen over the peak to provide a measure of the characteristic line intensity.  For the ideal, isolated peak the channels are chosen symmetrically about the peak centroid.  The sum of the counts recorded in the $\eta_P$ channels will be designated $N_T$. The $N_T$ counts contain a contribution of B counts from the background underlying the peak.  Thus the net counts above background which can be attributed to the peak is given by

$$P = N_T - B \tag{4.201}$$

In order to estimate B two background regions are selected, one on each side of the peak.  The background regions are symmetrically located with respect to the peak centroid, and each contains $\eta_B/2$ channels.  The background within the selected peak region is estimated as

$$B = \frac{\eta_P}{\eta_B} (N_{B1} + N_{B2}) \tag{4.202}$$

where $N_{B1}$ is the total counts in the $\eta_B/2$ channels in the lower energy background region, and $N_{B2}$ is the total counts in the $\eta_B/2$ channels in the higher energy background region.  Consequently, the net peak counts are estimated by

$$P = N_T - \frac{\eta_P}{\eta_B} (N_{B1} + N_{B2}) \tag{4.203}$$

Using the results in Sec. 4.5 the random error in P due to counting statistics can be computed to be

$$
\begin{aligned}
\sigma_P &= \sqrt{N_T + \left(\frac{\eta_P}{\eta_B}\right)^2 (N_{B1} + N_{B2})} \\
&= \sqrt{P + B + \left(\frac{\eta_P}{\eta_B}\right) B} \\
&= \sqrt{P + B\left(1 + \frac{\eta_P}{\eta_B}\right)}
\end{aligned}
\tag{4.204}
$$

It is evident from Eq. (4.204) that the random error in P can be minimized by utilizing a large number of channels $\eta_B$ to estimate the background. This is not always possible since neighboring peaks may restrict the width of the available background regions. Generally, one should strive for the number of background channels to be at least equal to the number of peak channels.

The net peak intensity is obtained by dividing the net peak counts by the recorded livetime

$$I_{p-b} = \frac{P}{t_\ell} \qquad (4.205)$$

Consequently the relative random error in the net peak intensity is equal to the relative random error in the net peak counts.

$$\frac{\sigma_{I_{p-b}}}{I_{p-b}} = \frac{\sigma_P}{P} = \frac{\sqrt{P + B(1 + \eta_p/\eta_B)}}{P}$$

$$= \frac{1}{\sqrt{P}} \sqrt{1 + \frac{B}{P}\left(1 + \frac{\eta_p}{\eta_B}\right)} \qquad (4.206)$$

If P is large compared to B, the error contribution from the background will be unimportant. On the other hand, when the net peak count is close to zero, the statistical error from the background dominates.

It can also be shown that there is an optimum choice for the number of channels $\eta_p$ summed in the peak. The results in Fig. 4.65 have been computed by assuming a gaussian shaped peak of height $\ell$ situated on an average background level b. The peak has a full width at half maximum height denoted by $\Gamma$, and has a total area of A counts. The resolution $\Gamma$ is expressed in the same units as $\eta_p$, i.e., channels, for convenience. If the background level were zero, then the optimum precision in Eq. (4.206) would be obtained by summing all the A counts in the peak. That is, the $\eta_p$ channels would cover the entire peak. The optimum relative precision in the case of zero background would be

$$\frac{\sigma_P}{P} = \frac{1}{\sqrt{A}} \qquad\qquad (4.207)$$

All cases of finite background in Fig. 4.65 have been referenced to the zero background case by multiplying the obtained value of $\sigma_p/P$ by $\sqrt{A}$. This normalized relative precision has been plotted in Fig. 4.65 as a function of the ratio $\eta_p/\Gamma$, which is the ratio of the number of summed peak channels to the peak resolution. The four curves plotted correspond to different values of the parameter $(\ell/b)/(1 + \eta_p/\eta_B)$. There are two ways of interpreting this parameter. If the ratio $\eta_p/\eta_B$ is considered to be fixed then the four curves correspond to different values of the line-to-background ratio $\ell/b$. The graph shows that a high line-to-background ratio provides better relative precision, as might have been expected. If the line-to-background ratio is considered to be fixed, the four curves demonstrate the effect of the ratio $\eta_p/\eta_B$ on the relative precision. Best precision is obtained when the number of background channels is large compared to the number of peak channels utilized.

For low line-to-background ratios, the curve labeled "0.1" indicates that the best relative precision is obtained when $\eta_p = 1.17\Gamma$. For high line to background ratios the value of $\eta_p$ should be increased slightly; although, a value of $\eta_p = 1.17\Gamma$ is not far from the optimum precision, even for an infinite line to background ratio. In general, the number of channels chosen on the peak should correspond to a width which is slightly greater than the FWHM resolution $\Gamma$. Where the width cannot be set precisely, it is better to set it too wide than too narrow. In fact, wider windows can be useful in reducing the sensitivity to peak shift. Figure 4.66 shows similar information for the effect of the peak and background integration regions on the minimum detection limit $C_{MDL}$. The results are based on Eqs. (11.37) and (11.39) in Chap. 11. An optimum setting of $\eta_p = 1.17\Gamma$ is also reflected in the detection limit curves for the normal situation of isolated windows.

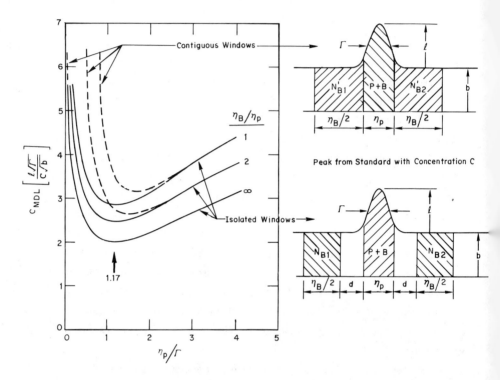

Figure 4.66  The effect of peak and background integration regions on the minimum detection limit $C_{MDL}$.  Reprinted by courtesy of EG&G ORTEC.

In some cases the background subtraction will not be as ideal as has been depicted in Fig. 4.65.  Neighboring peaks will lie so close to the peak of interest that accurate background estimation from the measured spectrum is not possible.  In such cases it is often possible to estimate the background through the quantitative regression analysis models.  When the calibration curve for concentration as a function of measured intensity is calculated, a background term can be included to account for the lack of accurate background subtraction in the spectrum.  A simple background constant can be used if the background does not exhibit strong variations as a result of changing specimen composition.  Where the peak-to-background ratio is low and the specimen composition varies widely, it may be necessary to include corrections to the background which depend on specimen composition.

In cases where the peak of interest is not isolated, but suffers an interference from an overlapping peak the peak integration region should be shifted slightly off center, away from the interfering peak.  This will tend to reduce the contribution from the overlapping peak.

## 4.10   SENSITIVITY OF THE X-RAY FLUORESCENCE SPECTROMETER

One of the most important parameters by which the performance of a spectrometer can be judged is the sensitivity.  Sensitivity is defined as the fractional change in instrument response per unit change in concentration of the measured element.  If a calibration curve is plotted with x-ray intensity on the y axis and element concentration on the x axis, the slope of the curve dy/dx is a measure of the sensitivity of the technique.  As far as bremsstrahlung sources are concerned, sensitivity may change by several orders of magnitude over the usual wavelength range of the spectrometer. The upper diagram in Fig. 4.67 shows a curve of sensitivity as a function of wavelength, and the lower curve relates wavelength with atomic number for different series lines.  The sensitivity curve is of the form of a smooth curve with a maximum at around 1 $\overset{\text{o}}{\text{A}}$.  At long wavelengths the sensitivity is reduced by the combined effects of a low fluorescence yield and strong absorption of the low-energy x-rays.  Several points in the instrument contribute to the absorption of the long-wavelength x-rays.  These include (1) self-absorption of low-energy x-rays in the x-ray tube anode, (2) absorption in the x-ray tube window, (3) absorption in the specimen, and (4) absorption in the detector window.  For wavelengths shorter than 1 $\overset{\text{o}}{\text{A}}$ the fluorescence yield is high and absorption losses are negligible.  However, the absorption-edge energies for the elements become a significant fraction of the x-ray tube voltage for wavelengths shorter than 1 $\overset{\text{o}}{\text{A}}$.  This causes the

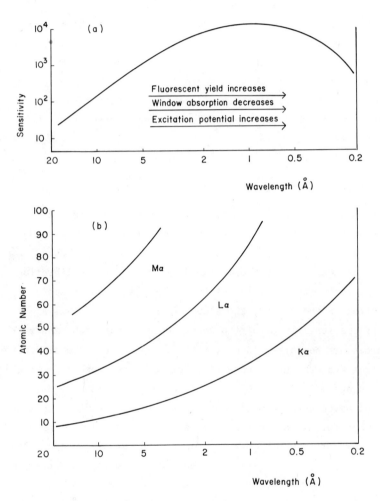

Figure 4.67    (a) The variation in sensitivity of a fluorescence
spectrometer employing broadband or bremsstrahlung excitation.
(b) Wavelengths of the major lines as a function of atomic number.

sensitivity to decrease at short wavelengths.

Study of the two curves in Fig. 4.67 shows that there is a
region of high sensitivity for elements between atomic numbers 20
and 50 using the K series, and between 70 and 90 using the L
series.   It must also be remembered, however, that the fluorescent
yield at given wavelengths drops as one goes from the K to the L

to the M series.  Thus, whereas a wavelength of 1 $\overset{o}{A}$ would correspond to a fluorescent yield of about 0.6 in the K series, it would
only correspond to about 0.1 in the L series.  This in turn means
that sensitivities obtainable with the K series are generally
better than those obtainable with the L series, which are in turn
superior to those obtainable with the M series.

## REFERENCES

1.  G. A. Brown, Applications of a x-ray pulse spectroscope in
    *Siemens Rev., 34* (Special Issue):69 (1967).
2.  Jan Krugers (ed.), *Instrumentation in Applied Nuclear Chemistry,*
    Plenum Press, New York/London, 1973.
3.  L. V. Sutfin and R. E. Ogilvie, in *Energy Dispersion X-ray
    Analysis:  X-ray and Electron Probe Analysis,* American Society
    for Testing and Materials Special Publication STP 485,
    Philadelphia, 1971, p. 197.
4.  A. J. Burek and R. L. Blake, *Advan. X-ray Anal., 16:*37 (1973).
5.  Robert W. Hendricks, *Rev. Sci. Instrum., 40:*1216 (1969).
6.  R. Jenkins, *The Gas Flow Proportional Counter in X-ray
    Spectrometry,* Philips Scientific Report FS6, N. V. Philips'
    Gloeilampenfabrieken, Scientific and Analytical Equipment
    Department, Eindhoven, The Netherlands, 1970.
7.  N. Spielberg and D. I. Tsarnas, *Rev. Sci. Instrum., 46:*1086
    (1975).
8.  W. Bambynek, *Nucl. Instrum. Methods, 112:*103 (1973).
9.  Neven Karlovac and T. V. Blalock, *IEEE Trans. Nucl. Sci.,
    NS-22:*457 (1975).
10. D. R. Beaman, J. A. Isasi, H. K. Birnbaum, and R. Lewis,
    *J. Phys. E: Sci. Instrum., 5:*767 (1972).
11. Robley D. Evans, *The Atomic Nucleus,* McGraw-Hill, New York,
    1955.
12. D. Edwards, Jr., *Rev. Sci. Instrum., 47:*1186 (1976).
13. Neven Karlovac, *Rev. Sci. Instrum., 43:*1540 (1972).
14. N. Karlovac, in *Operating and Service Manual for the ORTEC
    Model 449 Log/Linear Ratemeter,* ORTEC, Inc., Oak Ridge,
    Tennessee, 1971.
15. D. A. Landis, F. S. Goulding, and B. V. Jarrett, *Nucl. Instrum.
    Methods, 101:*127 (1972).
16. J. M. Jaklevic and F. S. Goulding, *IEEE Trans. Nucl. Sci.,
    19:*384 (1972).
17. Rolf Woldseth, *Amer. Lab., 4:*79 (1972).
18. D. A. Gedcke, E. Elad, and G. R. Dyer, *Proceedings of the
    Sixth National Conference on Electron Probe Analysis,* Electron
    Probe Analysis Society of America, New York, 1971, p. 5A.

19.  D. A. Gedcke and E. Elad, *Proceedings of the Sixth International Conference on X-ray Optics and Microanalysis,* University of Tokyo Press, 1972, p. 253.

20.  D. A. Gedcke, *X-ray Spectrom., 1*:129 (1972).

21.  D. A. Gedcke, in *Quantitative Scanning Electron Microscopy* (D. B. Holt, M. D. Muir, P. R. Grant, and I. M. Boswarva, eds.), Academic Press, London, 1974, p. 403.

22.  K. Kandiah, A. Stirling, D. L. Trotman, and G. White, *International Symposium on Nuclear Electronics,* vol. 1, Versailles, 1968, p. 69-1.

23.  K. Kandiah, *Semiconductor Nuclear Particle Detectors and Circuits* (L. Brown, ed.), National Academy of Sciences Publication 1593, Washington, D.C., 1969, p. 495.

24.  K. Kandiah, *Nucl. Instrum. Methods, 95*:289 (1971).

25.  D. A. Landis, F. S. Goulding, R. H. Pehl, and J. T. Walton, *IEEE Trans. Nucl. Sci, 18*:115 (1971).

26.  N. Karlovac and D. A. Gedcke, *Proceedings of the Eighth National Conference on Electron Probe Analysis,* New Orleans, paper 17, 1973.

27.  E. Fairstein and J. Hahn, *Nucleonics, 23*:56 (1965); *23*:81 (1965); *23*:50 (1965); *24*:54 (1966); *24*:68 (1966).

28.  F. S. Goulding, *Nucl. Instrum. Methods, 100*:493 (1972).

29.  F. J. Walter, in *Energy Dispersion X-ray Analysis: X-ray and Electron Probe Analysis,* ASTM Special Publication STP 485, American Society for Testing and Materials, Philadelphia, 1971, p. 82.

30.  R. L. Heath, *Advan. X-ray Anal., 15*:1 (1972).

31.  D. H. Wilkinson, *Proc. Cambridge Phil. Soc., 46*:508 (1950).

32.  R. W. Schumann and J. D. McMahon, *Rev. Sci. Instrum., 27*:675 (1956).

33.  G. Williams, in *Energy Dispersion X-ray Analysis: X-ray and Electron Probe Analysis,* ASTM Special Publication STP 485, American Society for Testing and Materials, Philadelphia, 1971, p. 125.

34.  S. J. B. Reed, *J. Phys. E., Sci. Instrum., 5*:994, 997 (1972).

35.  H. D. Keith and T. C. Loomis, *X-ray Spectrom., 5*:93 (1976).

36.  S. J. B. Reed and N. G. Ware, *J. Phys. E: Sci. Instrum., 5*:582 (1972).

37.  N. A. Dyson, *Nucl. Instrum. Methods, 114*:131 (1974).

38.  P. A. Ross, *J. Opt. Soc. Amer., 16*:433 (1928).

39.  J. R. Rhodes, in *Energy Dispersion X-ray Analysis: X-ray and Electron Probe Analysis,* American Society for Testing and Materials Special Publication STP 485, Philadelphia, 1971, p. 243.

40.  I. S. Sokolnikoff and R. M. Redheffer, *Mathematics of Physics and Modern Engineering,* McGraw-Hill, New York, 1958, Chap. 8.

41.  Philip R. Bevington, *Data Reduction and Error Analysis for the Physical Sciences,* McGraw-Hill, New York, 1969.

42.  Stuart L. Meyer, *Data Analysis for Scientists and Engineers,* Wiley, New York, 1975.

43. Jörg W. Müller, *Nucl. Instrum. Methods, 112*:47 (1973).
44. T. Inouye, *Nucl. Instrum. Methods, 104*:541 (1972).
45. J. E. Stewart, H. R. Zulliger, and W. E. Drummond, *Advan. X-ray Anal., 19*:153 (1976).
46. Jörg W. Müller, *Nucl. Instrum. Methods, 117*:401 (1974).
47. A. Foglio Para and M. Mandelli Bettoni, *Nucl. Instrum. Methods, 70*:52 (1969).
48. W. Feller, *On Probability Problems in the Theory of Counters*, Courant Anniversary Volume, Interscience, New York, 1948, p. 105.
49. D. A. Gedcke, *Proceedings of the Tenth Annual Conference of the Microbeam Analysis Society*, Las Vegas, 1975, paper 25.
50. J. Bartosek, J. Masek, F. Adams, and J. Hoste, *Nucl. Instrum. Methods, 104*:221 (1972).
51. A. R. Lowes, U. S. Patent No. 3,814,937 (June 4, 1974).
52. ORTEC Model 6200 Instruction Manual and Schematics, 1972; ORTEC Model 6230 Instruction Manual and Schematics, 1975.
53. J. Harms, *Nucl. Instrum. Methods, 53*:192 (1967).
54. C. F. Masters and L. V. East, *IEEE Trans. Nucl. Sci., 17*:383 (1970).
55. M. W. Barnhart, U. S. Patent No. 3,896,296 (July 22, 1975).
56. B. E. Warren, *X-ray Diffraction*, Addison-Wesley, Reading, Mass., 1969.

# 5

## Statistics

## 5.1  INTRODUCTION

X-ray spectrometry, like other instrumental analytical procedures,
is capable of producing large quantities of data which are assumed
to relate to the chemical composition of the specimen being analyzed.  In this chapter, the application of statistical concepts to
these numerical results will be examined without considering in any
detail the various factors that affect the proportionality between
x-ray intensity and analyte concentration.

A simple numerical result, such as 30.35% copper, for example,
may convey the wrong imporession to the reader if the analytical
precision is not included.  The precision *must always* be given whenever any analytical results are presented.  The mere fact that the
answer (30.35% copper) is written with numerical information in the
hundredths place in *no way* assures the reader that such precision
is obtainable nor does it reassure the reader that a subsequent
measurement will produce the same results.

In addition to estimating the precision, the statistical
method can be used to discover the various sources of error found in
a particular analytical technique.  This "analysis of variance" can

Table 5.1  Definition of Some Basic Terminology Relating to the
Statistical Method

---

*Accuracy:*  *The degree of agreement* of a measurement made from a
   specimen being analyzed with the "true result" obtained from an
   accepted reference standard.

*Precision:*  The degree of mutual agreement between repeated individual measurements made on the same sample.  The precision of a
   method may be excellent but its accuracy may be very poor.  (See
   Fig. 5.1.)  The terms accuracy and precision are often misused
   and frequently interchange.  These definitions have been taken
   from recommended ASTM standards [1].

*Bias or systematic error:*  A consistent deviation of the experimental results from the accepted reference level.  The sources of
   systematic errors such as environmental factors, etc., will be
   discussed more fully in subsequent paragraphs.

*True value:*  The value of a characteristic obtained from an accepted reference standard.

*Confidence limits:*  The confidence limits are upper and lower
   bounds (on any statistic) between which the specimen values will
   fall with a given probability P.

*Random error:*  Nonsystematic fluctuations in experimental conditions and measurement methods.  These may arise from the machine
   or the operator.

*Specimen and sample:*  As used in this chapter, the term *specimen*
   will refer to the actual small object that is placed into the
   spectrometer.  A *sample* is the as-received larger object from
   which an aliquot has been taken and prepared to give a specimen.
   For example, in a fertilizer analysis, the sample might consist
   of a 100-lb bag of fertilizer, whereas the specimen might be a
   5-g aliquot pressed into a 1 1/4-in diameter pellet.

---

be used to obtain a better experimental design since the analyst's

attention can be focused on the error-producing steps in the

technique.

   A subject as broad and far reaching as statistics cannot be

covered in a single chapter, and no attempt will be made to do so.

It is hoped that this chapter will provide a basic understanding

and appreciation of the statistical approach and thus stimulate

interest in further investigation on the part of the reader.  In

order to provide a uniform base upon which to build this subject,

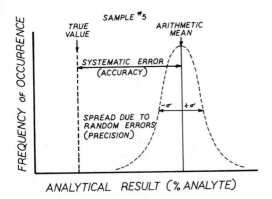

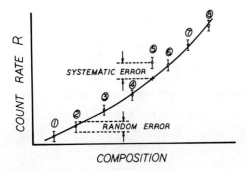

Figure 5.1  A schematic illustration of the definition of several important statistical terms.

it is wise to define some of the key terms at the start, and these are listed in Table 5.1 (see also Fig. 5.1).  Note that other terms will also be defined at appropriate locations in this text.

## 5.2  STATISTICAL THEORY

It will be assumed in this section that the reader is familiar with the general field of statistics but perhaps has not applied the statistical technique to specific problems in x-ray spectrometry. Unless otherwise specified, it will also be assumed that all variables discussed in this chapter are normally (gaussian) distributed so that only one frequency function need be discussed.  As we

proceed, it will be necessary to introduce and define a few additional terms.  These will be defined with a minimum of mathematical
rigor.  The reader who is interested in the derivations should consult the Refs. [2-4] listed at the end of this chapter.

As defined in Table 5.1, a *specimen* is a small "representation"
of a larger sample.  In general, x-ray spectrometry, like other
analytical techniques, will measure only a few *specimens* taken and
prepared from a large sample.  The attempt is then made to draw some
meaningful conclusions about the entire sample, i.e., (*the total or
parent population of specimens*) based upon the results obtained from
a small number of specimens.  The statistical methods developed in
this chapter particularly apply to such small (in number) specimen[*]
techniques.

### 5.2.1  The Arithmetic Mean (A Measure of Central Tendency)

A simple statistic that is representative of a set of n measurements
of $x_i$ is called the *arithmetic mean* $(\bar{x})$.  Note the mean of the
entire population of possible specimens is designated $\mu_x$.  The
arithmetic mean $(\bar{x})$ is expressed in Eq. (5.1)

$$\bar{x} = \sum_{i=1}^{n} \frac{x_i}{n} \qquad\qquad (5.1)$$

### 5.2.2  Dispersion about the Mean (the Variance)

The *variance* of a set of observations is defined to be the sum of
the squares of the deviations of the observations from the arithmetic mean.  Thus, the small specimen variance $S_x^2$ (the population
variance is given by $\sigma_x^2$) is defined as

$$S_x^2 = \sum_{i=1}^{n} \frac{(x_i - \bar{x})^2}{n - 1} \qquad\qquad (5.2)$$

Notice that the true average square deviation from the mean would
be obtained by dividing by n rather than n - 1 as shown in Eq.

---

[*] When the expression *small specimen* is used, it will refer to the
number of specimens rather than the physical size of the specimens.

Table 5.2  Symbols used for Small Sample and Population Parameters

| Small Specimen Parameter | | Population Parameter |
|---|---|---|
| $\bar{x}$ | Mean | $\mu_x$ |
| $S_x^2$ | Variance | $\sigma_x^2$ |
| $S_x$ | Standard deviation | $\sigma_x$ |

(5.2). Statistical theory [2] will show that if division is by n, the value of $S_x^2$ so obtained *will not* be an unbiased estimate of the true population variance $\sigma_x^2$. The distinction is particularly important when only a small number (around 10) of measurements are being made.

5.2.3  The Standard Deviation

The *standard deviation* $(S_x)$ of a set of measurements also measures the dispersion about the mean and it is shown in Eq. (5.3)

$$S_x = \sqrt{S_x^2} = \sqrt{\frac{\sum_{i=1}^{n} \{x_i - \bar{x}\}^2}{n - 1}} \tag{5.3}$$

Table 5.2 summarizes these three parameters and the population parameters of which they are an estimate.

5.3  DISTRIBUTION FUNCTIONS

If the results of a large number of replicate determinations are plotted, it is generally the case that they will form a smooth distribution of values. Some values near the arithmetic mean will occur with high frequency, while others (those near the tails of the curve) occur far less frequently. The emission of x-rays from a sample placed in an x-ray spectrometer (photon counting) has been shown to closely approximate the normal (or gaussian) distribution function if the number of counts N is large. The normal

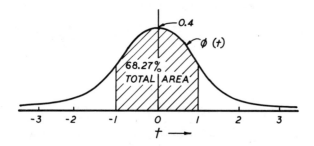

Figure 5.2  A graphical representation of Eq. (5.5) showing a normal distribution function having a mean of zero and a variance of unity.

distribution function f(x) is shown in Eq. (5.4).[*]

$$f(x) = c \exp\left[ -\frac{1}{2}\left(\frac{x - a}{b}\right)^2 \right] \qquad -\infty < x < \infty \qquad (5.4)$$

where $a = \mu_x$, $b = \sigma_x$, and $c = I/(\sigma\sqrt{2\pi})$. Equation (5.4) may also be expressed in terms of another variable t as shown in Eq. (5.5)

$$\phi(t) = \frac{1}{\sqrt{2\pi}} \exp\left[ -\frac{t^2}{2} \right] \qquad -\infty < t < \infty \qquad (5.5)$$

where $t = (x - \mu_x)/\sigma_x$. Equation (5.5) is the so-called standard form where the variable is normally distributed having a mean of zero and a variance of unity. Figure 5.2 illustrates the graph of Eq. (5.5). Recall that the area under this curve is unity.

It was shown by Rutherford and Geiger as early as 1910 that radioactive counting measurements, for large values of N, are normally distributed having a mean $\overline{N}$ and a standard deviation $\sqrt{\overline{N}}$, where $\overline{N}$ is the average number of counts measured from a series of replicate measurements $N_1$, $N_2$, $N_3$. As stated above, this has also been shown to be valid for normal x-ray intensity counting so that one can speak of a standard counting error $s_c = \sqrt{N}$. (It should be noted that this expression is only valid when a negligible background intensity is present (see Chap. 4). Thus for any counting

---

[*] This does not automatically imply that replicate x-ray spectrometric analytical results are normally distributed since systematic biases may exist.

measurement there will always be associated a *minimum* error $s_c$ whose value depends *only* upon the total number of events counted.

## 5.4 STATISTICAL TESTING OF HYPOTHESIS

At this point, it is useful to introduce the concept of a *random sample*. Let f(x) be the distribution function for some continuous random variable x and take multiple samples from this distribution each of size n. A property, such as composition, is measured on each of the n specimens from each of the multiple samples. The sampling is said to be random if the frequency distributions of the measured property from each set of n specimens are equal and, in fact, equal f(x). The "normal distribution theorem" therein states that if x is normally distributed[*] with a mean $\mu_x$ and variance $\sigma_x^2$ and a random sample of size n* is taken, then the average $\bar{x}$ is also normally distributed having a mean $\mu_x$ and variance $\sigma_x^2/n$.

This theorem clearly shows the increasing precision of the sample mean ($\bar{x}$) as an estimate of the population mean ($\mu_x$) as the sample size n is increased.

It also provides a mechanism for testing hypotheses concerning sample means as described below.

### 5.4.1 Application of the Normal Distribution Theorem: Testing Equality of Sample Means

A large sample was removed from a production process and 30 random specimens were prepared and examined in an x-ray spectrometer in order to determine the concentration of element A in the mill concentrate. A mean value of 15.5% A and a standard deviation of 1.8% were obtained.

A new chemical process was initiated in the manufacturing operation and the experiment was repeated, resulting in a new mean value of 14.8% A. Can it be concluded that the new process produces a different result?

---

[*] Henceforth, this statement will be abbreviated using $N(\mu_x, \sigma_x^2)$ also the term *random sample of size* n will be abbreviated by r.s.s. n.

The following two hypotheses can be proposed for the second experiment.

$H_0$:  $\mu_x$ = 15.5, i.e., there is no significant difference between $\mu_x$ and 15.5% A.

$H_1$:  $\mu_x$ < 15.5.

We want to reject the null hypothesis ($H_0$ is commonly referred to as the *null hypothesis*) that $\mu_x$ = 15.5, *if* $\mu_x$ is not 15.5 as indicated by $\overline{x}$ significantly larger than 15.5% A, i.e., by large absolute values of the variable t, where t was defined in Eq. (5.5) and Fig. 5.2.

For our problem, x is N (15.5, $(1.8)^2$); thus $\mu_x$ = 15.5 and $\sigma_{\overline{x}}$ = 1.8/$\sqrt{30}$ = 0.33.  The value of $\overline{x}$ as obtained from the second experiment from a r.s.s. 30 was 14.8% A.  Thus t can be calculated as follows:

$$t = \frac{\overline{x} - \mu_x}{\sigma_{\overline{x}}} = \frac{14.8 - 15.5}{0.33} = 2.12$$

From Fig. 5.2 or a table of normal distribution areas and ordinates [2,3,5], the probability of obtaining such a value of t is P ($|t|$ > 2.12) = 0.5000 - 0.48300 = 0.017.  This probability value 0.017 is highly unlikely because it lies outside of two standard deviations from the mean; thus we reject the null hypothesis $H_0$ that $\mu$ = 15.5 for our second experiment, and it is concluded that the new chemical process produces a new value for the concentration of A which might be detrimental to the manufacturing process.

5.4.2  Small Sampling Theory, n < 30

*The Chi-Squared $(\chi^2)$ Distribution*

In the example above, statistical tests have been used to evaluate the equality of sample means.  Similar test procedures are avail-able that enable one to statistically compare the standard devia-tions measured from two or more sets of small specimens.  This test of variance equality requires an understanding of another mathematical distribution which is called the *chi-squared distribution*. The $\chi^2$ distribution provides a measurement of the discrepancy between expected and observed frequencies. For example, when throwing a single die 60 times, each face would be expected to show up 10 times. In general, this does not occur and the $\chi^2$ test could be used, for example, to locate a bias if it existed.

Here the example of throwing dice is used simply to explain the nature of the $\chi^2$ distribution function shown in Eq. (5.6).

$$\chi^2 = \sum_{J=1}^{K} \frac{(0_j - e_j)^2}{e_j} \tag{5.6}$$

where

> $K$ = the number of frequency pairs to be compared (for a die $K = 6$)
>
> $e_j$ = expected frequency of occurrence of each event (for a die rolled 60 times $e_j = 10$)
>
> $0_j$ = observed frequency of occurrence of each event

A value of $\chi^2 = 0$ signifies exact agreement with expectation, and as $\chi^2$ becomes larger, the agreement becomes poorer. For random samples of size n (mean = $\bar{x}$) drawn from a normally distributed population whose variance is $\sigma^2$, the statistic $\chi^2$ can be defined as follows:

$$\chi^2 = \frac{(x_1 - \bar{x})^2 + (x_2 - \bar{x})^2 + (x_3 - \bar{x})^2 + \cdots + (x_n - \bar{x})^2}{\sigma^2}$$

If other random samples of size n are drawn from this population and $\chi^2$ computed for each, a sampling distribution of $\chi^2$ can be obtained. This continuous $\chi^2$ distribution function is given by Eq. (5.7).

$$f(\chi^2) = C\chi^{\nu-2} \exp\left(-\frac{1}{2}\chi^2\right) \tag{5.7}$$

where $\nu$ is called the *number of degrees of freedom* (in this case $\nu = n - 1$, the number of independent observations on the sample minus one) and c is a constant (depending on $\nu$) to make the area

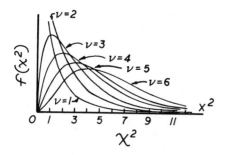

Figure 5.3   The $\chi^2$ distribution for various degrees of freedom.

under the curve equal 1.   Figure 5.3 illustrates the appearance of
the $\chi^2$ distribution for various values of $\nu$.

### 5.4.3   Testing Sample Variances

Theorem

If x is normally distributed with variance $\sigma^2$, and if $s^2$ is the
sample variance based on a r.s.s. n, then $\nu s^2/\sigma^2$ has a $\chi^2$
distribution with n - 1 degrees of freedom.

The simple example below will illustrate the application of
this theorem to the problem in which standard deviations from two
samples are being compared.

Example

A new and faster sample preparation procedure is being evaluated and compared with an older, slow procedure.   The older
procedure has been used extensively and has a population standard deviation $\sigma$ = 1.5%.   Using the newer procedure, a random
sample of size 25 produced a standard deviation s = 1.7%.   Is
the new procedure inferior, i.e., has the variability truly
increased?

We begin as before by testing two hypotheses:

$H_0$:   $\sigma$ = 1.5%, i.e., there is no significant difference between the standard deviation and 1.5%.
$H_1$:   $\sigma$ > 1.5%.

We want to reject the null hypothesis $H_0$: $\sigma = 1.5\%$ if $\sigma > 1.5\%$, as indicated by values of s significantly larger than $\sigma$, that is, by large positive values of $\chi^2$.  For this problem, $vs^2/\sigma^2$ possesses a $\chi^2$ distribution with 24 degrees of freedom. $\chi^2$ may be calculated as shown.

$$\chi^2 = \frac{vs^2}{\sigma^2} = \frac{24(1.7)^2}{(1.5)^2} = 30.8$$

Consulting a table of $\chi^2$ distribution for $v = 24$ [2,5], we find that the probability is $\pm$ 0.1 of having such a large value of $\chi^2$.  Thus we reject the null hypothesis that $\sigma = 1.5$ and must conclude that the new procedure will indeed have a larger variability.

### 5.4.4   Confidence Limits for $\sigma^2$

Two values of $\chi^2$ can be found from a table of $\chi^2$ for n - 1 degrees of freedom, namely $\chi_1^2$, and $\chi_2^2$, such that the probability is 0.025 that $\chi^2$ observed > $\chi_2^2$ and is 0.975 that $\chi^2$ observed > $\chi_1$.  This defines the 95% confidence limits for $\sigma^2$ as shown below:

$$P(\chi_1^2 < \chi^2 \text{ observed} < \chi_2^2) = P\;\chi_{0.0975}^2 < \frac{vs^2}{\sigma^2} < \chi_{0.025}^2 = 0.95$$

$$\text{and}\quad P\!\left(\frac{vs^2}{\chi_{0.925}^2} < \sigma^2 < \frac{vs^2}{\chi_{0.975}^2}\right) = 0.95 \tag{5.8}$$

For the previous example, the 95% confidence limits for $\sigma$ when s = 1.7% and $v$ = 24 may be calculated using Eq. (5.8).  Values of $\chi_{0.975}^2$ and $\chi_{0.025}^2$ are obtained from $\chi^2$ tables [2,5].

$$\frac{24(1.7)^2}{39.4} < \sigma^2 < \frac{24(1.7)^2}{12.4}$$

$$1.76 < \sigma^2 < 5.59 \qquad\qquad 1.33\% < \sigma < 2.37\%$$

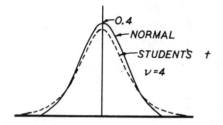

Figure 5.4  The Student's t distribution.

5.4.5  Small Sample Approximations to the Normal Distribution

When the sample size n is greater than 30, the sampling distributions of many statistics are approximately normal (gaussian), and
the approximation becomes increasingly better as n increases.  However, as n decreases below 30, the normal approximation becomes
progressively more invalid.  A distribution function is needed to
enable statistical testing (like those described in the paragraphs
above) to be performed precisely when n is small.  The so-called
Student's t distribution can be used for these cases and will be
described briefly below.

As Fig. 5.4 shows, the Student's t distribution closely approxi-
mates the normal distribution as $\nu$, the number of degrees of freedom, increases.  Equation (5.9) gives the distribution function for
the Student's distribution curve shown in Fig. 5.4.  Equation (5.9)
is valid under the provisions of the following theorem.

> Theorem
> If a variable u is normally distributed with mean zero and a
> variance of unity and a variable $V^2$ has a $\chi^2$ distribution with
> u degrees of freedom (u and V must be independently distri-
> buted, then the variable $t = u\sqrt{\nu}/V$ has a Student's distribu-
> tion with $\nu$ degrees of freedom given by Eq. (5.9).

The confidence limits for the mean of a small sample where x
is normally distributed with mean $\mu$ and variance $\sigma^2$ (with $\bar{x}$ and $s^2$
as the sample estimates based on a r.s.s. n) can be determined
using the Student's distribution.  Let

$$u = \frac{\bar{x} - \mu}{\sigma^2/\sqrt{n}} \qquad\qquad v^2 = \frac{ns^2}{2}$$

These selections of u and v meet the conditions of the theorem above, and the variable t given by Eq. (5.10) possess a t distribution with n - 1 degrees of freedom.

$$f(t) = \frac{C}{(1 + t^2/v)^{(v+1)/2}} \qquad\qquad v = n - 1 \qquad\qquad (5.9)$$

$$t = \frac{\bar{x} - \mu}{s_x} \sqrt{n} \qquad\qquad\qquad\qquad (5.10)$$

Small sample hypothesis testing follows the same pattern described in previous paragraphs except that a table of Student distribution ordinates and areas would be used rather than a normal distribution table [2,3,5].

### 5.4.6  Variance Testing:  The F Distribution

Frequently it is required to isolate each individual source of error in an analytical procedure or simply to compare variances from two sets of samples.  The following theorem defines the F test and the F *distribution*.

Theorem

If u and v possess independent $\chi^2$ distributions having $v_1$ and $v_2$ degrees of freedom, respectively, then

$$F = \frac{u/v_1}{v/v_2}$$

has the F distribution with $v_1$ and $v_2$ degrees of freedom, given by the probability density function

$$f(F) = \frac{\Gamma[(v_1 + v_2)/2]}{\Gamma(v_1/2)\Gamma(v_2/2)} \left(\frac{v_1}{v_2}\right)^{v_1/2} F^{(v_1-2)/2}$$

$$\times \left(1 + \frac{v_1}{v_2} F\right)^{-(v_1+v_2)/2}$$

Tables of the F function are found in most statistics texts as well
as in many mathematical handbooks [2,3,5]. Note for
$F = (u/\nu_1)/(v/\nu_2)$ with $u = \nu_1 s_1^2/\sigma_1^2$ and $v = \nu_2 s_2^2/\sigma_2^2$, both u and v
possess $\chi^2$ distributions. It can be shown that

$$F = \frac{s_1^2/\sigma_1^2}{s_2^2/\sigma_2^2} = \frac{s_1^2}{s_2^2}\left(\frac{\sigma_2^2}{\sigma_1^2}\right)$$

When testing the typical null hypothesis, $H_0: \sigma_1^2 = \sigma_2^2$, F becomes

$$F = \frac{s_1^2}{s_2^2}$$

In using the F test ratio, the larger of the two unbiased variance
estimates is usually placed in the numerator.

Since the F function depends upon two parameters $\nu_1$ and $\nu_2$, a
three-dimensional table would be needed to tabulate all possible
values of F corresponding to $\nu_1$ and $\nu_2$ plus the various probabilities. Generally, only the 5% and 1% right-tailed F test is tabulated.

While this procedure can be used to test the equivalence of
sample variances (as shown below), it has much greater importance
because of its application to what is known as the *analysis of
variance technique* used in experimental design. This will be discussed later.

5.4.7 Testing Variances: Example Problem

When testing sample means, using the normal distribution or
Student's test, it is a good procedure to first test the assumption
generally made that $\sigma_x = \sigma_y$.

For example, two sets of specimens ($n_x = 17$, $n_y = 26$) are
measured using two different procedures, and the standard deviations
are $s_x = 1.8\%$ and $s_y = 3.5\%$, respectively. Can it be assumed that
$\sigma_x^2 = \sigma_y^2$? What if $s_y = 2.5\%$?

$$F = \frac{s_y^2}{s_x^2} = \frac{12.25}{3.24} = 3.78$$

In a table of F values where $\nu_1$ = 16 and $\nu_2$ = 25, the F value associated with the probability of 10% is 2.06.  Since the calculated value of F (3.78) is greater than 2.06, the hypothesis that $\sigma_x$ = $\sigma_y$ must be rejected.  As expected, the method giving $s_y$ does have a lower precision.  If $s_y$ were lowered to 2.5, then the calculated F = 1.97, which is lower than 2.06, there would be a 90% confidence that the variance associated with each procedure is the same (i.e., $\sigma_x^2 = \sigma_y^2$).

## 5.5   LEAST-SQUARES CURVE FITTINGS

When correlations exist between observed data (such as composition and intensity on a calibration curve, or intensity and time as in equipment stability analysis), it becomes possible to formulate equations relating the independent variables with one or more dependent variables.  The simplest case is that of data which when plotted form a straight line

$$y = a_o + a_1 x \tag{5.11}$$

where y is the dependent variable and x is the independent variable. This linear equation is frequently written as y = c + mx, where c and m are defined as in Fig. 5.5.

Generally, in x-ray spectrometry, the relationship between the independent (intensity) variable y and the dependent (composition) variable x cannot be expressed by a simple linear equation such as y = c + mx when x ranges over wide values of composition.

In these cases, the relationship between y and x may be expressed by any number of mathematical functions although a polynomial of low degree will generally be very satisfactory.  Table 5.3 lists several common polynomial forms used for curve fitting nonlinear data.

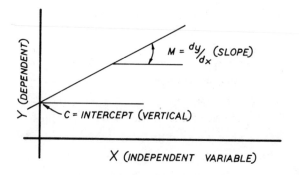

Figure 5.5  Graph of the linear equation y = mx + c.

Table 5.3  Polynomials Used for Curve Fitting

---

$y = a_0 + a_1x$                                           Straight line

$y = a_0 + a_1x + a_2x^2$                                   Parabolic curve

$y = a_0 + a_1x + a_2x^2 + a_3x^3$                          Cubic curve

$y = a_0 + a_1x + a_2x^2 + \cdots + a_nx^n$                 nth degree polynomial

---

5.5.1  Criterion for Determining Best Fit:  The Least-Squares Method

Regardless of the curve to which the data are to be fit, a method
must be used that will achieve the "best" curve through the data
points.  The curve drawn in Fig. 5.6 shows that the data points
$(x_i, y_i)$ all have small $\delta_i$ differences (in the y direction) from the
curve.  The best curve that can be drawn through these data points
is the one for which the sum of the $\delta_i^2$ is a minimum.  A best curve
obtained in such a manner is said to fit the data in a *least-square*
*sense*.  The least-square method will then provide the best values
of the coefficient $a_0$, $a_1$, ..., $a_n$ for the various curves found in
Table 5.3, depending upon which degree polynomial is chosen.

The equations which provide the best values of $a_i$ are called
the *normal* equations.  A set of normal equations exists for each
polynomial shown in Table 5.3.  For the simple linear case, these
equations are shown below.

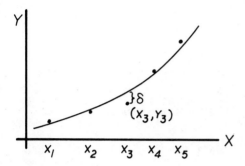

Figure 5.6  Least squares data fitting to a curve showing $\delta$ for a data point.

$$y = a_0 + a_1 x$$

$$\sum_i y_i = a_0 N + a_1 \sum_i x_i$$

$$\sum_i x_i y_i = a_0 \sum_i x_i + a_1 \sum_i x_i^2 \qquad (5.12)$$

where N is the number of sets of data $(x_i, y_i)$.

When these equations are solved simultaneously, the best values of $a_0$ and $a_1$ will be obtained.  For details of this derivation, the reader is directed to Refs. 2 and 6.  For the parabolic case such normal equations are obtained whose simultaneous solutions will provide the best values of $a_0$, $a_1$, and $a_2$.  These "best fit" least-squares curves are also called *linear* (or *curvilinear*) *regression curves* of the variable y on x since y is estimated from x.  Computer programs are available for easy calculation of such best fit polynomial regression curves.

5.5.2  Multiple Regression

The dependent variable y (the intensity) often depends upon more than one independent variable x, such as in complex alloys where the intensity from element A depends upon the concentration of element A *and* the concentrations of elements B, C, D, etc.  In this case, we may have an equation expressing *linear multiple regression* as in Eq. (5.13).

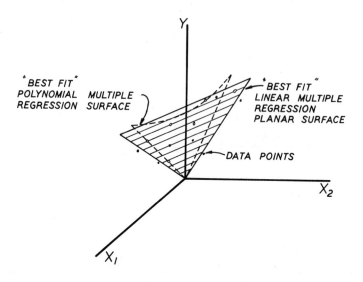

Figure 5.7  A sketch of the best fit polynomial regression surface.

$$y = a_0 + a_1 x_1 + a_2 x_2 + \cdots + a_n x_n \tag{5.13}$$

If $y$ is a function of more than two variables (e.g., quaternary alloy), then geometric intuition is lost even for the simple case of *multiple linear regression* since more than three dimensions are required to plot the data.

If the linear multiple regression model above does not fit the data, a *polynomial multiple regression* technique will probably work [9]; see Eq. (5.14).  For a ternary alloy (two independent variables), a multiple regression nonplanar surface can be defined that is a best fit to the observed data (see Fig. 5.7).

$$y = a_0 + a_1 x_1 + a_2 x_1^2 + a_3 x_1^3 + a_4 x_2 + a_5 x_2^2 + a_6 x_2^3 \tag{5.14}$$

The best fit criterion is a least-squares technique as described previously, and *normal* equations can be defined for multiple regression planes and surfaces.  Computer programs are essential and are available for these types of analyses.

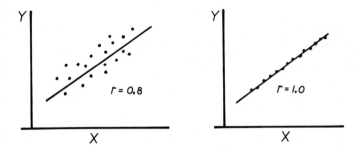

Figure 5.8  The definition of the correlation coefficient r when the
independent and the dependent variables are assumed to be linearly
correlated.

### 5.5.3  Goodness of Fit:  The Correlation Coefficient

A necessary but not sufficient proof of correlation between any two
variables[*] is defined by the *correlation coefficient* r.  In Eq.
(5.15), r is defined in a general fashion to be valid for both
linear and nonlinear simple correlation between *two* variables y
and x.

$$r = 1 - \frac{s_{y,x}^2}{s_y^2} \qquad\qquad (5.15)$$

where

$$s_y = \sqrt{\frac{\sum_i (y_i - \bar{y})^2}{N - 1}}$$

the standard deviation of the y values and

$$s_{y,x} = \sqrt{\frac{\sum_i (y - y_{curve})_i^2}{N - g}}$$

the standard error of estimate of y on x for $N - g = \nu$ (number of
degrees of freedom).  The term $(y - y_{curve})_i = \delta_i$, is defined in
Fig. 5.6.

_____

[*] Techniques are available for the calculation of nonlinear multiple
correlation coefficients, but these go beyond the scope of this
chapter.  For more details, see Refs. 2 and 3 at the end of the
chapter.

A value of r = 1 indicates perfect correlation of the variables
y and x, whereas a value of r = 0 indicates no correlation whatso-
ever.  Figure 5.8 shows the correlation coefficient r calculated for
two scatter diagrams when the data are *assumed* to be *linearly*
correlated.


5.6  ANALYSIS OF VARIANCE AND EXPERIMENTAL DESIGN

Errors are a common denominator of all analytical procedures and
x-ray spectrometry is no exception.  The aim of most analysts should
be to identify and quantify the error-producing steps in their
particular analytical procedures.  Statistics can prove to be an
invaluable aid for this type of analysis as will be seen in the
paragraph that follows.

The novice in the field of x-ray spectrometry often feels that
his analytical precision is largely governed by the easiest-to-
calculate error source, the counting error.  In practice, however,
this is frequently the least important contribution to total error
in the system.  The next section will discuss sources of error in
x-ray spectrometry and some possible solutions.  In this section,
it will be assumed that many sources of error exist and an attempt
to locate them will be made using a statistical tool called
*analysis of variance*.

A basic assumption of the components of variance models in
statistics is that

$$\sigma^2_{total} = \sigma^2_A + \sigma^2_B + \sigma^2_C + \cdots + \sigma^2_N \tag{5.16}$$

Equation (5.16) states that the total variance of any procedure involving steps A, B, C, etc., equals the sum of the variances
of each of the steps.  This assumes the parameters in each step are
independent and normally distributed.  In x-ray spectrometry, typical steps might include sampling, grinding, weighing, pelletizing
locating the peak, etc.  If the variance of each of these individual
steps is calculated, they can be compared to the total variance to

Table 5.4  Analysis of Variance Method Applied to Sample Preparation Methods

| | | $GT_L$ | | $GT_H$ | | Averages |
|---|---|---|---|---|---|---|
| | | $MT_L$ | $MT_H$ | $MT_L$ | $MT_H$ | |
| $W_L$ | $BP_L$ | 15.1-14.8 | 15.0-14.9 | 15.5-15.5 | 15.3-15.7 | 15.23 |
| | $BP_H$ | 15.5-15.7 | 15.4-15.7 | 15.6-15.8 | 15.5-15.9 | 15.64 |
| $W_H$ | $BP_L$ | 15.0-15.2 | 14.7-14.5 | 15.6-15.5 | 15.1-15.3 | 15.14 |
| | $BP_H$ | 15.3-15.4 | 14.9-15.2 | 15.8-15.6 | 15.4-15.6 | 15.59 |
| | | 15.25 | 15.04 | 15.61 | 15.48 | Grand average = $\bar{\bar{X}}$ = 15.34 |

GT = grinding time.
MT = mixing time.
W = weight of powder.
BP = briquetting pressure.
Subscripts: H = high level; L = low level.

estimate their individual importance.  The measurement of their
variance components can be very time consuming unless some statistical experimental design is used.  Without prior knowledge of experimental design, it would be necessary to run many replicate tests of
each procedure keeping all other variables constant.  To test the
importance of grinding time, for example, it would be necessary to
grind a single well-mixed sample for a series of times $t_1$, $t_2$,
$t_3$, ..., $t_n$ and then prepare pellets under identical conditions of
weight, briquetting, pressure, and measurement conditions.  A
properly conceived experimental design can test all of these variables simultaneously using a minimum of time and effort.

Consider the problem of locating the major sources of analytical error [7] from among the four sample preparation steps; GT =
grinding time, MT = mixing time, W = weight of powder, and BP =
briquetting pressure.  Instead of using a series of times, weights,
and pressures, only two levels for each variable were used, namely
a high value (H) and a low value (L).  This is called a *factorial
design experiment,* in this case a 2n factorial, where n = 4.  The
general procedure will be to compare the overall variance obtained
from all the sample combinations with the variances of the individual steps.  The factorial design evaluates not only the independent effects of the several variables but also their interaction-
combination effects.  Note that other simpler experimental designs
exist and can be used when it is known that interaction effects are
not significant.  Since two levels of four variables are being
tested, we have $2(2)(2)(2) = 16$ samples to prepare.  Table 5.4
shows the factorial design for determining the effects of the four
variables mentioned above.  The numbers that appear in the various
columns are *hypothetical* values for duplicate determination of the
percentage of element x in the samples being tested.

The data from Table 5.4 can be reduced to several two-factor
tables for ease of calculation.  Table 5.5 lists the data for the
two variables mixing time and grinding time.  The sum of squares
of grinding time can be calculated from

Table 5.5   Analysis of Variance[a]

|          | $GT_L$ | $GT_H$ | Averages |
|----------|--------|--------|----------|
| $MT_L$   | 15.25  | 15.61  | 15.43    |
| $MT_H$   | 15.04  | 15.48  | 15.26    |
| Average  | 15.15  | 15.55  |          |

[a]Reduced table for two variables of sample preparation.
MT = Mixing time.
GT = Grinding time.

Table 5.6   Variance Analysis Summary for Hypothetical Data Shown
on Table 5.4

| Source of Variation | Sum of Squares | Degrees of Freedom | Variance | F Ratio |
|---------------------|----------------|--------------------|----------|---------|
| Grinding time       | 0.64           | 1                  | 0.80     | 2.66    |
| Mixing time         | 0.13           | 1                  | 0.36     |         |
| Weight of sample    | 0.12           | 1                  | 0.35     |         |
| Briquetting pressure| 0.49           | 1                  | 0.70     | 2.33    |
| Duplicates          | 0.45           | 16                 | 0.17     |         |
| All interactions[a] | 1.00           | 24                 | 0.20     |         |
| Total               | 2.88           | 31                 | 0.30     |         |

$\overline{\overline{x}}$ = 15.34

[a]Calculated by difference assuming second-order interactions are
zero.

$$\text{sum of squares} = \sum_{i=1}^{n} N(x_i - \overline{\overline{x}})^2 \tag{5.17}$$

where $\overline{x}_1$ = 15.15, $\overline{x}_2$ = 15.55, N = 8, and $\overline{\overline{x}}$ (the grand average) =
15.34 from Table 5.4.  Thus, for grinding time we have sum of
squares equals

$$8(15.15 - 15.34)^2 + (15.55 - 15.34)^2 = 0.64 \text{ (grinding time)}$$

Similar values for the other variables are calculated and
shown in Table 5.6.  The duplicate variance was calculated by

Table 5.7   Designed Experiments for Analysis of Sample Changer Errors

| Test No. | Description | Error Designation[a] | Repeat Cycles | Total Test Duration (h) |
|---|---|---|---|---|
| 1a | Total long-term equipment error excluding changer, single specimen, not removed | ET1L | 150 | 4.5 |
| 1b | Total short-term error as above | ET1S | 42 | 1.25 |
| 2 | Single specimen not removed insert, count, retract | ET2 | 25 | 1.25 |
| 3 | Same as test 2 except that specimen removed and re-placed between each measurement | ET3 | 25 | 1.25 |
| 5 | Same as test 3 except that 10 different holders are used with the *same* specimen | ET5 | 3 | 2 |
| 6 | Same as test 3 except that 10 similar specimens are used in the *same* holder | ET6 | 3 | 1.75 |
| 7a | Total long-term equipment error, single wavelength, 10 similar specimens in 10 different positions | ET7A | 10 | 4.5 |
| 7b | As for test 7a but with more cycles | ET7B | 22 | 10 |
| 8 | Total long-term equipment error but using two wavelengths ratiod one to the other to remove long-term drift | ET8 | 3 | 4.5 |

[a]Experimentally determined errors for the seven sets of experimental conditions. Conditions are designated ET1, ET2, ..., ET7. Other errors are designated as follows:

    Counting error = EC.
    Sample changer error = ES.
    Short-term equipment error = E1S.
    Long-term equipment error = E1L.

subtracting from each analysis the average of the two duplicates
and summing the squares of these differences. The total sum of
squares for all individual measurements was calculated using Eq.
(5.2) and the interaction sum of squares was calculated by summing
all other sum-of-squares terms and subtracting from the total sum
of squares.

Among other things, Table 5.6 demonstrates that the hypothetical data are rather poor. Note that the F test can be used to compare these variances. The F values for grinding time and briquetting pressure are not highly significant (the values are smaller
than the 5% one-tailed F values found in F distribution tables
$v_1$ = 1 and $v_2$ = 30), owing to the overall poor quality of the data
(i.e., the total variance is large), but do indicate that these are
the two most troublesome areas. For details of the calculation
involved, the reader is referred to Ref. 7.

## 5.6.1 Isolation of Equipment Errors

An experiment run by one of the authors (Jenkins) was designed to
isolate and evaluate the magnitude of errors from an experimental
sample changer plus other equipment errors. The experiments are
described on Table 5.7. Table 5.8 shows the derivation of the individual errors described in Table 5.7. The experimental results
are summarized on Table 5.9 and the errors are summarized on
Table 5.10.

The results in Table 5.10 show that the dominant sources of
error for this application, in order of significance, are (1)
specimen holder differences 0.29%, (2) long-term equipment error
0.2%, (3) specimen difference error 0.17%, (4) placement of specimen
error 0.077%, and (5) counting error 0.71%. Note that the counting
error is fifth in importance in this particular instance.

## 5.6.2 Statistical Testing for Equipment Stability

Statistical methods can be used to locate errors induced by
ment instability. In this section, we will be concerned with using
statistical methods to identify short- and long-term unidirectional

Table 5.8   Derivation of Individual Errors from the Designed
Experiments

---

$$EC = 100/\sqrt{N}$$

$$E1S = \sqrt{ET1S^2 - EC^2}$$

$$E1L = \sqrt{ET1L^2 - EC^2}$$

$$E2 = \sqrt{ET2^2 - ET1S^2}$$

$$E3 = \sqrt{ET3^2 - ET2^2}$$

$$E5 = \sqrt{ET5^2 - ET3^2}$$

$$E6 = \sqrt{ET6^2 - ET3^2}$$

$$E7 = \sqrt{ET7^2 - E6^2}$$

$$E8 = \sqrt{ET8^2 - E6^2}$$

$$ES = \sqrt{E2^2 + E3^2 + E5^2}$$

*Assuming* that all errors are independent errors the total error ET
is given by

$$ET = \sqrt{\sum_n (ETN)^2}$$

Thus for two experiments giving two errors Ei and Ej the total
error will be

$$ET = \sqrt{Ei^2 + Ej^2}$$

and   $$Ei = \sqrt{ET^2 - Ej^2}$$

this gives the means of isolating a given individual error, in this
case Ei.

---

drift (unidirectional changes in intensity) and oscillation errors
(short-term intensity fluctuations).  This subject has been discussed by Hooten and Parsons [8] in great detail.

A typical x-ray spectrometer stability test requires repeated
measurements of a single intensity without deliberate changes in
any other variable.  This is basically a statistical time series

Table 5.9   Experimental Data

| Test No. | Error Designation | Actual Number of Repeats | $\bar{N}$ | σ% | Standard error $\sigma\%/\sqrt{2}(n-1)$ |
|---|---|---|---|---|---|
| 1a | ET1L | 150 | 1971035 | 0.2128 | ± 0.0123 |
| 1b | ET1S | 42 | 1975410 | 0.0885 | ± 0.0098 |
| 2 | ET2 | 25 | 2002526 | 0.1040 | ± 0.0150 |
| 3 | ET3 | 25 | 1942105 | 0.1296 | ± 0.0187 |
| 5 | ET5 | 30[a] | 1999806 | 0.3213 | ± 0.0422 |
| 6 | ET6 | 30 | 1992148 | 0.2132 | ± 0.0280 |
| 7a | ET7A | 100 | 1958383 | 0.6072 | ± 0.0432 |
| 7b | ET7B | 219 | 1926704 | 1.8295 | ± 0.0876 |
| 8 | ET8 | 30 | --[b] | 0.4497 | ± 0.0590 |

[a]First set of data rejected from statistical analysis.

[b]Test 8 was recorded as a ratio; thus a correlation with $\bar{N}$ is irrelevent.

Table 5.10   Errors due to Specimen Changer Device

| Source of Error | Designation | Value(σ%) | Standard Error |
|---|---|---|---|
| a. Typical total system error | | | |
|   i. Multiple element (ratio method) | E8 | 0.4146 | ±0.0659 |
|   ii. Single element (absolute method) | E7 | 0.5831 | ±0.0463 |
| b. Equipment error (less specimen changer) | | | |
|   i. Short term (1.25 h) | E1S | 0.0532 | ±0.0163 |
|   ii. Long term (4 h) | E1L | 0.2007 | ±0.0130 |
| c. Counting error | EC | 0.0707 | |
| d. Specimen changer error (total) | ES | 0.3089 | ±0.0459 |
|   i. Retraction/insertion | E2 | 0.0546 | ±0.0327 |
|   ii. Placement of specimen | E3 | 0.0773 | ±0.0373 |
|   iii. Specimen holder differences | E5 | 0.2940 | ±0.0468 |
| e. Specimen difference error | E6 | 0.1693 | ±0.0381 |

problem where the independent variable X is time. Generally, the intensity is counted for a fixed time increment and sufficient counts are accumulated for each measurement so that the counting variance will be significantly lower than any other source of variance. It has been shown that for photon counting with a stable source intensity, the population variance $\sigma^2$ is given by

$$\sigma = \sqrt{\mu} \qquad (5.18)$$

where $\mu$ = the population mean. For a stable instrument, a good estimate of $\mu$ is given by $\overline{X}$, the mean observed number of photons; thus

$$\sigma^2 = \overline{X} \qquad (5.19)$$

For photon counting, the variance $\sigma^2$ can be estimated from a single measurement of N total counts by

$$\sigma^2 \simeq N \qquad (5.20)$$

from which the familiar standard counting error expression is derived

$$\sigma = \sqrt{N} \tag{5.21}$$

or  $\sigma\% = \dfrac{\sqrt{N}}{N} \times 100 = \dfrac{100}{\sqrt{N}}$ \hfill (5.22)

## Various Criterion for Equipment Stability [8]

1.  If 99.7% of observations lie within 3 standard deviations ($\overline{X} \pm 3\sqrt{\overline{X}}$) of the mean, the equipment is considered stable.
2.  The proportions of observation within the two ranges $\overline{X} \pm 2\sqrt{\overline{X}}$ and $\overline{X} \pm \sqrt{\overline{X}}$ are close to 95.5% and 68.3%, respectively.
3.  Perform a $\chi^2$ test ratioing $s^2$ from subsets of the total population to the population variance $\sigma^2$. If n random observations are taken having a sample variance $s^2$ from a population having a variance $\sigma^2$, then

$$\frac{(n - 1)s^2}{\sigma} = \chi^2 \tag{5.23}$$

has a $\chi^2$ distribution with $\nu = n - 1$ degrees of freedom. Thus, we may use a $\chi^2$ two-tail 0.05 significance test (see Sec. 5.4.3). If $\chi^2_{0.975} < \chi^2 < \chi^2_{0.025}$, then there is a 95% probability that $s^2$ *is* from a normal population of variance $\sigma^2$. If the above inequality does not hold, i.e., $\chi^2_{0.975} > \chi^2 > \chi^2_{0.995}$ or $\chi^2_{0.025} < \chi^2 < \chi^2_{0.005}$, then $s^2$ is significantly different than $\sigma^2$ and stability is doubtful. In the extreme cases of highly significant differences between $s^2$ and $\sigma^2$, $\chi^2$ would be such that $\chi^2 < \chi^2_{0.995}$ or $\chi^2 > \chi^2_{005}$.

4.  Perform a linear regression analysis on the time series data. Since $y = c + mx$, for perfectly stable equipment, c must equal $\overline{X}$ and the slope m must equal zero. The correlation coefficient r (see Sec. 5.5.3) will indicate the degree of correlation of the intensity variable with the time. A value of r = 0 would indicate no correlation between time and specimen intensity (i.e., no unidirectional drift in the intensity).
5.  A newly developed approach is called the *mean square successive difference* (MSSD) approach which can differentiate between short- and long-term drifts. The MSSD technique considers the order of the observations as well as their individual values. A statistic $\delta^2$ is calculated where

$$\delta^2 = \sum_{i=1}^{n-1} \frac{X_{i+1} - X_i}{n - 1} \qquad (5.24)$$

For a random sample of size n,

$$\delta^2 \rightarrow 2\sigma^2 \qquad (5.25)$$

$\delta^2$ will be less affected by drift than $s^2$ but it more sensitive to
oscillations or short-term drift. The ratio $\delta^2/s^2 = \eta$ should approach 2, and values of $\eta$ significantly < 2 indicate long-term drift
and values of $\eta$ significantly > 2 indicate short-term oscillations.
Tables of critical $\eta$ values are available [10] and are used like $\chi^2$
tables to test the significance of the difference between $\eta$ and 2.
If the number of observations $\eta$ is greater than 25, a more convenient statistic $\varepsilon$ is given:

$$\varepsilon = 1 - \frac{\eta}{2} \qquad (5.26)$$

$\varepsilon$ will thus have an average value of zero and is nearly normally
distributed with variance

$$\sigma_\varepsilon^2 = \frac{n - 2}{(n - 1)(n + 1)}$$

Critical values of Student's t (see Sec. 5.4.4) can be used where

$$t_s^2 = \frac{\varepsilon}{\sigma_\varepsilon}$$

If $t_s^2$ is positive and $> t_{0.005}^2$, then slow drift is significantly
indicated. On the other hand, if $t_s^2$ is negative, oscillations
(short-term drift) have occurred. Since $\bar{x}$ is a good estimate of $s^2$,
the variables $\eta_{\bar{x}}$ and $t_{\bar{x}}$ are defined

$$\eta_{\bar{x}} = \frac{\delta^2}{\bar{x}}$$

$$t_{\bar{x}} = \frac{1 - \eta_{\bar{x}}/2}{\sigma_\varepsilon}$$

For stable equipment, all the statistics mentioned above, $\chi^2$, $\eta_s^2$,
$\eta_{\bar{x}}$, $t_s^2$, $t_{\bar{x}}$, will be statistically insignificant. Table 5.11 taken

Table 5.11 Values of the Statistics Subject to Various Instability Conditions

| Instability conditions | $\chi^2$ | $n_s^2$ | $n_{\bar{x}}$ | $t_s^2$ | $t_{\bar{x}}$ |
|---|---|---|---|---|---|
| Long-term drift | Very significant | Low value, very significant | Low value, very significant or not significant | Large (+) value, very significant | Large (+) value, very significant or not significant |
| Short-term drift | Very significant | High value, very significant | High value, very significant | Large (+) value, very significant | Large (-) value, very significant |
| Drift plus oscillation | Very significant | Low value, very significant | High value, very significant | Large (+) value, very significant | Large (-) value, very significant |
| Unstable no drift, no oscillation | Very significant | Not significant or very significant | High value, very significant | Not significant | Large (-) value, very significant |

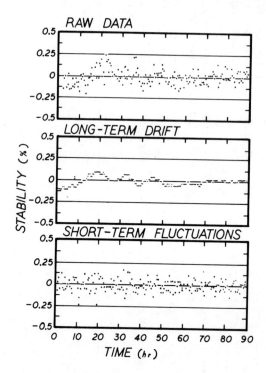

Figure 5.9  An illustration of short-term fluctuations and long-term drift obtained from an energy-dispersive fluorescence spectrometer.  Reprinted by courtesy of EG&G ORTEC.

from Ref. 8 summarizes the drift information obtainable with the statistics listed above.

Hooten and Parsons [8] have calculated these statistics for several series of drift tests and the interested reader should consult their paper.

Figure 5.9 represents the results obtained by Gedcke for the calculation of short- and long-term drift of an energy dispersive fluorescence analyzer.

In this example, each data point represents 1000 s of livetime (approximately 25 min of real time).  Successive counting intervals were run at constant excitation conditions for approximately 88 h. The long-term drift curve was generated by an 11-point smooth on the

raw data. The resulting curve contains the drift components for
periods of the order of 5.5 h, or longer. The long-term drift is
computed from

$$\pm\% \text{ Drift} = \frac{\text{max} - \text{min}}{\text{max} + \text{min}} \times 100\%$$

where max and min are the maximum and minimum intensities recorded
in the long-term drift curve.

The short-term fluctuations are extracted by subtracting the
long-term drift curve from the raw data and adding back the mean.
The short-term fluctuation curve contains components with periods
of the order of 5.5 h down to 25 min and includes most of the counting statistics. The excess short-term fluctuations beyond counting
statistics are reported as

$$XS = (\text{variance in data} - \overline{N})^{1/2} \times 100\%/\overline{N}$$

where the mean $\overline{N}$ is subtracted from the variance in the short-term
fluctuation data.

In the example run, the long-term drift was ±0.10% and the
short-term fluctuations beyond counting statistics were 0.00%.


## 5.7 SOURCES OF ERROR IN X-RAY SPECTROCHEMICAL ANALYSIS

The basic act of counting a steady source of photons with any
radiation detector has been shown to have associated with it a
"standard counting error" $\sigma = \sqrt{N}$, where N can be increased at the
discretion of the operator, thus minimizing the percent counting
error. The entire subject of counting strategy is discussed in
more detail in Chaps. 4 and 9. In general, all other errors will
be in excess of the standard counting error if N has been chosen
sufficiently large and the peak-to-background ratio is sufficiently
high. We have mentioned several of these sources of error in
previous sections of this chapter, particularly in regard to the
analysis of variance techniques. In this section, we will consider

Table 5.12  Sources of Error in X-ray Spectrometric Analysis

| Sample | Equipment | Other |
|---|---|---|
|  | Systematic |  |
| Absorption | Short-term drift | Preparation of working curves |
| Enhancement | Long-term drift | Estimation of composition |
| Heterogeneity | Sample heating |  |
| Segregation | Resetting goniometer |  |
| Particle size distribution | Sample repositioning |  |
| Contamination | Resetting operating conditions |  |
| Mixing | Crystal deterioration |  |
| Weighing | Tube contamination |  |
| Briquetting |  |  |
| Surface Preparation |  |  |
| Polishing |  |  |
| Etching |  |  |
| Microstructure |  |  |
| Chemical shifts |  |  |
| Spectral line interference |  |  |
| Deterioration of standards |  |  |
|  | Random |  |
| Counting errors |  |  |
| Tube and generator stability |  |  |
| Electronic instability |  |  |
| Random sample selection |  |  |

sources of error that affect the accuracy and precision of x-ray
spectrometric analysis.

The two terms *random* and *systematic* error are often used when
discussing the sources of error in analytical procedures.  Both
types are prevalent in x-ray spectrometric measurements.  Random
errors are like those associated with photon counting from a radioactive source.  Even though the photon emission occurs at a specific
mean rate, the instantaneous rate fluctuates yielding a distribution of rates having R as the mean value.  This spread implies that
no two measurements are likely to be exactly the same, and thus a
*random* error results.  On the other hand, if during the course of
this photon measurement experiment the detector is moved slightly
away from the source after each measurement, the resulting intensity
change would be caused by a *systematic error* or *imposed bias* on
the results.

Table 5.12 lists the major sources of random and systematic
errors encountered in x-ray spectrometry.  *Methods* for handling
these error sources are discussed in subsequent chapters.  As Table
5.12 shows, the major sources of *random* error arise from counting
errors and equipment instability.  Equipment random error has improved to the point where this source of error is generally of very
low order, i.e., approximately 0.05 to 0.1%.  Thus, in order to increase analytical precision, the major *systematic* errors must be
recognized and eliminated or reduced.

Jenkins [15] has shown that the general analytical precision
can be approximated by the empirical relationship

$$\sigma = K \sqrt{C}$$

where

C = concentration in percent

K = figure of merit for precision

For x-ray spectrometry, K has been observed to have an approximate minimum value of 0.01 compared to a range of 0.008 to 0.05 for
round-robin wet chemical analysis of metals.  Currie [11] has

discussed the limits of precision in analytical chemistry from a statistical basis and Harvey [12] has described a method for identifying "wild" or potentially rejectable data points.  Buchanan and Tsai [13] and Camp and Rhodes [14] have also recently investigated the precision and accuracy of the x-ray spectrometer method.

REFERENCES

1.  American Society of Testing and Materials Standard E-177-71, Part 30, *Annual ASTM Standards,* 1972, p. 360.
2.  P. G. Hoel, *Introduction to Mathematical Statistics,* Wiley, New York, 1954.
3.  G. W. Snedecor, *Statistical Methods,* Collegiate Press, Iowa, 1946.
4.  W. L. Gore, *Statistical Methods for Chemical Experimentation,* Interscience, New York, 1952.
5.  W. C. Beyer, *CRC Handbook of Tables for Probability and Statistics,* 2nd Ed., CRC Press, 1974.
6.  E. T. Whittaker and G. Robinson, *The Calculus of Observations,* Blackie and Sons, Limited, London, 1929.
7.  B. J. Mitchell, *Advan. X-ray Anal., 11:*129 (1968).
8.  K. A. H. Hooten and M. L. Parsons, *Anal. Chem., 45:*2218 (1974).
9.  B. J. Alley and H. Myers, *Anal. Chem., 37:*1685 (1965).
10. C. A. Bennett and N. L. Franklin, *Statistical Analysis in Chemistry and the Chemical Industry,* Wiley, New York, 1954, p. 678.
11. L. A. Currie, *Anal. Chem., 40:*586 (1968).
12. P. K. Harvey, *X-ray Spectrom., 2:*147 (1973).
13. E. B. Buchanan, Jr. and Foo-Chong Tsai, *Anal. Chem., 46:*1701 (1974).
14. D. C. Camp and J. R. Rhodes, *X-ray Spectrom., 3:*47 (1974).
15. R. Jenkins, *An Introduction to X-Ray Spectrometry,* Heyden, London, 1974, p. 118.

# 6

## General Computer Applications and Quantitative Spectrum Analysis as Applied to Energy-Dispersive Analysis

### 6.1 INTRODUCTION

The problems encountered in x-ray fluorescence spectrometry are well suited to the use of digital computers. Many of the quantitative models used to convert measured intensities to concentrations are mathematically complicated and difficult to solve by hand. In energy-dispersive spectrometers peak overlaps often require mathematical fitting techniques to separate the elemental intensities. Digital computers provide the computational power to make these tasks manageable. In many applications the fluorescence spectrometer is required to measure the composition of a large number of specimens of the same type on a repetitive basis. For such applications computer automation provides efficient control of the calibration procedures, specimen presentation, and selection of excitation and measurement conditions.

The list of potential beneficial applications of digital computers to x-ray fluorescence spectrometry is almost endless. In Table 6.1 a few of the more important applications are summarized. Since a comprehensive treatment of all these topics would require a complete book, this chapter will represent only a brief introduction.

Table 6.1  Important Applications of Digital Computers to X-ray
Spectrometry

---

A.  Spectrum analysis

  1.  Background fitting and subtraction
  2.  Fitting and separating overlapping peaks
  3.  Computation of net peak intensity
  4.  Reporting of characteristic line energies or wavelengths
      for rapid qualitative analysis
  5.  Deadtime correction
  6.  Computation of the statistical error in the net peak inten-
      sity

B.  Concentration calculations

  1.  Determination of calibration curves and interelement cor-
      rection coefficients
  2.  Computation of concentration from measured intensities

C.  Mathematical computations not associated with specimen analysis

D.  Automation of specimen analysis

  1.  Control of x-ray tube voltage and current and other excita-
      tion conditions
  2.  Selection of counting time
  3.  Automatic specimen sequencing
  4.  Analyzing crystal selection
  5.  Control of $2\theta$ setting for peak and background measurements
  6.  Control of PHS window setting
  7.  Selection of amplifier and proportional counter gain

E.  Process control feedback

  1.  Product quality monitoring
  2.  Defective product composition alarm
  3.  Process adjustment recommendations
  4.  Incoming materials inspection
  5.  Closed-loop process control

---

For those seriously pursuing the use of a computer, Refs. 1 and 2
are recommended for further reading.  These two books cover the
mathematical techniques commonly encountered in computer applications.  Further references will be given for the specific topics
discussed.  The latter part of this chapter will concentrate on
spectrum analysis, a topic most likely to be of direct interest to
the energy dispersive spectroscopist.  Discussion of concentration
computation algorithms is provided in the chapters on quantitative
analysis.

Table 6.2   Types of Computer Systems

---

1.  Hosted satellite system
2.  Remote terminal, timeshared computer
3.  Large batch processing computer
4.  On-line minicomputer
5.  Microprocessor
6.  Hardwired automator

---

## 6.2  TYPES OF COMPUTER SYSTEMS

Table 6.2 summarizes the types of computer systems available in
approximate order of computational power.  Although the purchaser
of a new fluorescence spectrometer is likely to opt for the on-line
minicomputer, the operator of older equipment may find significant
utility in either the large batch processing computer or the timeshared computer.

The large batch processing computer is generally readily accessible in universities, large organizations, and scientific
institutes.  This type of computer is installed in a central location, and is designed to be efficient in handling complicated,
computational problems.  Its memory is large, and it is fast.  All
the operating system software exists and has been developed to a
sophisticated state.  The programmer need only learn to work in
high-level languages such as Fortran.  In dealing with one of these
computing centers, one quickly develops the feeling that his computing power is unlimited.  The program and perhaps a magnetic data
tape are handed into the center in the morning, and by afternoon
the answer to even a very complicated problem is available, neatly
printed out.  If one of these computers is available, it offers the
ideal means for a novice to learn how to benefit from a computer.
Fortran can be learned in a surprisingly short time, and often
within a week the novice can start to solve his problems on the
computer.  This type of facility offers the greatest power for

computing concentrations and fitting the most complicated spectra.
However, it does have drawbacks. Most computing centers are reluc-
tant to handle data tapes. Even when they do accept magnetic tapes,
generation and transportation of the data tape is awkward at best.
Scientific computing centers are usually not set up for reading in
large amounts of data and reducing it to answers. Their forté lies
in generating a wealth of numbers derived from loops of mathematical
algorithms. Even though the computing speed is high, turn-around
time is often slow. Centers usually accumulate programs until the
start of the next batch or shift. This results in a turn-around
delay of four to twelve hours. Nevertheless the batch processing
computing center can be the ideal solution for the analyst whose
computing time requirements are light. In fact, even if an on-line
computer is acquired, the computing center will still be invaluable
for the really complex problems involving extensive computations.

Some of the turn-around time delays can be diminished by
turning to the large timeshared computer with a remote terminal.
In most locations this type of service is as available as the nearest telephone line. For a monthly fee and a computing time charge,
a remote input/output terminal can be installed in the laboratory,
coupled to telephone lines. Due to the timesharing feature, access
to the computer will appear to be instantaneous. This system offers
most of the benefits of the batch processing computer, but with
better access. It is an excellent solution for small operations
unable to justify either a batch processing computer or a dedicated
on-line computer. Most timesharing companies offer a variety of
programs developed by other users. Sometimes the solution to the
problem at hand is already in their library. Unfortunately, this
system is still lacking when it comes to efficiently transferring
experimental spectra to the computer for analysis. Also, it has
negligible potential for computer control of the fluorescence
analyzer.

Most of the tasks outlined in Table 6.1 can be handled by a
16-bit minicomputer with a memory size between 8,000 and 16,000

words, a fast paper tape punch and reader, a hard copy input/output
terminal, and for the larger problems, a disk. The cost of the
average system is comparable to the price of a low-cost fluorescence
analyzer. The on-line minicomputer offers fast, dedicated access.
Since the fluorescence analyzer can be interfaced directly to the
computer, data transfer is extremely efficient. This system provides
the minimum turn-around time for data acquisition and analysis. By
using standard CAMAC modules [3-7] or custom designed circuits, the
computer can be interfaced to almost any instrument or process to
provide computer readout and control. Several companies offer
higher level languages for minicomputer systems. Hence, most programming tasks can be rather simple. Several fluorescence analyzer
manufacturers offer complete computerized systems based on minicomputers, including all of the necessary problem-solving software.
This type of system offers the most efficient solution for the
laboratory that has a heavy analysis load. On the other hand, the
minicomputer does have some drawbacks. Its computing speed using
higher level languages is usually slow. For the ultimate in computing speed the operator may want to write some of his programs
in assembly language. Assembly language programming requires a
little more skill and more effort.

It might appear that buying quick access to a dedicated on-line
computer entails giving up the sophisticated computing power of a
large computer. Conversely, opting for the heavy computing capability of the large computer would seem to rule out the fast data
transfer and turn around time provided by the on-line minicomputer.
There is a solution to this dilemma which has been employed in
several larger institutes [8-10]. A minicomputer is interfaced
directly to the measuring instrumentation (the spectrometer in the
case of fluorescence analysis). The memory size of this "satellite"
computer is chosen to be just large enough to handle data acquisition, computer control of the instrumentation, and very simple
computational problems. To achieve sophisticated computing capability the minicomputer is connected to a large, central, data

processing computer via a suitable data link. The large "host" computer is a timeshared computer which may be located at a great distance from the satellite minicomputer. The minicomputer provides dedicated accessibility, and fast turn-around time where and when it is needed. When complicated data analysis must be performed, the minicomputer efficiently transfers the appropriate data to the timeshared computer. The timeshared computer is very quick in calling up the correct program and passing back the answers to the operator via the minicomputer. This system represents the ultimate in analytical capability. In systems incorporating an on-line mini-computer an interactive display is usually provided to let the operator see his data and select further data manipulation based on the displayed information.

Hardwired computers and controllers have been available since well before the advent of the minicomputer. They represent a very satisfying and economical solution where a specific, dedicated, and unchanging set of functions is required. Some examples of hard-wired computers are the multichannel pulse-height analyzer, axis positioners, programmed angle selectors ($2\theta$), and data acquisition sequencers for multielement measurements, and other programmed spectrometer controllers. This type of product is sometimes easier to operate than a software computer because interaction is by hard-ware controls logically arranged to suit the desired function. There are many applications where this type of instrument provides a satisfactory solution. Hardwired automators are particularly applicable to x-ray spectrometers which are used infrequently or to routine quality control operations. Hardwired computers and con-trollers generally sacrifice flexibility and the ability to make complicated concentration calculations.

In between the computational capabilities of the minicomputer and the rigidity of the hardwired controller lies the microprocessor. These instruments emulate the hardwired programs of the controller via a read-only memory (ROM). This memory is preprogrammed by the manufacturer rather than the operator. To gain some operator

flexibility microprocessors frequently include some programmable
read-only memory (PROM). The microprocessor is more capable than
the hardwired controller and with little or no increase in cost.
Compared to the minicomputer, the microprocessor is less expensive.
However, the microprocessor is efficient in handling only the very
simple mathematical algorithms.

## 6.3 COMPUTER PROGRAMMING LANGUAGES

There are basically four types of computer languages: machine language, assembly language, compiler languages, and interpretive
languages. The only way that information can be understood by the
computer is by storing instructions and data as binary words.
Therefore, all the computer "sees" is words made up of 0s and 1s.
This is machine language. Very few programs are written in machine
language because it is difficult to use even for experienced pro-
grammers.

The next higher level language is assembly language. This is
a symbolic language where there is nearly a one-to-one conversion
from symbolic statements to machine language. The programs are
written with mnemonic symbols representing machine language instructions. The assembler is a program which looks at the mnemonic-
coded program and translates it into the appropriate machine language statements. Many systems programs and hardware control programs are written in assembly language.

Because programming in assembly language can become very
tedious, high-level languages have been developed. These languages
enable even an inexperienced user to write computational programs.
Instead of writing several assembly language instructions to do a
task, the programmer writes single statements which can be translated into large blocks of machine language statements. Table 6.3
gives an example of the relationship between the types of languages.

There are two main types of high-level languages. One is
compiler-based languages such as Fortran, Algol, or Cobol. The

Table 6.3   The Same Operation Compared Among Different Languages

| High-Level Languages | Assembly Language | Machine Language (octal) |
|---|---|---|
| A = B + C - D | ADD RØ, R1 | 060001 |
|  | SUB R2, R1 | 160201 |
|  | MOV R1, R3 | 010103 |

second is *interpreter-based* languages, such as ORACL, Basic, and
Focal.  A compiler-based language requires the use of a translating
program called a *compiler*, which takes a complete program expressed
in high-level statements, such as SET C = A*B + D↑2, and translates
it into a unique and self-contained machine language-type program.
The resultant compiled program can then be executed after it is
loaded into the computer.  The important point is that the compil-
ing and executing processes occur in two distinctly separate steps.

The interpreter-based languages have the characteristic that
translation and execution are simultaneous processes.  The interpreter program, unlike the compiler, remains core-resident during
execution.  The interpreter translates and executes the program
line by line.  Because the interpreter is core-resident, on-line
editing and immediate execution of programs is possible.  However,
because each program statement must be translated, as well as executed, during the execution state, execution time is relatively
slow.  In addition, memory space (typically 4000 to 5000 words) must
be allocated to the interpreter itself at all times.  Although compiler languages are more core-efficient than interpreter languages
during execution, the actual program compilation requires a great
deal of hardware overhead.  Programs written in an interpretive
language whose execution times are too long can be made to run
faster by rewriting the slow segments of code into assembly language and combining them with the rest of the program.

Higher level languages are rather easily learned.  For example,
the book by McCracken [11] is recommended as an introduction to
Fortran.  For other languages the reader can consult computer

manufacturer's instruction manuals.  Most universities offer intro-
ductory courses on programming and can recommend suitable textbooks.


6.4   INTRODUCTION TO QUANTITATIVE SPECTRUM ANALYSIS

Quantitative spectrum analysis encompasses the techniques for ex-
tracting from a recorded spectrum the most accurate information
possible about characteristic line energies and intensities.  Although simple methods are often very practical, the more sophisticated treatments based on information theory yield the ultimate
precision.  The following sections are aimed at energy dispersive
analysis but contain many concepts applicable to wavelength dispersive spectrum analysis.

The recorded x-ray spectrum will be considered to be a histogram with the x axis representing wavelength intervals or energy
intervals.  These intervals are sometimes referred to as *channels*
(hence, channel number).  The y axis represents the number of counts
recorded in the time t in each interval or channel.  For the purpose
of this section it will be assumed that the system has negligible
deadtime or is operating with a proper livetime clock so that the
standard deviation in the observed counts in each channel is the
square root of the number of counts in that channel.  Furthermore,
the x-y coordinates described above will be used in a general
fashion, without reference to the specific energy or wavelength
calibration   The x coordinate will be considered to have nonnega-
tive integer values (0, 1, 2, 3, ...).


6.5   SMOOTHING

Smoothing the spectrum to reduce statistical fluctuations prior to
quantitative analysis is not recommended.  Smoothing provides
little improvement in the statistical precision obtainable with
proper integration of isolated peaks, and is of no advantage prior
to least-squares fitting of peaks.  It is of benefit, however, for

qualitative analysis, particularly for identifying weak peaks near the lower limit of detection.

The simplest smoothing method uses a uniform weighting of adjacent channels. That is, y', the smoothed value for the counts in channel x, is obtained from adjacent channels by

$$y'(x) = \frac{1}{2n + 1} \sum_{i=-n}^{+n} y(x + i) \qquad (6.1)$$

Here the smoothing takes place over 2n + 1 adjacent channels. If the values of y(x + i) are all approximately the same over the 2n + 1 channels, it is easy to show that the standard deviation in the smoothed data σ' is reduced by the factor $\sqrt{2n + 1}$ relative to the standard deviation of the original data σ. That is,

$$\sigma' = \frac{\sigma}{\sqrt{2n + 1}} \qquad (6.2)$$

Consequently, smoothing can reduce the statistical fluctuations in the background and make it easier to notice a small peak whose amplitude is slightly greater than the standard deviation in the original background data.

The deficiency of uniform weighting becomes apparent when smoothing over a peak. In this case the expected counts should vary rapidly with x. The counts in channels adjacent to channel x are poor representations of the expected counts in channel x. Consequently, the counts in channel x + i should receive less weight the larger the absolute value of i. Thus the smoothing algorithm becomes

$$y'(x) = \sum_{i=-n}^{+n} h(i)y(x + i) \qquad (6.3)$$

where h(i) is a weighting function such that

$$\sum_{i=-n}^{+n} h(i) = 1 \qquad (6.4)$$

A commonly employed weighting function is the binominal distribution [1]. For a three-channel smooth, the values of the function are

$$h(+1) = \frac{1}{4}$$
$$h(0) \ = \frac{1}{2}$$  \hfill (6.5)
$$h(-1) = \frac{1}{4}$$

For a five-point smooth, the function is

$$h(+2) = \frac{1}{16}$$
$$h(+1) = \frac{1}{4}$$
$$h(0) \ = \frac{3}{8}$$  \hfill (6.6)
$$h(-1) = \frac{1}{4}$$
$$h(-2) = \frac{1}{16}$$

A most interesting case is to consider a small peak superimposed on a high, flat background. If the peak is near the detection limits, information theory shows that there is an optimum weighting function to produce the best minimum detection limit [12-15]. For the case considered the optimum function is

$$h(i) = s(x - x_o) \hfill (6.7)$$

where $s(x - x_o)$ is the function describing the expected peak shape centered at channel $x_o$. In other words, the optimum weighting function has the same shape as the peak. In information theory the weighting function is also known as a *filter function* or a *correlator function*. The effect of the optimum filter is to concentrate all the usable information from the peak in the channel at the center of the peak to permit the best minimum detection limit using only the center channel.

A gaussian function is a fairly good representation of peak shape, i.e.,

$$f(x) = \frac{A}{\sigma \sqrt{2\pi}} \exp\left(\frac{-(x - \bar{x})^2}{2\sigma^2}\right) \tag{6.8}$$

where A is the area under the peak, $\bar{x}$ is its centroid, and $\sigma$ is its standard deviation. The full width at half maximum (FWHM) of the peak is given by

$$\Gamma = 2(2 \ln 2)^{1/2}\sigma = 2.35\sigma \tag{6.9}$$

Since Eq. (6.8) is a continuous function, the observed histogram function will be

$$s(x) = \int_{x-1/2}^{x+1/2} f(x)\ dx \tag{6.10}$$

where the width of a channel is unity and the channel number x represents the center of the channel. Consequently, Eq. (6.10) represents the optimum filter function for the detection of a weak peak on a high, flat background.

It is interesting to compare the uniform weighting function of Eq. (6.1) with the optimum function. For continuous functions it can be shown that the best width for the uniform weighting function is 1.17$\Gamma$ and the resulting detection limit is only about 6% worse than with the optimum filter. The uniform weighting function (rectangular filter) is equivalent to a simple integration of the counts over 2n + 1 channels about the center of the peak. If the region of integration extends over approximately 1.17$\Gamma$ at the center of the peak, it will provide detection limits within 6% of the best that can be obtained. Figures 4.65 and 4.66 show that it is not important to chose exactly 1.17$\Gamma$. Clearly, it is not necessary to smooth the spectrum to obtain good detection limits. A smoothing will be useful primarily for visual inspection. Furthermore, a simple rectangular filter is adequate and convenient since it is faster and simpler to program.

Note that smoothing increases the width of peaks and makes resolution of overlapping peaks more difficult. The optimum filter described above broadens the peak by a factor of $\sqrt{2}$.  In general, the optimum filter will depend on the background shape as well as the peak shape and will be affected by the peak-to-background ratio.

## 6.6  SIMPLE BACKGROUND SUBTRACTION METHODS

Where simple methods suffice, they should be used.  In many cases sophisticated techniques add considerably to the mathematical complexity and computational time without significant improvement in the accuracy of the answer.  Where severe peak overlap is not encountered and well-defined background regions exist on both sides of the peak, a simple linear background interpolation can be used. This method is discussed in Chap. 11 (Sec. 11.2.2) and Chap. 4 (Sec. 4.9.3).

A number of channels $\eta_P$ located symmetrically about the centroid of the peak is chosen for integration (Fig. 6.1).  Two background regions, each containing $\eta_B/2$ channels, are selected on either side of the peak, equidistant from the peak centroid.  The background included in the peak channels $\eta_P$ is estimated from

$$B = \frac{\eta_P}{\eta_B} (B_U + B_L) \tag{6.11}$$

where $B_U$ is the counts in the $\eta_B/2$ background channels above the peak and $B_L$ is the counts in the corresponding background region below the peak.  The net peak counts above background are

$$P = N_t - B = N_t - \frac{\eta_P}{\eta_B} (B_U + B_L) \tag{6.12}$$

where $N_t$ is the total counts in the $\eta_P$ channels.  The expected statistical error in B is

$$\sigma_B = \frac{\eta_P}{\eta_B} \sqrt{B_U + B_L} \tag{6.13}$$

Consequently, the expected relative standard deviation in P is

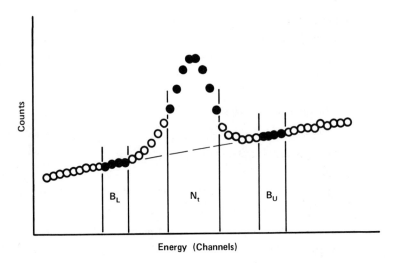

Figure 6.1  A simple linear background interpolation.  Reprinted by courtesy of EG&G ORTEC.

$$\frac{\sigma_p}{P} = \frac{[P + B(1 + \eta_p/\eta_B)]^{1/2}}{P} \tag{6.14}$$

The statistical error can be minimized by choosing $\eta_B \gg \eta_p$ where sufficient background regions exist.  For $P \gg B$ the peak integration region should be chosen to be approximately twice the peak FWHM to minimize Eq. (6.14).  For peak intensities near the detection limit $\eta_p \approx 1.17\Gamma$ should be chosen.  It should be noted, however, that $\eta_p \gg \Gamma$ provides less sensitivity to peak shifts caused by gain changes in the spectrometer [16].


6.7  LEAST-SQUARES FITTING OF THE BACKGROUND

When the assumption of a linear background, inherent in the method of Sec. 6.6, is not valid, then linear least-squares fitting of more representative functions can be employed.  If g(x) represents the general expected functional form of the background, then the specific function g(x) providing the best representation of the background is obtained by minimizing $\chi^2$ in Eq. (6.15).

$$\chi^2 = \sum_i \frac{1}{\sigma_i^2} [y_i - g(x_i)]^2$$

$$= \sum_i \frac{[y_i - g(x_i)]^2}{y_i} \qquad\qquad (6.15)$$

For example, if g(x) is a linear equation

$$g(x) = a + bx \qquad\qquad (6.16)$$

then the particular values of a and b providing the most probable estimate of the true function sampled by the data is obtained by determining the values of a and b which minimize Eq. (6.15) [1]. The solution is obtained from

$$\frac{\partial \chi^2}{\partial a} = 0 \qquad\qquad \frac{\partial \chi^2}{\partial b} = 0 \qquad\qquad (6.17)$$

and for a linear function is

$$a = \frac{1}{\Delta} \left( \sum \frac{x_i^2}{\sigma_i^2} \sum \frac{y_i}{\sigma_i^2} - \sum \frac{x_i}{\sigma_i^2} \sum \frac{x_i y_i}{\sigma_i^2} \right)$$

$$b = \frac{1}{\Delta} \left( \sum \frac{1}{\sigma_i^2} \sum \frac{x_i y_i}{\sigma_i^2} - \sum \frac{x_i}{\sigma_i^2} \sum \frac{y_i}{\sigma_i^2} \right) \qquad (6.18)$$

$$\Delta = \sum \frac{1}{\sigma_i^2} \sum \frac{x_i^2}{\sigma_i^2} - \left( \sum \frac{x_i}{\sigma_i^2} \right)^2$$

The $x_i$ represent the channel numbers used in the fit, and the $y_i$ are the counts in the corresponding channels. The weighting factors are $\sigma_i^2 = y_i$ for fitting spectra, where $y_i$ is the number of counts in a channel. The summation over i covers all the channels to be utilized in estimating the background. Once the specific function describing the background under the peak is determined, the interpolated contribution due to background can be subtracted from the total peak counts $N_t$.

The method can be applied in a similar fashion with a polynominal of n-th order

$$g(x) = \sum_{j=0}^{n} a_j x^j \qquad\qquad (6.19)$$

or with a sum of functions

$$g(x) = \sum_{j=0}^{n} a_j f_j(x) \qquad\qquad (6.20)$$

where each $f_j(x)$ is a different predetermined function of x.  In
both cases, $a_j$ is the coefficient of the j-th term to be determined
by fitting to the data.

The benefit of the least-squares fitting technique lies in the
fact that the solution represents the most probable value for the
background provided

1.  The general function truly describes the shape of the
    background spectrum.
2.  The data are distributed about the expected value accord-
    ing to the "normal" or gaussian probability function.

The latter criterion is valid where $y_i$ > 10 counts, since the
Poisson distribution describing counting statistics approaches the
normal distribution for large counts.  The validity of assuming both
criteria are met can be tested by comparing the $\chi^2$ actually obtained
with the value to be expected [1].  Abnormally high or low $\chi^2$ values
indicate violation of at least one of the above criteria.  Frequently, an abnormally high $\chi^2$ is encountered in spectrum fitting,
indicating that the choice of g(x) is wrong.

Usually background can be fitted with a first- or second-
degree polynomial.  Occasionally a third-degree polynominal is necessary.  The higher order polynominals should be avoided whenever
possible.


6.8  BACKGROUND CORRELATOR FUNCTIONS

The problem with least-squares fitting is the requirement to
specify the shape of the background.  In many cases it is more convenient to have a method which is completely general and requires

( a )

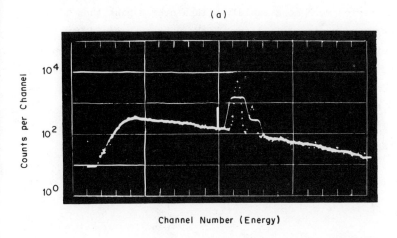

( b )

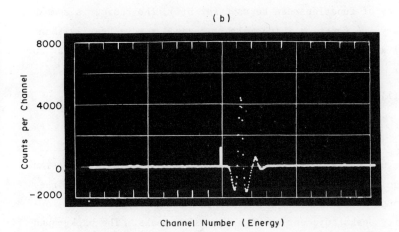

Channel Number ( Energy )

Figure 6.2  Background subtraction using the simple rectangular
correlator.  (a) The original spectrum with the estimated background
superimposed and intensified.  (b) The spectrum after background
subtraction.  Note the distortion of the small Cr Kβ peak caused by
overestimation of the background in the vicinity of the larger
Cr Kα peak.  Reprinted by courtesy of EG&G ORTEC.

no prior knowledge of the shape of the background.  Correlator functions, and Fourier filtering are equivalent techniques in this
general class of methods [12-15,17].  They are based on the fact

that y(x) varies much more rapidly with x over a peak than it does
in a background region.  A filter which suppresses rapid variations
of y(x) with x but retains slowly varying components will suppress
peaks and retain the background information.  If the smoothing function in Eq. (6.3) is extended over a large number of channels compared to the width of a peak, then the smoothed spectrum is a
reasonably good estimate of the background.  That is, the background is given by

$$g(x) = \sum_{i=-n}^{+n} h(i)y(x + i) \qquad\qquad (6.21)$$

provided

$$2n + 1 \gg 2\Gamma \qquad\qquad (6.22)$$

A variety of functions can be used for h(i), the choice is somewhat
arbitrary.  A function which weights channel x more heavily than
channel x + i is generally desirable.  However the rectangular correlator

$$h(i) = (2n + 1)^{-1} \qquad\text{for } -n \leq i \leq n$$
$$= 0 \qquad\qquad\text{for } -n > i > n \qquad (6.23)$$

is easier to apply and provides satisfactory results [12].

Figure 6.2 illustrates background subtraction using the simple
rectangular correlator.  Note that this method overestimates the
background in the vicinity of a peak, causing suppression of low-
intensity peaks adjacent to high-intensity peaks.  The background
estimate is particularly faulty in a wide region of severe peak
overlap.

By adding several refinements to the simple background correlator method, performance in the vicinity of peaks can be improved
substantially [15,17-19].  Where the correlator predicts a background higher than the original data, the original data are retained as the better background estimate [18].  The change in slope
of the background can also be monitored to discard regions identified as belonging to peaks because the slope changes too rapidly

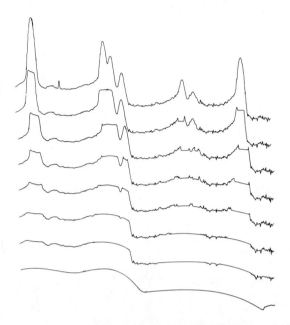

Figure 6.3  Background estimation with a correlator algorithm designed to reduce the perturbation due to peaks.  The top line is the
original spectrum.  The seven steps in arriving at the final estimate on the bottom line are illustrated.  The vertical scale is
logarithmic, with each spectrum displaced vertically for clarity.
Reprinted by courtesy of EG&G ORTEC.

[15,18,19].  The spectrum can be broken up into smaller regions and
the minimum intensity for each section identified as a background
sample [15,19].  A slowly varying function is used to interconnect
these points and a slope test is applied.

    Figure 6.3 illustrates the selection of background using a
refined correlator technique.  The vertical (intensity) scale is
logarithmic.  The top spectrum is the raw data.  Seven smoothing
operations are carried out with a correlator.  In each case the resulting spectrum is shown, displaced downward from the previous
result.  In each operation the new smoothed spectrum and the
previous spectrum are compared.  Where the original data is more
than three standard deviations below the smoothed data, the original
data is retained as the best background estimate.

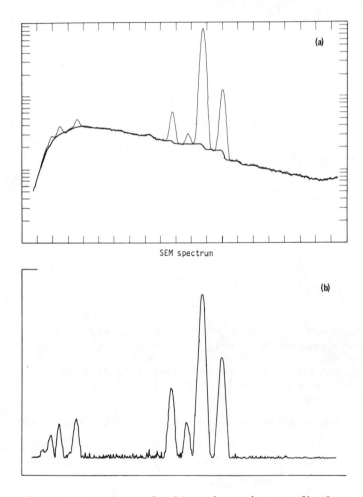

SEM spectrum

Figure 6.4  Background subtraction using a refined correlator technique.  (a) The original spectrum with the background estimate
superimposed and intensified.  (b) The spectrum with background
subtracted.  Reprinted by courtesy of EG&G ORTEC.

Figure 6.4 shows the application of this technique to an x-ray
spectrum obtained with a Si(Li) detector system on a scanning electron microscope.  In the top spectrum the background is superimposed
on the raw spectrum.  In the lower spectrum the background has been
subtracted, leaving only the peaks.  In this case an additional
refinement has been added to account for the step in background

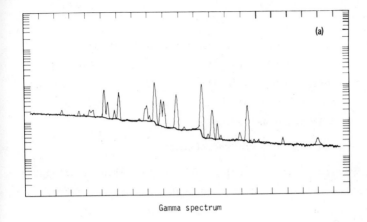

Gamma spectrum

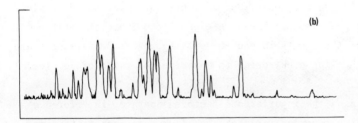

Figure 6.5  The refined correlator background subtraction with a
complicated γ-ray spectrum for a Ge(Li) detector.  (a) The original
spectrum with background superimposed.  (b) The spectrum after
background subtraction.  Reprinted by courtesy of EG&G ORTEC.

level across a peak due to incomplete charge collection in the detector [18] (see Sec. 11.3 and Fig. 11.8).  The complicated γ-ray
spectrum in Fig. 6.5 illustrates the general power of the correlator method in estimating the background without prior knowledge of
the detailed shape of the background spectrum.

By reference to Eq. (6.14) it can be concluded that the standard deviation in the net peak counts above background will lie
somewhere between $\sigma_p = (P + B)^{1/2}$ and $\sigma_p = (P + 2B)^{1/2}$, depending
on the smoothing interval and the details of the background correlator algorithm.

## 6.9   SIMPLE TREATMENT OF PEAK OVERLAPS

Although least-squares fitting of overlapping peaks theoretically
provides the optimum solution, it is mathematically complicated and
often consumes excessive computational time.   In repetitive measurements, where a high throughput of specimens is important, simpler
and faster solutions are advantageous.

### 6.9.1   Predetermined Peak Interference Factors

One of the simplest and most useful methods for handling overlapping
peaks involves the use of predetermined peak interference factors.
Except for severe peak overlap cases, this method can produce results nearly as good as least-squares fitting, but is considerably
faster.   Figure 6.6 illustrates the method.

   In Fig. 6.6(c) two peaks (dashed lines) are overlapped to yield
the composite peak (solid line).   Regions 1 and 2 are chosen for
integration of the counts in each underlying peak.   Best results are
obtained by choosing each region to maximize the ratio of Eq. (6.26)
to Eq. (6.27).

   To determine the interference factors separate, interference-
free spectra are recorded for each peak from specimens containing
only one of the elements.   These reference spectra are accumulated
for a time sufficient to render the statistical errors in the measured interference factors negligible.   Figures 6.6(a) and (b) show
the two reference spectra after background subtraction.   The peak
areas in regions 1 and 2 are measured and noted as $Q_1$ and $Q_2$ for
peak Q and $R_1$ and $R_2$ for peak R.   The peak interference factors are
defined by

$$q = \frac{Q_2}{Q_1} \qquad\qquad r = \frac{R_1}{R_2} \qquad\qquad\qquad (6.24)$$

   In the composite spectrum [Fig. 6.6(c)] an appropriate method
is used to estimate the background contributions $B_1$ and $B_2$ for
regions 1 and 2, respectively.   The total counts in regions 1 and 2
can be expressed as

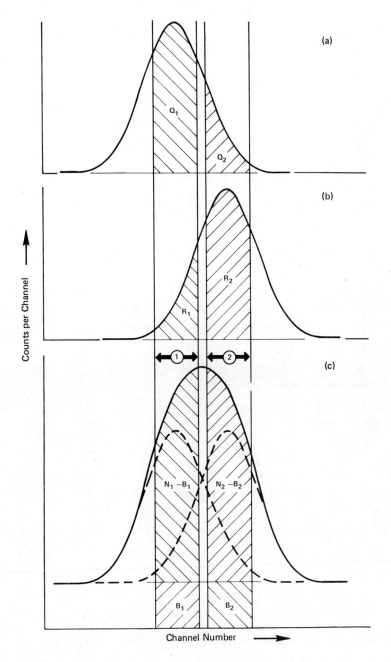

Figure 6.6  Illustration of doublet deconvolution using peak interference factors.  See text for details.  Reprinted by courtesy of
EG&G ORTEC.

$$N_1 = N_Q + B_1 + rN_R$$
$$N_2 = N_R + B_2 + qN_Q$$
$$\text{(6.25)}$$

where $N_Q$ is the contribution to region 1 due to peak Q and $N_R$ is the contribution to region 2 due to peak R. Equations (6.25) can be solved for the net peak areas of interest, $N_Q$ and $N_R$.

$$N_Q = \frac{N_1 - B_1 - r(N_2 - B_2)}{1 - qr}$$
$$N_R = \frac{N_2 - B_2 - q(N_1 - B_1)}{1 - qr}$$
$$\text{(6.26)}$$

Thus the interfering peaks can be resolved into interference-free peak areas by integrating the counts in regions 1 and 2 in the composite spectrum, estimating the backgrounds $B_1$ and $B_2$, and applying the previously determined interference factors q and r.

The statistical errors in the resolved areas $N_Q$ and $N_R$ can be shown to be

$$\sigma_Q = \frac{[N_1 + \sigma_{B1}^2 + r^2(N_2 + \sigma_{B2}^2)]^{1/2}}{1 - qr}$$
$$\sigma_R = \frac{[N_2 + \sigma_{B2}^2 + q^2(N_1 + \sigma_{B1}^2)]^{1/2}}{1 - qr}$$
$$\text{(6.27)}$$

providing the errors in q and r are made negligible. The contributions due to the background will depend on the method of background estimation but will generally fall between the limits $0 < \sigma_{B1}^2 \leq B_1$ and $0 < \sigma_{B2}^2 \leq B_2$.

Consider the case where $N_R$ is large compared with $N_Q$. For simplicity assume that $\sigma_{B1}$ and $\sigma_{B2}$ are zero. The error in the weak peak area becomes

$$\sigma_Q = \frac{[(N_Q + rN_R + B_1) + r^2(N_R + qN_Q + B_2)]^{1/2}}{1 - qr}$$

Since q and r are both less than 1, $B_1 \approx B_2$, and $N_Q \ll N_R$, the expression simplifies to

$$\sigma_Q \approx \frac{[N_Q + rN_R + B_1]^{1/2}}{1 - qr} \tag{6.28}$$

From Eq. (6.28) it is evident that the intense peak significantly contributes to the statistical error in determining the weak peak area.  In fact if $rN_R$ is large compared with $N_Q + B_1$ the statistical error in the weak peak is totally due to the intense peak.  This is the major limitation in measuring a weak peak in the presence of an intense interfering peak, and is basic to *all* the methods for resolving interfering peaks.

Because Eq. (6.26) is very simple, an on-line minicomputer can solve for $N_Q$ and $N_R$ very quickly.  Providing the peak shapes and positions in the reference spectra are identical to those in the composite spectrum, the method is relatively free of systematic biases in estimating the net intensities.  The technique can also be applied to overlap between two line series.  For example, the Mn Kβ lies under the Fe Kα on energy dispersive spectrometers.  The Mn Kα peak area can be used to subtract the Mn Kβ interference from the Fe Kα line.  However, a word of caution is in order when using Kα/Kβ or Lα/Lβ ratios for resolving interferences.  Occasionally the composite specimen contains an absorption edge between the two lines, which is not present in the pure element reference spectra. Thus the Kα/Kβ or Lα/Lβ ratios are different in the composite specimen than in the pure element reference standards.

## 6.9.2  Sequential Peak Stripping

The removal of the Mn Kβ interference from the Fe Kα using the interference-free Mn Kα intensity suggests another method.  First the composite spectrum is recorded on the unknown specimen.  Background is subtracted using one of the general methods previously outlined, and a full qualitative analysis is performed.  Once the peaks are all identified the spectrum is "stripped" starting with the most interference-free and most prominant peak.  For example, suppose the Mn Kα is interference free, but the Mn Kβ interferes with the Fe Kα.  The Mn peaks would be stripped first, by recording

a reference spectrum from a pure manganese specimen.  The same
background subtraction method is applied to the reference spectrum.
Next the reference spectrum intensity is multiplied by an appropriate constant and subtracted from the composite spectrum.  The constant is adjusted so that the subtraction cancels the Mn Kα peak in
the composite spectrum.  As a result, the Mn Kβ interference is also
removed from the Fe Kα peak.  This process is repeated for each
element in the spectrum, moving from the most interference-free and
intense peaks toward the weakest peak.  At the end of the process
the operator is left with a spectrum containing no significant
residual peaks and a list of constants for the elements found.
These constants represent the ratio of the intensity in the unknown
to the pure element intensity for each element.  This information
forms the input for a quantitative concentration calculation.

To avoid error contributions from the reference spectra due to
counting statistics, each pure element standard should be counted
long enough to have an intensity at least twice the intensity contained in the composite spectrum.  If this involves using a different counting time, the appropriate correction must be made to
the intensity ratio.  This is most easily achieved by dividing the
counts by the counting time for all spectra.  Care must be taken to
ensure that peak shapes and positions in the reference spectra are
identical in the unknown spectrum.  Gain and baseline shifts due to
counting rate changes can cause problems in this respect.  Note
also that this method will not properly strip the pileup spectrum.

The major shortcoming of the method is the fact that most of
the residual error is assigned to the last peak stripped in a
region of peak overlaps.  Typically this last peak is also the
smallest and can least tolerate the additional intensity error.
In addition, the method is slow where many peaks must be stripped.
Some relief can be gained by using calculated spectra rather than
measured reference spectra.  This is particularly helpful where a
pure element standard is unavailable.  Here again it is important
to check for the absorption edge problem described in Sec. 6.9.1.

## 6.10  LEAST-SQUARES FITTING OF PEAKS

Providing peak shapes are accurately depicted, least-squares fitting
provides the most accurate measurement of component peak intensities
for overlapping peaks.

### 6.10.1  Linear Least-Squares Fitting

For a region of a spectrum containing a background and n overlapping
peaks, the fit is performed with the function

$$g(x) = f_o(x) + \sum_{j=1}^{n} f_j(x) \tag{6.29}$$

The background is described by $f_o(x)$, while each $f_j(x)$ describes one
of the peaks.  For simplicity it will be assumed that the peak
shape is gaussian, i.e.,

$$f_j(x) = \frac{A_j}{\sigma_j \sqrt{2\pi}} \exp\left(\frac{-(x - \bar{x}_j)^2}{2\sigma_j^2}\right) \tag{6.30}$$

with an area $A_j$, a centroid $\bar{x}_j$, and a full width at half maximum
height.

$$\Gamma_j = 2(2 \ln 2)^{1/2} \sigma_j k \tag{6.31}$$

Here k is the electronvolt-per-channel calibration, i.e., $k = dE/dx$,
$\Gamma_j$ is in electron volts, and $\sigma_j$ is in units of "channel number."
If each of the overlapping peaks has been identified, then its correct centroid can be read from tables of characteristic line ener-
gies.  If $\bar{x}_j$ is known, then $\Gamma_j$ and $\sigma_j$ can be computed from the
relation

$$\Gamma_j = \left[(\Gamma_{noise})^2 + (2.35 \sqrt{\varepsilon F E_j})^2\right]^{1/2} \tag{6.32}$$

Where $E_j$ is the energy corresponding to channel number $\bar{x}_j$.  With $\bar{x}_j$
and $\sigma_j$ determined, the only free parameter remaining in $f_j(x)$ is
$A_j$.

The background function will typically be a quadratic, i.e.,

$$f_o(x) = a_0 + a_1 x + a_2 x^2 \tag{6.33}$$

Consequently, the function to be fitted to the spectrum is of the form

$$g(x) = \sum_k a_k f_k(x)$$

where the $f_k(x)$ are predetermined functions and the $a_k$ are the parameters to be determined by optimizing the fit. This means that a linear least-squares fitting procedure can be used. As before, the $\chi^2$ to be minimized is

$$\chi^2 = \sum_i \frac{[y_i - g(x_i)]^2}{y_i} \tag{6.34}$$

and the solution is obtained from the set of linear equations.

$$\frac{\partial \chi^2}{\partial a_0} = 0 \qquad\qquad \frac{\partial \chi^2}{\partial a_1} = 0$$

$$\frac{\partial \chi^2}{\partial a_2} = 0 \tag{6.35}$$

$$\frac{\partial \chi^2}{\partial A_j} = 0 \qquad\qquad j = 1 \text{ to } n$$

Note that the solution does not require iteration. From Eqs. (6.35) the most probable values of $a_0$, $a_1$, $a_2$, and $A_j$ are obtained. These values substituted into Eq. (6.34) yield a value of $\chi^2$ which can be compared with the theoretically expected value. Abnormal values of $\chi^2$ indicate an improper choice of fitting functions or data which do not follow the normal distribution law. The former is usually the cause of the poor fit. Frequently an abnormally large $\chi^2$ value indicates an overlooked peak.

6.10.2. Nonlinear Least-Squares Fitting

If the $\bar{x}_j$ are not known, and must be determined by the fit, then an additional set of equations must be included in Eqs. (6.35), i.e.,

$$\frac{\partial \chi^2}{\partial \overline{x}_j} = 0 \qquad\qquad j = 1 \text{ to } n \qquad\qquad\qquad (6.36)$$

Unfortunately Eqs. (6.36) are not linear in $\overline{x}_j$, and the set of Eqs.
(6.35) and (6.36) must be solved by iteration. This is called a
*nonlinear least-squares fit*. A detailed description of this method
can be found in Ref. 1. Sometimes it is also necessary to include
the $\sigma_j$ as free parameters, making the fit even more complicated.
The problem with nonlinear least-squares fitting is the extensive
time consumed in searching for the optimum set of parameters.
Therefore, it is preferable to determine $\overline{x}_j$ and $\sigma_j$ prior to the fit
in order to use a linear least-squares fit.

## 6.11  DETERMINING PEAK SHAPES

Except when fitting peaks containing very few counts, the assumption
of a gaussian peak shape is clearly inadequate. Figure 6.7 illustrates the principle effects contributing to the peak shape. If
the resolution broadening caused by the x-ray spectrometer is de-
convoluted from the spectrum, the characteristic x-ray line approaches a delta function with a width limited only by the natural
line width. This narrow line is superimposed on the background
from the specimen. Occasionally, when the characteristic photon is
detected in the Si(Li) detector, not all of the charge produced is
collected. This incomplete charge collection leads to a detector-
produced background shelf below each characteristic line. Although
this shelf has been depicted as flat in Fig. 6.7 it usually has
some slow variation with pulse height. The sharp edge of the background shelf and the characteristic line are both broadened by the
detector resolution. The broadening is due to preamplifier noise
and detector ionization statistics. These processes ideally produce a gaussian broadening. However a combination of bulk trapping
and incomplete charge collection in the Si(Li) diode causes a

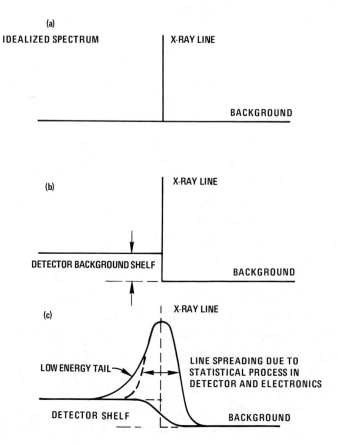

Figure 6.7  Contributions to the peak shape.  (a) The spectrum from the specimen.  (b) The background contribution due to the detector. (c) Broadening of all sharp features due to spectrometer resolution. Reprinted by courtesy of EG&G ORTEC.

significant deviation from a gaussian in the form of a low-energy tail.

The best description of the actual peak shape has been developed by Gunnick and Niday [18].  They describe the peak shape by the function

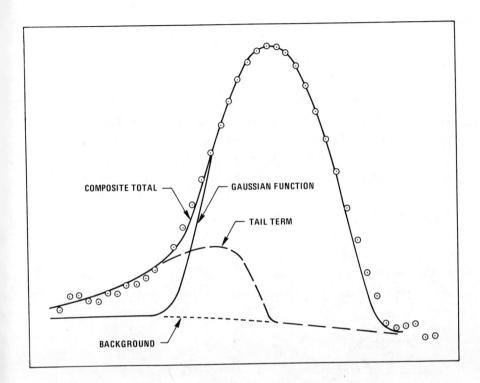

Figure 6.8  The GAMANAL peak shape.  Reprinted by courtesy of
EG&G ORTEC.

$$f(x) = y_0 \exp\left[\frac{-(x - \overline{x})^2}{2\sigma^2}\right]$$

$$+ ay_0\left\{\exp[b(x - \overline{x})]\right\}\left[1 - \exp\left(\frac{-c(x-\overline{x})^2}{2\sigma^2}\right)\right]\delta \qquad (6.37)$$

The first term is the standard gaussian, and the second term describes the low-energy tailing.  The peak height at $x = \overline{x}$ is $y_0$; a,
b, and c are parameters used to describe the low-energy tail, and

$$\delta = 1 \qquad\qquad \text{for } (x - \overline{x}) \leq 0$$

$$= 0 \qquad\qquad \text{for } (x - \overline{x}) > 0 \qquad\qquad (6.38)$$

The function is illustrated in Fig. 6.8. The shape contains six parameters: $y_o$, $\bar{x}$, $\sigma$, a, b, and c. Usually the last four parameters are predetermined from calibration spectra and expressed as functions of $\bar{x}$. This leaves two free parameters, $y_o$ and $\bar{x}$, and the fit is nonlinear. If $\bar{x}$ can be predetermined by qualitative analysis, only one free parameter remains, and the fit is linear.

Equation (6.37) describes a continuous function, whereas the spectrum is a histogram. Therefore, the algorithm in Eq. (6.10) must be applied to convert Eq. (6.37) to a histogram representation.

An alternative method is to define the peak shapes by measuring reference spectra on pure element standards. These stored spectra are called up as the functions $f_j(x)$ [Eqs. (6.29) and (6.30)] when the linear least-squares fit is to be performed. Some of the shortcomings listed in Sec. 6.9.2 are also applicable to the use of pure element reference spectra in least-squares fitting.


## 6.12   OTHER TECHNIQUES

The science of spectrum analysis is extensive and comprehensive. Therefore, the  analyst who is interested in skillfully applying these techniques is advised to consult the complete list of references at the end of this chapter. Several interesting topics not covered here, such as Fourier transformation and resolution enhancement, are treated by several of the references.


REFERENCES

1.  Philip R. Bevington, *Data Reduction and Error Analysis for the Physical Sciences,* McGraw-Hill, New York, 1969.
2.  P. Quittner, *Gamma-Ray Spectroscopy,* Akademiai Kiado, Budapest, 1972.
3.  *CAMAC: A Modular Communication Bridge,* ORTEC, Oak Ridge, Tenn., 1972.
4.  A. Senator, I. N. Hooton, G. L. Miller, H. P. Lie, and E. A. Gere, in *Proceedings for the Skytop Conference on Computer Systems in Experimental Nuclear Physics,* Clearing House for Federal Scientific and Technical Information, National Bureau of Standards, U.S. Department of Commerce, Springfield, Va., 1969, p. 394.

5. I. N. Hooton, in *Proceedings for the Skytop Conference on Computer Systems in Experimental Nuclear Physics,* Clearing House for Federal Scientific and Technical Information, National Bureau of Standards, U.S. Department of Commerce, Springfield, Va., 1969, p. 466.

6. *CAMAC Tutorial Papers,* in *IEEE Trans. Nucl. Sci.,* NS-20, 2:1-71 (1973).

7. AEC Reports TID-25876 (March 1972), TID-25875 (July 1972), and TID-25877 (December 1972), Superintendent of Documents, U.S. Government Printing Office, Washington, D.C.

8. D. S. Gemmell, in *Proceedings for the Skytop Conference on Computer Systems in Experimental Nuclear Physics,* Clearing House for Federal Scientific and Technical Information, National Bureau of Standards, U.S. Department of Commerce, Springfield, Va., 1969, p. 37.

9. D. G. Kyser, G. L. Ayers, and D. E. Horne, *Proceedings of the Eighth National Conference on Electron Probe Analysis,* Electron Probe Society of America, New Orleans, 44A (August 1973).

10. P. M. Grant, T. R. Lusebrink, and D. G. Taupin, *Industrial Research, 50* (November 1972).

11. Daniel D. McCracken, *A Guide to Fortran IV Programming,* Wiley, New York, 1972.

12. A. Robertson, W. V. Prestwich, and T. J. Kennett, *Nucl. Instrum. Methods, 100:*317 (1972).

13. M. A. Mariscotti, *Nucl. Instrum. Methods, 50:*309 (1967).

14. W. W. Black, *Nucl. Instrum. Methods, 71:*317 (1969); A. L. Connelly and W. W. Black, *Nucl. Instrum. Methods, 82:*141 (1970).

15. T. Inouye and N. C. Rasmussen, *Trans. ANS, 10:*38 (1967); T. Inouye, *Nucl. Instrum. Methods, 30:*224 (1964); T. Inouye, T. Harper, and N. C. Rasmussen, *Nucl. Instrum. Methods, 67:*125 (1969); T. Inouye, *Nucl. Instrum. Methods, 104:*541 (1972).

16. I. L. Fairweather, and D. A. Gedcke, *Nucl. Instrum. Methods, 49:*1 (1967).

17. G. K. Wertheim, *J. Electron Spectroscopy and Related Phenomena, 6:*239 (1975).

18. R. Gunnink and J. B. Niday, *Computerized Quantitative Analysis by Gamma-Ray Spectrometry,* Vols. I to V, Lawrence Livermore Laboratory Report UCRL-51061, TID-4500, University of California, 1971.

19. E. Achterberg, F. C. Iglesias, A. E. Jech, J. A. Moragues, M. Pérez, J. J. Rossi, W. Scheuer, and J. F. Suárez, *IEEE Trans. Nucl. Sci., NS-19:*3 (1972).

20. Gary Horlick, *Anal. Chem., 43:*61A (1971); *44:*943 (1972).

21. J. W. Colby, in *Proceedings of the Sixth International Conference on X-Ray Optics and Microanalysis,* (G. Shinoda, K. Kohra, and T. Ichionikawa, eds.), University of Tokyo Press, 1972, pp. 247-251.

22. K. J. Blinowska and E. F. Wessner, *Nucl. Instrum. Methods, 118:*597 (1974).

23. M. Dojo, *Nucl. Instrum. Methods, 115:*425 (1974).

24.  Marcel J. E. Golay, *IEEE Trans. Computers, 299* (March 1972).
25.  N. G. Volkov, *Nucl. Instrum. Methods, 113*:483 (1973).
26.  Kanji Tasaka, *Nucl. Instrum. Methods, 109*:547 (1973).
27.  J. Libert, *Nucl. Instrum. Methods, 109*:609 (1973).
28.  W. Teoh, *Nucl. Instrum. Methods, 109*:509 (1973).
29.  J. W. Tepel, *Nucl. Instrum. Methods, 40*:100 (1966).
30.  F. H. Schamber, *Proceedings of the Eighth National Conference on Electron Probe Analysis,* Electron Probe Society of America, New Orleans, 44A (August 1973), p. 85A.

# 7

## Specimen Preparation

### 7.1 INTRODUCTION

The intensity of a spectral line can be strongly influenced by two
major sources of error, namely selective absorption of the primary
beam and the absorption and enhancement of the fluorescent radiation. These effects may be further complicated by physical phenomena which are already present, or introduced, as a result of
specimen preparation procedures. The dependence of x-ray spectral
intensity upon the physical state of the specimen is well known.
Effects such as surface roughness, particle shape, particle size,
and size distribution can all lead to nonproportional relationships
between spectral intensity and elemental composition. This chapter
will discuss the problems associated with specimen preparation.
The basic techniques will be covered briefly and special attention
will be devoted to several methods in common use today.

The guiding principles for specimen preparation techniques are
reproducibility, accuracy, simplicity, low cost, and rapidity of
preparation. Often these are mutually exclusive goals, and it becomes the job of the analyst to choose from the wide variety of
available methods or to design a new method to fit his particular

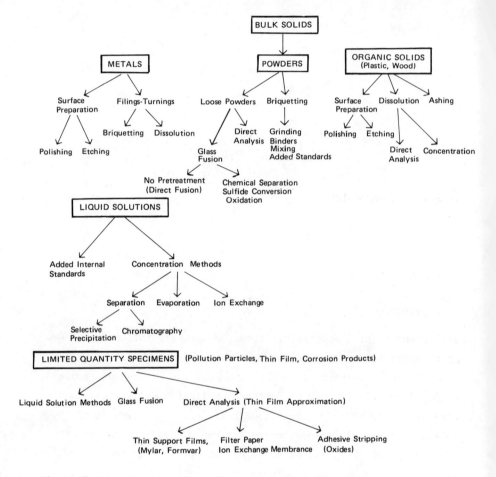

Figure 7.1  A summary of sample preparation procedures.

problem.  Generally, it is not simply a matter of deciding which of
the many procedures will work, but a consideration of the precision,
accuracy, and timecost factors involved.

A broad general outline of specimen preparation procedures are
shown in Fig. 7.1 and Table 7.1.  These show the wide variety of
preparation procedures available to the x-ray analyst.  Details of
these specific preparation methods may be found by consulting the
list of references at the end of this chapter.  Several of the most
common techniques will however be discussed in this chapter, owing
to their broad application to a wide variety of sample types.

Table 7.1 Advantages and Disadvantages of Several General Specimen
Preparation Procedures Used in X-ray Spectrochemical Analysis

---

*Bulk Solids Such as Metals and Ceramics* [1-4]

Advantages: Specimen preparation generally rapid, high fluorescent
x-ray intensities; if identical comparison standards are available, good analytical accuracy can be obtained.

Disadvantages: Heterogeneity problem can be large; results very
sensitive to surface preparation method used; no standard can be
added; identical standards are not always available.

*Liquids* [5,6]

Advantages: Specimen inhomogeneity eliminated; easy preparation of
standards; dilution to reduce or eliminate interelement effects;
chemical separation or concentration possible; various sample
forms can be analyzed; no surface effects.

Disadvantages: Sample must be destroyed; dissolution may be slow
and tedious or very difficult; high dilution results in decreased
sensitivity; very low intensity from low-atomic-number elements
(high dilution) window absorption, high background scatter; out-
gasing and heating of liquid can seriously affect analysis,
especially in vacuo (splatter, bubbles, expansion, induced precipitation).

*Fusion (Glass)* [7,8]

Advantages: Specimen inhomogeneity eliminated; matrix effects can
be minimized; easy preparation of standards, matrix modification
possible, surface effects minimal for atomic numbers > 15, moderately rapid specimen preparation now possible, capable of highest
precision and accuracy of any sample preparation method.

Disadvantages: Not all samples will fuse without prior treatment
(metals, sulphides, organics); detection limit of very low atomic
number elements is high ($Na_2O$, 500-700 ppm); large dilution factor results in lower count rates; ignition losses must be considered.

*Powders* [1,2,9-11]

Advantages: Rapid convenient specimen preparation; high x-ray intensity; standards can be added; matrix modification (dilution or
heavy absorber) is possible; permanent standards can be prepared
by briquetting.

Disadvantages: Can have severe heterogeneity problems; trace-level
grinding contamination; pressure-density variations in
briquettes; grinding, weighing, and mixing errors probable; surface effects will strongly influence low-atomic-number analysis.

---

Table 7.2   General Approach to Specimen Preparation for X-ray
Spectrochemical Analysis

---

1.  Have analytical goals clearly in mind (precision, accuracy,
    concentration range, cost, speed).
2.  Anticipate the expected interelement effects by considering
    absorption edges and emission lines.
3.  Consult the available literature.
4.  Select several possible procedures that have the potential to
    meet the conditions set forth in the analytical goals.
5.  Evaluate each possible procedure to determine its strengths and
    weaknesses.   Some laboratory measurements may be necessary.
    Narrow your choices.
6.  Return to the literature to obtain more details on the particu-
    lar method or methods selected in step 5.
7.  Choose a procedure, modify it to meet your needs, and perform a
    laboratory and statistical evaluation.

---

As a rule, the literature should always be consulted prior to
starting any analytical method development.   It is a common situation to find that somebody else has faced your problem and has made
a good start toward a solution.

## 7.2   GENERAL APPROACH TO SPECIMEN PREPARATION

Table 7.2 presents a general approach to the specimen preparation
problem in x-ray spectrochemical analysis.   The various steps shown
in Table 7.2 are discussed in more detail in the paragraphs that
follow.

As stated above, the analytical goals must be clearly in mind
before beginning a specimen preparation method.   To provide a basis
for the discussion that follows, a typical specimen preparation
problem is outlined below.   Consider the analysis of a mine tailings
sample containing approximately 90 to 95% $SiO_2$.   This sample contains minor amounts of $Fe_2O_3$, $TiO_2$, $Cr_2O_3$, $Al_2O_3$, and several other
elements in the parts per million range.   Assume that the sample
arrives in the form of a coarse powder having an average particle
size of approximately 100 mesh.   Further assume that the analyst

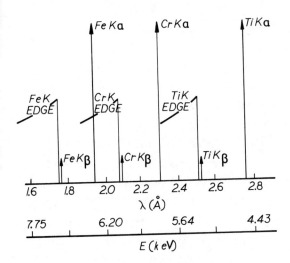

Figure 7.2   Absorption edges and emission lines used to locate potential interelement effects.

has been requested to determine the $TiO_2$ and $Cr_2O_3$ content within a standard deviation of less than ± 0.03% absolute.   Figure 7.2 illustrates a general *first step* to be taken in any x-ray spectrochemical analysis prior to the selection of a sample preparation method.   This figure is a visual presentation of the expected interelement effects (see Chaps. 2 and 9), it allows the investigator to anticipate the magnitude of the matrix effects before attempting to select his specimen preparation method.   From Fig. 7.2 it can be seen that moderate to severe interelement effects can be expected, owing to the relative position of the Cr Kα emission line and the titanium absorption edge.   It will be assumed for this example that the analyst has a chromium target x-ray tube at his disposal.   The *second step* requires the analyst to carefully consult the available literature for his particular analytical problem.   This consultation should not be limited to available textbooks but must include current journals, various abstracting services such as the *X-ray Fluorescent Spectroscopy Abstracts* [12] and the biennial (even Years) "X-ray Absorption and Emission Reviews found in *Analytical Chemistry* [13].

For the mine tailings analysis presented above, one can immediately locate 18 references for titanium analysis in ores and
minerals in a recent textbook by Liebhafsky [14].  In the "1970
X-ray Absorption and Emission Analytical Review" issue of *Analytical
Chemistry,* there are five specific references pertinent to the
analysis of titanium in minerals and ores.  Reference to the previous years of this journal indicates the following number of
specific references:  1968, 4; 1966, 9; 1964, 4; 1962, 10.  Thus,
before selecting any particular specimen preparation technique
shown in Fig. 7.1, the analyst has available a minimum of some 50
references to consider.  Having scanned the references discussed
above, decisions can be made regarding the various possibilities
commensurate with the analytical goals requested of the analyst,
the interelement effects described in Fig. 7.2, and the cost factors
involved in the analysis.  Because of the required accuracy, assume
that the analyst has narrowed his choice of methods down to
(1) grinding and briquetting, (2) liquid solution, and (3) glass
fusion.  A study of the literature indicates that the desired accuracy could not be obtained with briquetted powders unless the
samples are ground such that all particles are less than 20 micrometers (elimination of particle size-induced heterogeneity effects).
This grinding was found to be excessively time consuming, and thus
the powder method was abandoned.  The liquid solution technique
is especially unsatisfactory for low-atomic-number elements.  The
technique that appears to have the best promise of success would be
the glass fusion technique, and this will be discussed in more
detail in subsequent sections of this chapter.

Having made the decision to consider the glass fusion procedure, the analyst should again turn to the literature.  Table 7.2 is
a listing of some of the available papers in the area of glass
fusion sample preparation methods.  Many of these references were
first located using the analytical review issue of *Analytical
Chemistry,* available textbooks [1,11] and commerical literature
[15].  From several of the references listed in Table 7.3, a glass

Table 7.3 Bibliography of the Glass Fusion Method of Specimen
Preparation

1. I. Adler, in *X-Ray Emission Spectroscopy in Geology*, Elsevier,
   Amsterdam, 1966, pp. 124-126.
2. A. D. Ambrose, R. Rutherford, and S. Muir, The analysis of
   oxide systems by x-ray fluorescence -- a method of wide application, *Metallurgia, 491:*119-124 (1970).
3. G. Anderman and J. D. Allen, The evaluation and improvement
   x-ray emission analysis of raw mix and finished cements,
   *Advan. X-ray Anal., 4:*414-432 (1961).
4. D. G. Ashley and K. W. Andrews, Analysis of aluminosilicate
   materials by x-ray spectrometry, *Analyst, 97:*841-847 (1970).
5. The X-ray analysis of metallic elements in nonmetallic material, *A. R. L. Spectrographers Newsletter, 7:*1-3 (1954).
6. L. Bean, A method of producing sturdy specimens of pressed
   powders for use in x-ray spectrochemical analysis, *Appl.
   Spectrosc., 20:*191 (1966).
7. W. Becker and G. Wronka, Mechanised sample preparation for
   x-ray fluorescence analysis of oxide materials, *Siemens Rev.,
   XXXIX:*24-27 (1972).
8. D. F. G. Brown, A. M. MacKay and A. Turek, Preparation of
   stable silica standard solutions in rock analysis using
   lithium tetraborate, *Anal. Chem., 41:*2091 (1969).
9. K. G. Carr-Brion, A rapid method of casting fused beads for
   using x-ray fluorescence analysis, *Analyst, 89:*556-557 (1964).
10. F. Claisse, Accurate x-ray fluorescence analysis without
    internal standards, *Norelco Reporter, 4(1):*3-7 (1957).
11. T. J. Cullen, Potassium pyrosulphate fusion technique in
    determination of copper in mattes and slags by x-ray spectro-
    scopy, *Anal. Chem., 32:*516-517 (1960).
12. B. P. Fabbi, A refined fusion x-ray fluorescence technique and
    determination of major and minor elements in silicate stan-
    dards, *Amer. Mineral, 57:*237-245 (1972).
13. R. J. Fredericks and N. Longbottom, Determination of iron,
    zinc, and calcium in boiler fireside deposits using x-ray
    spectroscopy, *Norelco Reporter, 9:*44-45 (1962).
14. S. Sarian and H. W. Weart, X-ray fluorescence analysis of
    polyphase metals. An application of the borax fusion tech-
    nique, *Anal. Chem., 35:*115 (1963).
15. R. K. Harvey, An accurate fusion method for the analysis of
    rocks and chemically related materials by x-ray fluorescence
    spectrometry, *X-ray Spectrom., 2:*33-44 (1973).
16. R. Jenkins and J. L. de Vries, in *Practical X-Ray Spectro-
    metry*, 2nd Ed., MacMillan, London, 1969.
17. O. I. Joensuu and N. H. Suhr, Spectrochemical analysis of
    rocks, minerals, and related materials, *Appl. Spectrosc.,
    16:*101-104 (1962).

Table 7.3 (cont.)

18. J. O. Larson, R. A. Windler, and J. C. Guffy, A glass fusion method for x-ray fluorescence analysis, *Advan. X-ray Anal.*, *10*:489 (1966).
19. R. LeHoutllier and S. Turmel, Bead homogeneity in the fusion technique for x-ray spectrochemical analysis, *Anal. Chem.*, *46*:734 (1974).
20. W. Lodding and D. W. Rhett, Sample preparation for x-ray fluorescence analysis:  Li borate glass disks, *Amer. Mineral*, *57*:281 (1972).
21. C. L. Luke, Trace analysis of metals by borax disk x-ray spectrometry, *Anal. Chem.*, *35*:155 (1963).
22. F. J. M. J. Maessen and P. W. J. M. Boumans, Critical examination of the borate fusion technique for spectrochemical trace analysis of geological materials using the dc arc, *Spectrochem. Acta*, *22B*:739 (1968).
23. K. Norrish and J. T. Hutton, Preparation of samples for analysis by x-ray fluorescent spectrography, *Div. Rep. Div. Soils* (1954), CSIRO 3/64.
24. K. Norrish and J. T. Hutton, An accurate x-ray spectrographic method for the analysis of a wide range of geological samples, *Geochim. Cosmochim. Acta*, *33*:431 (1969).
25. C. Plug and J. N. Van Niekerk, A universal x-ray fluorescent method for the quantitative analysis of nonmetallics, *J. S. African Chem. Inst.*, *18*:71 (1965).
26. F. S. Rinaldi and P. E. Aguzzi, A simple technique for casting glass disks for x-ray fluorescence analysis, *Spectrochem. Acta*, *23B*:15 (1969).
27. H. J. Rose, I. Adler, and F. J. Flanagan, X-ray fluorescence analysis of the light elements in rocks and minerals, *Appl. Spectrosc.*, *17*:81 (1963).
28. J. A. T. Smellie, Preparation of glass standards for the use in x-ray microanalysis, *Mineral Mag.*, *38*:614 (1972).
29. D. A. Stephenson, An improved flux-fusion technique for x-ray emission analysis, *Anal. Chem.*, *41*:966 (1969).
30. D. A. Stephenson, Theoretical analysis of quantitative x-ray emission data:  glasses, rocks, and metals, *Anal. Chem.*, *43*:176 (1971).
31. A. Strasheim and M. P. Brandt, A quantitative x-ray fluorescence method of analysis for gelogical samples using a correction technique for the matrix effect, *Spectrochem. Acta*, *23B*:183 (1967).
32. R. Tertian, A rapid and accurate x-ray determination of rare earth elements in solid solution on liquid materials using the double-dilution method, *Advan. X-ray Anal.*, *12*:546 (1969).
33. R. Tertian, Quantitative chemical analysis with x-ray fluorescence spectrometry -- an accurate and general mathematical correction method for the interelement effect, *Spectrochem. Acta*, *24B*:447 (1969).

Table 7.3 (cont.)

---

34.  R. Tertian, Fluorescence x theorie et pratique de l'analyse
     des solutions liquides ou solides par une methode double
     mesure, Thesis (Doctoral) A L'Universite De Paris VI No.
     d'enregistrement au C.N.R.S.A.O. 7410, June 7, 1972.
35.  J. E. Townsend, X-ray spectrographic analysis of silica and
     alumina base catalyst by a fusion-cast disc technique, *Appl.
     Spectrosc., 17:*37 (1963).
36.  M. S. Wang, Rapid sample fusion with lithium tetraborate for
     emission spectroscopy, *Appl. Spectrosc., 16:*141 (1962).
37.  E. E. Welday, Silicate sample preparation for light-element
     analyses by x-ray spectrography, *Amer. Mineral, 49:*889 (1964).
38.  H. J. Rose, I. Adler, and F. J. Flanagan, Use of $La_2O_3$ as a
     heavy absorber in x-ray fluorescence analysis of silicate
     rocks, *U.S. Geol. Surv., Prof. Pap., 345-B:*80-82.
39.  C. O. Ingamells, Lithium metaborate flux in silicate analysis,
     *Anal. Chem. Acta, 52* (1970).
40.  E. Kunkel, Sample preparation of elastomers and other polymers
     for x-ray fluorescence analysis by borax fusion, *A. Anal.
     Chem., 270:*126 (1974).

---

fusion specimen preparation procedure was selected and the analysis
was performed to meet the requirement set forth in the analytical
goals.

The analysis described above is typical of many faced by the
x-ray analyst.  The time spent searching the literature (Table 7.2,
steps 4 and 6) will pay many dividends beyond the solution of the
particular problem at hand.  An acceptable literature search can be
made in one day (if adequate library facilities are available) and
will often save many days of fruitless laboratory studies.

## 7.3  PREPARATION OF SOLID SPECIMENS

### 7.3.1  Bulk Solids

When bulk solid specimens are analyzed directly, a major cause of
error arises from improper surface preparation.  This effect is
more severe when longer wavelength (low-energy) x-rays are being
measured.  The penetration depth of x-rays into a solid surface

Table 7.4  Mass Absorption Coefficients[a]

| Measured Radiation | Average Matrix Composition | $\mu$(matrix) $(cm^2/g)$ | $\rho$ | $x_{90\%}$ ($\mu m$) |
|---|---|---|---|---|
| Mg K$\alpha$ | Zn | 9500 | 7.13 | 0.012 |
| S  K$\alpha$ | Fe | 1158 | 7.87 | 0.12 |
| Fe K$\alpha$ | Al | 93 | 2.70 | 4.2 |
| Cr K$\alpha$ | Fe | 113 | 7.87 | 1.2 |
| Zr K$\alpha$ | U | 65 | 18.7 | 0.86 |
| Fe K$\alpha$ | Cr | 475 | 7.19 | 0.31 |

[a]Mass absorption coefficients found in the x-ray literature can vary significantly from one set of tables to another, particularly for the longer wavelength x-ray region. (Variations of ± 10% are not uncommon.) The values represented here are average values selected from several sources.

(such as a metal) is very shallow. Minor surface irregularities act as heterogeneities in the path of the primary exciting x-rays and have even a greater effect on the secondary fluorescent x-rays. As pointed out by Birks [16], a general rule of thumb states that the surface roughness should not exceed the path length that would cause 10% absorption of the measured radiation. The path length x causing 10% absorption (90% transmission) can easily be calculated from the general absorption equation:

$$I_T = 0.90 = \exp(-\mu \rho x_{90\%})$$

where

$\mu$ = mass absorption coefficient of whole specimen

$\rho$ = density of specimen

$$\ln 0.90 = \mu \rho x_{90\%}$$

$$x_{90\%} = \frac{0.105}{\mu \rho} \text{ cm} \qquad\qquad 1 \text{ cm} = 10^4 \text{ } \mu m$$

Table 7.4 lists several 10% absorption path lengths for common analytical situations involving the analysis of metallic samples.

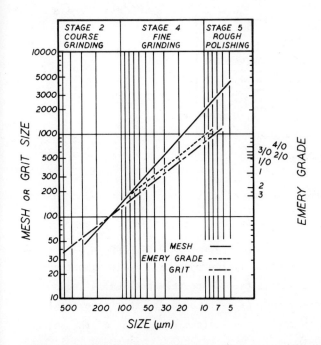

Figure 7.3 Relationship between grit size and micrometer size of abrasive particles. Reprinted with permission from *AB Metal Digest* *10(1)*, Buehler Ltd., Evanston, Illinois.

From Table 7.4 and the above, the amount of allowable surface roughness varies over a wide range. Thus, most metallic samples must be metallographically polished [17-19] prior to x-ray spectro-chemical analysis. Polishing generally requires rough grinding using a series of graded abrasives. Final polishing may require the use of very fine abrasives as shown in Figs. 7.3 and 7.4, possibly followed by chemical or electrochemical polishing [17,19].

Roughness is not the only criterion for a good polished sur-face. Smearing of a soft metal over a harder one will cause unusually large errors (a typical example is the analysis of silicon in aluminum casting alloys [3,4] or leaded copper alloys [38]). Thus, from Table 7.4, the measurement of iron in aluminum would re-quire a 4.2-μm surface finish which could not be obtained even with 600-grit paper. The smearing of the softer aluminum over the harder

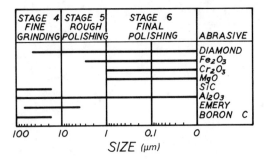

Figure 7.4 Micrometer size ranges of various abrasives and polishing stages. Reprinted with permission from *AB Metal Digest 10(1)*, Buehler Ltd., Evanston, Illinois.

iron-containing intermetallic particles would produce results which although they might be reasonably reproducible, would almost certainly lead to erroneous results. Since this problem is known to exist for metallic specimens, it is recommended that some metallography be performed to verify the surface conditions of the specimen preparation technique.

### 7.3.2 Powdered Specimens

Powered samples constitute one of the most frequently used specimen forms in x-ray spectrometry. Powders are analyzed in loose form but more often as a pressed pellet (briquette). While some samples arrive in a suitable powdered form, most powders must be obtained from larger bulk samples. Reduction generally takes the form of crushing and grinding in order to achieve a uniform small particle size. The analyst who selects the convenience of the powdered specimen form should be aware of some of the problems that may be encountered in its use.

*Particle Size Effects*

It is not unreasonable to expect a relationship between fluorescent x-ray intensity and particle size, since x-rays can only penetrate and emerge from a finite depth beneath the surface. Figure 7.5 indicates the layer thickness that emits 99.9% of the measured

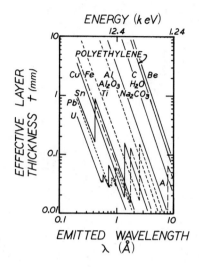

Figure 7.5  Effective layer thickness as a function of wavelength
for various matrices.   Reprinted with permission from *Siemens
Review*.

analyte line intensity as a function of the emission wavelength
and the matrix composition.   It is clear from this figure that wavelengths greater than 3 Å are only measured from a thin surface
layer in the thickness range of 1 to 100 μm.

There have been several theoretical treatments [9,10,20] of
the effect of particle size and size distribution upon x-ray
spectrochemical analysis results.   Particle size effects have even
been noted when using different size ion-exchange resins for trace
analysis procedures [21].   The qualitative effects of particle size
on the spectral line intensity are shown in Figs. 7.6 and 7.7.   As
can be seen, the effect can be very large and can be alleviated in
most cases by reduction of the entire sample to very small particle
sizes*.   Jenkins and de Vries [11] have shown (see Table 7.5) that
grinding to achieve sub-50-μm particles may require a considerable
expenditure of time and effort.   Selective particle size grinding

---

*This is not always true as shown by Claisse and Sampson [9] for
the analysis of aluminum in a $SiO_2$ matrix.

Table 7.5  Particle Size Distribution

| Type of Grinding Technique | Optimum Load | Particles > 150 μm | Microscopic Examination | Particles 80-150 μm | Microscopic Examination | Particles 50-80 μm | Microscopic Examination | Particles < 50 μm |
|---|---|---|---|---|---|---|---|---|
| Mechanical agate mortar 2 hr dry grinding | 20 g | 2% | Particle up to 300 μm 90% Mica | 15% | Rounded particles 30% Mica | 20% | | 54% |
| Mechanical agate mortar 2 h wet grinding | 20 g | 0.2% | Very large particles up to 600 μm | 1.7% | Rounded particles 90% Mica | 0.4% | 60% Mica | 97.7% |
| Tema disk mill 12 min dry grinding | 75 g | 0.1% | Particles up to 300 μm 95% Mica | 1.7% | Rounded particles 95% Mica | 3.3% | 40% Mica | 94.9% |
| Glen Creston 270 M special agate ball mill, 30 mins. dry grinding | 1.6 g | 0% | | <0.01% | Few particles of Mica | <0.01% | Few particles of Mica | 100% |

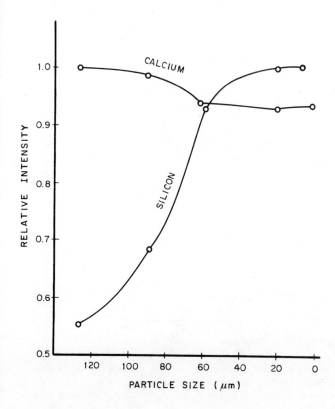

Figure 7.6 The effect of particle size on spectral line intensity in a system containing calcium and silicon. Reprinted from F. Bernstein, *Adv. X-Ray Anal.*, *6:*436 (1962), with permission from Plenum Publishing Corp.

[22] can result in a narrower particle size distribution (Fig. 7.8 and Table 7.6) with a resulting improvement in accuracy and precision for some elements (see particularly the chromium and iron intensities, Table 7.6).

*Some Common Errors Introduced When Using Powdered Specimens*

Most powdered specimens are prepared by some sort of comminution process. Thus, contamination with elements from the grinding containers and implements, particularly at the trace element level, is always present. If binders and/or matrix diluents are used, additional error can result from weighing and improper mixing.

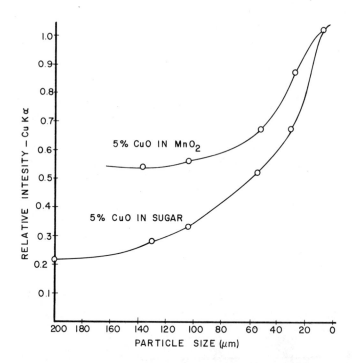

Figure 7.7 The effect of particle size on the intensity of a char-
acteristic copper line in two different matrices. Reprinted from
F. Bernstein, *Adv. X-Ray Anal.*, *6*:436 (1962), with permission from
Plenum Publishing Corp.

Differential particle size reduction occurs in most grinding procedures because the various constituents of the powder are reduced
in size at different rates, producing wide variation in particle
size and often creating severe segregation problems.

*Pressed Pellets*

Using loose powders as samples for x-ray spectrometric analysis
will not allow a quantitative analysis to be made. It is necessary
to eliminate or reduce the wide density and compositional variations
that occur naturally when a powder is poured into a specimen holder.
These variations in density can be reduced by pressing the powder
in a die, usually having a cylindrical shape, at pressures ranging
from 20,000 to 100,000 psi, preferably under vacuum to degas the

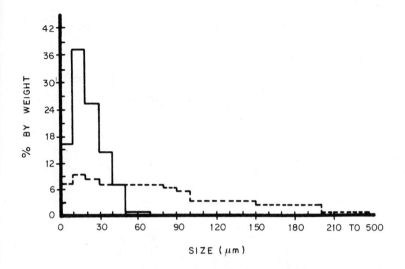

Figure 7.8  Particle size distribution of a ground sample of ferro-manganese [22].  Solid curve:  selectively ground using 400-mesh screens; dashed curve:  statically ground for 6 min.

powder before insertion into the vacuum spectrometer.  This process generally results in a flat pellet (briquette) that can be placed directly in the spectrometer.  Often binders and diluents are added prior to pressing and samples may also be pressed into plastically deformable metal caps [22,24] in order to form a durable pellet without the need of a binder.  Particle size effects must be considered as discussed in previous paragraphs and density gradients may often be accentuated by the pressing procedure.  Some pressing variables and the effect of briquetting pressure are shown in Fig. 7.9 and are further discussed quantitatively by Clark and Hench [25].

7.3.4  Glass Beads

The glass fusion technique was originally described by Claisse [7] in 1957 as a method to reduce or eliminate interelement and particle size effects by means of a high dilution in a borate glass matrix. Small quantities of metal oxides are readily soluble in sodium tetraborate ($Na_2B_4O_7$) and lithium tetraborate ($Li_2B_4O_7$) glasses.

Table 7.6  Effect of Selective Grinding on Intensity Ratios and
Composition Determination [22]

| | Fe Ratio | Fe % | Cr Ratio | Cr % | Si Ratio | Si % | Ca Ratio | Ca % |
|---|---|---|---|---|---|---|---|---|
| | | | | Basic Brick | | | | |
| | | | Statically Ground | | | | | |
| **Sample 1A** | | | | | | | | |
| Side T | 1007 | 8.45 | 1012 | 18.80 | 1044 | 4.85 | 832 | 0.88 |
| | 1002 | 8.41 | 1012 | 18.80 | 1054 | 4.92 | 830 | 0.88 |
| Side B | 998 | 8.38 | 1006 | 18.75 | 1051 | 4.90 | 826 | 0.87 |
| | 999 | 8.38 | 1005 | 18.75 | 1024 | 4.75 | 825 | 0.87 |
| **Sample 1B** | | | | | | | | |
| Side T | 1004 | 8.43 | 1014 | 18.86 | 1057 | 4.93 | 828 | 0.87 |
| | 1008 | 8.45 | 1018 | 18.95 | 1046 | 4.86 | 827 | 0.87 |
| Side B | 1005 | 8.44 | 1012 | 18.80 | 1068 | 5.00 | 829 | 0.88 |
| | 1007 | 8.45 | 1012 | 18.80 | 1049 | 4.90 | 828 | 0.87 |
| | | | Selectively ground through 325 mesh | | | | | |
| **Sample 2A** | | | | | | | | |
| Side T | 1068 | 9.17 | 1145 | 21.35 | 1047 | 4.87 | 814 | 0.86 |
| | 1072 | 9.18 | 1140 | 21.30 | 1038 | 4.84 | 809 | 0.85 |
| Side B | 1077 | 9.24 | 1139 | 21.30 | 1061 | 4.95 | 812 | 0.86 |
| | 1075 | 9.23 | 1147 | 21.36 | 1041 | 4.83 | 811 | 0.86 |
| **Sample 2B** | | | | | | | | |
| Side T | 1070 | 9.18 | 1136 | 21.25 | 1040 | 4.83 | 812 | 0.86 |
| | 1070 | 9.18 | 1132 | 21.22 | 1054 | 4.92 | 809 | 0.85 |
| Side B | 1070 | 9.18 | 1134 | 21.24 | 1050 | 4.90 | 816 | 0.86 |
| | 1072 | 9.18 | 1137 | 21.28 | 1049 | 4.90 | 814 | 0.86 |
| | | | Selectively ground through 400 mesh | | | | | |
| **Sample 3A** | | | | | | | | |
| Side T | 1083 | 9.28 | 1179 | 22.00 | 1027 | 4.76 | 781 | 0.82 |
| | 1080 | 9.27 | 1176 | 21.98 | 1024 | 4.75 | 781 | 0.82 |
| Side B | 1077 | 9.24 | 1176 | 21.98 | 1018 | 4.73 | 784 | 0.83 |
| | 1081 | 9.28 | 1175 | 21.97 | 1022 | 4.74 | 784 | 0.83 |
| **Sample 3B** | | | | | | | | |
| Side T | 1076 | 9.24 | 1168 | 21.92 | 1019 | 4.73 | 785 | 0.83 |
| | 1075 | 9.23 | 1168 | 21.92 | 1009 | 4.70 | 785 | 0.83 |
| Side B | 1082 | 9.27 | 1175 | 21.97 | 1036 | 4.82 | 782 | 0.82 |
| | 1081 | 9.27 | 1178 | 21.99 | 1021 | 4.74 | 783 | 0.83 |

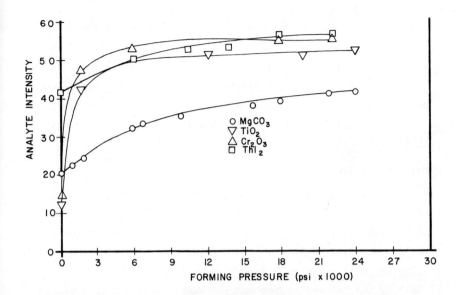

Figure 7.9  Effect of forming pressure on sample density and x-ray intensity [25].    Zero pressure was obtained by pressing the powders in the die until a very small positive pressure was observed on the gauge.

The method consists of weighing several hundred milligrams of the oxide sample and dissolving (fluxing) this in approximately 10 to 20 g of borate glass.  The borax dilution ratio is generally in excess of 10:1.  The glass is cast to form a disk having a smooth surface, thus also eliminating surface roughness effects* discussed in other preparation procedures.

Problems associated with fluxing, crucible sticking, casting, and bead explosion prevented widespread early acceptance of this technique.  Since 1957, many researchers (see annotated bibliography) have attacked the problems mentioned above, and today the method is becoming widely applied, even to nonoxide samples such as metals [26,27], sulfides [7,28], cements [29], and catalysts [30] and polymers [39].  Several automatic [29] and semiautomatic

---

* Surface grinding may be necessary for the analysis of elements having Z < 15 if the cast lower surface is not smooth.

[31,35] devices are currently being marketed which utilize special
nonwetting Pt-Au alloy crucibles RF induction heating and/or gas
flame burners.  In the Claisse apparatus the fusion is achieved
over gas burners in rotating platinum-gold crucibles designed to
provide thorough agitation [32] of the melt during dissolution.
At the end of fusion, the crucibles are inverted and the bead is
cast into the moldtop.

*Fluxing of Glasses*

Fusion in borate glass requires temperatures sufficiently in excess
of the melting point to provide a low-viscosity fluid melt.[*]  Radiation heating devices such as air-gas flames can produce temperatures
of approximately 1050 to $1150^{\circ}C$ which are sufficient to insure rapid
melting if small (~30 $cm^3$) crucibles of low heat content are used.
RF heating can achieve higher temperatures and $1300^{\circ}C$ is typically
used in the Philips apparatus [29].  This higher temperature results in more rapid dissolution since flux fusion is a diffusion
controlled process.  LeHouillier has shown the effectiveness of
agitation [32] which also accelerates diffusion by sweeping away
the concentration gradients adjacent to the dissolving particles.
A typical hand-agitated platinum crucible will generally require 15
min of frequent agitation to effect complete solution and release
of trapped gas bubbles.  Mechanical agitation [32] can achieve the
same result in several minutes.  Muffle furnace melting should be
avoided unless time is not a factor.

*Fluxing of Nonoxides*

Metal oxides are rapidly soluble in borate glasses at the 100:1
dilution factor.  Nonoxides require special treatment prior to, or
during, borate fusion.  Most metals may be roasted in air or oxygen
to induce oxidation.  Alternatively [7,26-28], the metal may be
heated in the presence of excess sulfur, dissolved in potassium

---

[*] Melting points as follows:  $Li_2B_4O_7$: $930^{\circ}C$, $Na_2B_4O_7$: $741^{\circ}C$,
$K_2S_2O_7$: $>300^{\circ}C$.

pyrosulfate, and then fused in borax.  Other techniques include
dissolving the metal in acid and evaporating to dryness prior to
borax fusion.  Claisse [7,28] has also used a $BaO_2$ oxidation of
sulfide minerals directly in the platinum crucible.  Approximately
1 g $BaO_2$ is added to a 100-mg sulfide sample in the bottom of the
crucible.  This is then covered with 10 to 20 g of borax.  Heating
the crucible disassociates the $BaO_2$ with subsequent oxidation of
the sulfides to sulfates which are soluble (in small amounts) in
the borax glass.

Many examples of the use of lower dilution ratios such as 1:1
are found in the literature [33,34].  Ratios such as these are used
for the analysis of trace elements and low atomic numbers such as
sodium.

*Fusion Problems*

Attack of the Pt-Au crucible will occur when certain easily reduced
oxides, such as those containing nickel and copper, are present.
Generally, this results in glass adhesion to the Pt-Au crucible
wall and solution-erosion of the smooth inner surface of the
crucible.  This problem can be largely avoided by the use of wetting
agents of which sodium iodide and sodium chloride are the most effective [35,36].  Platinum also forms a eutectic melt with elemental silicon and boron, and these materials should be avoided or
crucible cracking may result [35].  The problem of sample weight
losses (volatile evolution) has been treated quantitatively by
Tertian [8] and Harvey et al. [34], and various suggestions have
been made to handle this problem, which occurs frequently when car-
bonates, phosphates, and organic materials are present in the
sample.

*Casting of Glass Beads*

Many of the problems associated with fusion arise from improper
casting techniques.  Casting may be described as a controlled
cooling-rate process designed to avoid crystallization and trapped
freezing strains.  Too rapid a cooling rate will result in highly

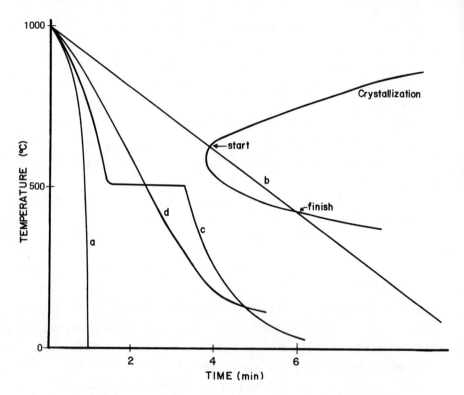

Figure 7.10  The cooling behavior of a glass fusion specimen:
(a) too rapidly cooled results in a highly strained fusion bead;
(b) very slow cooling from the melt causes crystalization to occur;
(c) ideal cooling behavior which includes a short thermal arrest at
$500^{\circ}C$ producing a tough, tempered fusion bead; (d) intermediate-
rate continuous cooling produces a satisfactory bead if conditions
are controlled properly.

unstable strained glasses that will shatter without warning.  If
cooled too slowly, crystallization may occur with resulting cracking
of the bead.  Even if cracking does not occur, crystallized beads
are unsuitable because they are no longer homogeneous glasses.  A
proper cooling rate will produce an annealed (tempered) glass having high mechanical strength and durability.  A properly cooled
glass fusion bead should easily withstand dropping on a hard floor
without shattering.  Figure 7.10 summarizes [28] the cooling behavior of glass fusion specimens and resembles a TTT curve used

in the steel industry for the thermal treatment of quenched and
annealed steels.

When casting beads by hand, the melt must be clear and free of
bubbles and undissolved material (thorough agitation).  Successful
beads can be cast onto the top of an aluminum hot plate held at
400-500°C.  The surface of the aluminum should be smooth but need
not be mirror polished.  The melt can be poured into a metal ring*
and allowed to solidify for 2 min before being pushed (using a
heated refractory pusher) onto a piece of thin refractory material
such as Transite (Johns Mansville Co.).  The temperature of the
sample and refractory must be very nearly the same to avoid thermal
shocks.  If the hot-plate temperature is too low, the bottom surface
of the button will be wavy and undulating in appearance.  If too
hot, the glass will adhere to the casting surface and rapid crystallization may take place.  The refractory and bead are then removed
from the hot plate and allowed to cool to room temperature.  At no
time during cooling should the hot button be touched by an object
cooler than itself.  This process is illustrated in Fig. 7.11(a)
to (g).

Modern fusion devices [31] are equipped with semiautomatic
casting accessories that use Pt-Au casting dishes.  Buttons made
from these devices are always smoother than those cast on aluminum.

Properly cast glass specimens generally require no surface
preparation if the analyte has an atomic number greater than 15.
Surface undulations will, however, affect the accuracy of the
analysis of low-atomic-number elements.  Figure 7.12 shows the effect of atomic number on the coefficient of variation (the percent
standard deviation) of the analysis and compares the precision of
the fusion method with the pressed-powder technique for the analysis of silicate rocks.  These results are summarized in Table 7.7.

---

* Oxidized 14-gage copper wire or machined aluminum will work satis-
factorily.

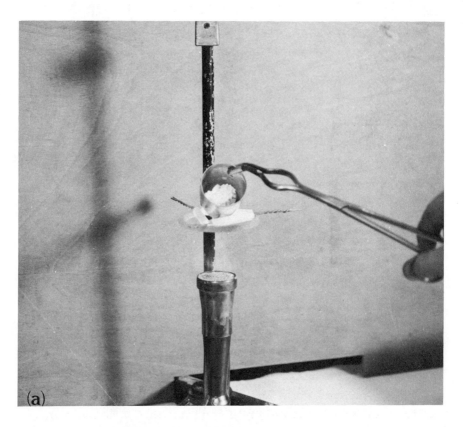

Figure 7.11  Melting and casting of fusion beads by hand using a
platinum/gold crucible, platinum-tipped tongs, a gas burner, and a
small hotplate.  (a)  A borax sample mixture is supported on a
ceramic triangle over a gas flame.  (b)  The crucible is agitated
during the melting operation.  (c)  The melt is ready to pour.
(d)  Pouring the melt into metal rings which have been placed on the
hot aluminum plate immediately prior to pouring.  (e)  The appearance of the upper surface of the glass disk.  (f)  The lower surface
of the disk cast onto aluminum.  (g)  The appearance of a disk which
was purposely allowed to crystallize during cooling.

*Preparation of Standards Using Flux Fusion*

Standards are prepared using carefully weighed quantities of pure

oxides which are fused in the same borate glass matrix as the un-

known samples.  If $BaO_2$ is used in the unknown bead, then the stand-

ard should also contain the same quantity of $BaO_2$.  Heavy oxides are

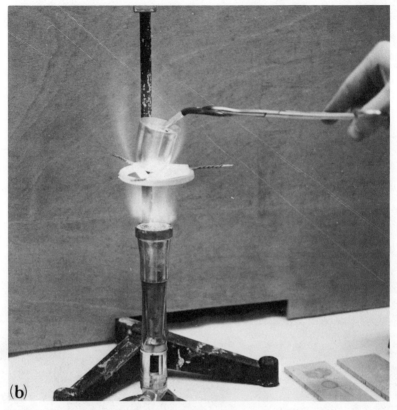

Figure 7.11 (cont.)

often purposely added to samples and standards to further aid in
matrix modification, $La_2O_3$ being a common example.  Standards can
contain several elements of interest if the dilution factor is
sufficiently high to eliminate interelement effects or if a matrix
modifier is purposely added.  Tertian [8] has described a very
general method (double dilution) which requires only two standards
for each element being analyzed rather than using the entire calibration curve procedure (see also Sec. 10.5.2).  This double-
dilution method is applicable to a wide variety of samples and can
achieve precisions equal to those obtained from the calibration
curve methods.  Table 7.8 illustrates the application of this method
to the analysis of silicate rocks [40].

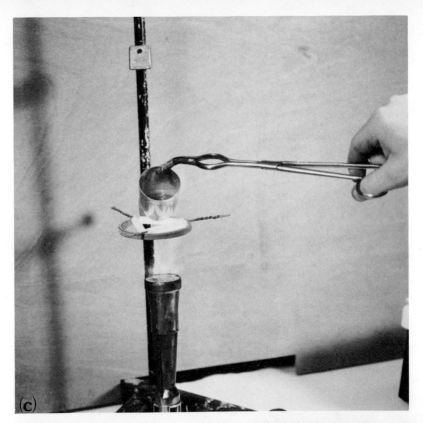

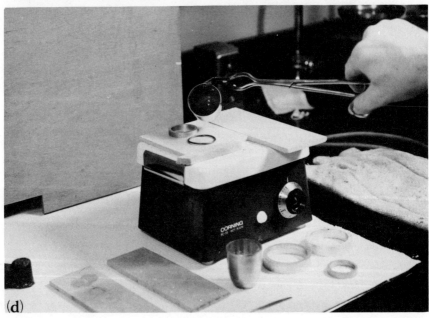

Figure 7.11 (cont.)

(e)

(f)

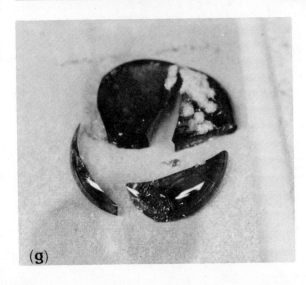

(g)

Table 7.7   Comparison of Briquetted and Fused Specimens for the Analysis of Silicate Rocks

| | Fusion and very high dilution using $Li_2B_4O_7$ and a heavy absorber. | Fusion and high dilution 1:10 borax glass. | Fusion and high dilution 1:2 $Li_2B_4O_7$. | Fusion and low dilution 1:2 $Li_2B_4O_7$. | Unfused powders. |
|---|---|---|---|---|---|
| Matrix effect | Virtually eliminated rocks of all compositions fall on straight line calibration curves. | Greatly reduced (with the exception of rocks of extreme composition) single straight-line calibration curves can be used for all elements. | Reduced. For accurate analysis standards a similar composition should be used or corrections made for absorption. | | A correction must be made for absorption either arithmetically or by using a standard of almost identical composition. |
| Particle size | Eliminated. | Eliminated. | Eliminated. | Eliminated. | Fine careful grinding is necessary. Sheet silicates create difficulties. |
| Sensitivity | Much reduced. Mg, Na, P and trace elements cannot be reached adequately. | Reduced. Suitable for all major elements down to Mg, but not trace elements. | Slightly reduced. Trace elements are still measurable usually down to 10 ppm. | | Very good. |
| Speed (1): sample preparation | A fusion preparation takes approximately twice the time of a normal unfused powder preparation. | | | | Very quick for fairly accurate results, but similar to fusion preparation times for very accurate results. |
| Speed (2): counting time (relatively insignificant) | Much slower. | One-tenth as fast. | One-third as fast. | | Very fast. |

Table 7.8  Comparative Analysis of Silicate Rocks [40]

| Samples | Concentrations % | | | |
|---------|------|-----|------|------|
|         | $Fe_2O_2$ | MnO | $SiO_2$ | $Al_2O_3$ |
| 01 | 2.78[a] | 0.09[a] | 69.71[a] | 14.61[a] |
|    | 2.83[b] | 0.083[b] | 69.71[b] | 14.61[b] |
| 02 | 1.36[a] | 0.05[a] | 75.58[a] | 12.63[a] |
|    | 1.34[b] | 0.047[b] | 76.29[b] | 12.61[b] |
| 03 | 12.92[a] | 0.20[a] | 38.49[a] | 10.31[a] |
|    | 12.92[b] | 0.20[b] | 38.10[b] | 9.96[b] |
| 04 | 9.75[a] | 0.21[a] | 52.65[a] | 17.42[a] |
|    | 9.67[b] | 0.21[b] | 52.02[b] | 17.64[b] |
| 05 | 8.36[a] | 0.11[a] | 39.10[a] | 3.06[a] |
|    | 8.38[b] | 0.12[b] | 39.03[b] | 2.87[b] |
| 06 | 23.22[a] | 0.04[a] | 7.36[a] | 54.30[a] |
|    | 23.36[b] | 0.045[b] | 7.18[b] | 55.03[b] |
| 07 | 0.69[a] | 0.008[a] | 36.45[a] | 59.03[a] |
|    | 0.62[b] | 0.0025[b] | 36.44[b] | 59.19[b] |
| 11 | 11.0[a] | 0.17[a] | 52.64[a] | 14.85[a] |
|    | 11.18[b] | 0.17[b] | 52.35[b] | 14.90[b] |
| 12 | 2.77[a] | 0.04[a] | 69.19[a] | 15.34[a] |
|    | 2.78[b] | 0.033[b] | 69.77[b] | 15.25[b] |
| 13 | 4.33[a] | 0.04[a] | 67.28[a] | 15.11[a] |
|    | 4.28[b] | 0.038[b] | 66.79[b] | 14.92[b] |
| 14 | 6.80[a] | 0.10[a] | 59.00[a] | 17.01[a] |
|    | 6.82[b] | 0.095[b] | 60.02[b] | 16.78[b] |
| 15 | 8.51[a] | 0.12[a] | 41.87[a] | 0.86[a] |
|    | 8.17[b] | 0.11[b] | 41.85[b] | 0.58[b] |
| 16 | 8.85[a] | 0.13[a] | 40.45[a] | 0.55[a] |
|    | 8.64[b] | 0.12[b] | 40.36[b] | 0.12[b] |
| 17 | 13.51[a] | 0.18[a] | 54.48[a] | 13.66[a] |
|    | 13.42[b] | 0.18[b] | 54.09[b] | 13.40[b] |

[a]Chemically analyzed value.

[b]X-ray fluorescence value using fusion double dilution.

## 7.4  LIQUIDS

In many respects, liquids represent a nearly ideal specimen type.
If all elements are in solution, the sample is truly homogeneous
and obviously no particle size effects are present.  Standards are

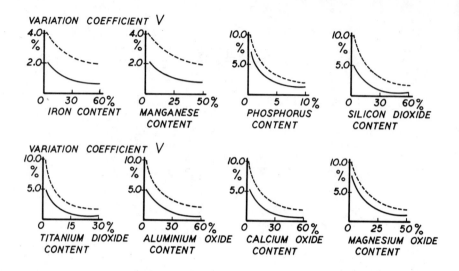

Figure 7.12  Comparison of the variation coefficients of various elements and compounds for pressed pellets (dashed line) and fusion pellets (solid line).

easily prepared and excellent quantitative results are possible. The major drawbacks arise from the dilution factor with resulting decreased intensities and the difficulty of analyzing low-atomic-number elements whose radiations must pass through the window of the sample cell.  On the other hand, it is a popular misconception that a dilution in the form of either a liquid or a fusion specimen will reduce the intensity by a factor equivalent to the dilution factor.  This is clearly not true, for example, a 100:1 dilution reduced the copper intensity by a factor of 2.5:1 [6,10].  Other more formidable difficulties are often associated with the chemical dissolution of many nearly insoluble samples.  Bertin [5,6] has reviewed the various solution techniques used in modern spectrochemical analysis and the reader is referred to this work for further information.

# REFERENCES

1. E. P. Bertin, in *Principles and Practice of X-Ray Spectrometric Analysis*, Plenum, New York, 1970, Chap. 16.
2. E. P. Bertin and R. J. Longobucco, *Norelco Reporter*, 9:31 (1962).
3. R. E. Michaelis and B. A. Kilday, *Advan. X-ray Anal.*, 5:405 (1961).
4. M. A. Kemper, *X-ray Spectrom.*, 3:111 (1974).
5. E. P. Bertin, *Norelco Reporter*, 12:15 (1965).
6. E. P. Bertin, *Advan. X-ray Anal.*, 11:1 (1968).
7. F. Claisse, *Norelco Reporter*, 4:3 (1957).
8. R. Tertian, *Spectrochem. Acta*, 24B:447 (1969).
9. F. Claisse and C. Sampson, *Advan. X-ray Anal.*, 5:335 (1962).
10. C. B. Hunter and J. R. Rhodes, *X-ray Spectrom.*, 1:107 (1972).
11. R. Jenkins and J. L. de Vries, in *Practical X-Ray Spectrometry*, 2nd Ed., Springer-Verlag, New York, 1970. Chap. 8.
12. *X-Ray Fluorescence Spectrometry Abstracts*, Science and Technology Agency, South Kensington, London (1970-present).
13. *X-Ray Absorption and Emission, Annual Review, 1949-52; Biennial 1954-present*, Analytical Chemistry.
14. H. A. Liebhafsky, *X-Rays, Electrons and Analytical Chemistry*, Wiley Interscience, New York, 1972.
15. *X-Ray Fusion Flux*, Bibliography, Chemplex Industries, Inc., Scarsdale, New York.
16. L. S. Birks, *X-Ray Spectrochemical Analysis*, 2nd Ed., Wiley Interscience, New York, 1969.
17. G. L. Kehl, *The Principles of Metallographic Laboratory Practice*, McGraw-Hill, New York, 1949.
18. L. E. Samuels, *Metallographic Polishing by Mechanical Means*, American Elsevier, New York, 1971.
19. *Metallography Structures and Phase Diagrams, Metals Handbook*, Vol. 8, American Society for Metals, 1973.
20. J. B. Rhodes and C. B. Hunter, *X-ray Spectrom.*, 1:107 (1972).
21. A. L. Allen and V. C. Rose, *Advan. X-ray Anal.*, 15:534 (1972).
22. A. H. Pitchford, *Norelco Reporter*, 7 (1960).
23. Spex Industries, Metuchen, N.J.
24. Somar Laboratories, New York.
25. D. E. Clark and L. L. Hench, *Norelco Reporter*, 17:25 (1970).
26. Sarian, Suren, and H. W. Weart, *Anal. Chem.*, 35:115 (1963).
27. C. L. Luke, *Anal. Chem.*, 35:1551 (1963).
28. F. Claisse, Notes SUNYA X-Ray Clinic, June 1973, Albany, N.Y.
29. P. R. Dijksterhuis, *Automatic X-Ray Analysis of Cement Materials Including Sample Preparation by Fusion or Pressing*, Scientific and Analytical Equipment Dept., N.V. Philips Gloeilampenfabrieken, Eindhoven, 1968.

30.  J. E. Townsend, *Appl. Spectrosc.*, *17*:37 (1963).
31.  *Claisse Stirrer Fusion Device,* Spex Industries, Metuchen, N.J.
32.  R. LeHouillier and S. Turmel, *Anal. Chem.*, *46*:734 (1974).
33.  B. P. Fabbi, *X-ray Spectrom.*, *2*:15 (1973).
34.  P. K, Harvey, D. M. Taylor, R. D. Hendry, and F. Buncroft, *X-ray Spectrom.*, *2*:33 (1973).
35.  K. Becker, K. Wilhelm, and G. Wronka, *Siemens Rev.*, *XXXIX*:24 (1972).
36.  R. W. Gould and J. T. Healey, *X-ray Spectrom.*, *3*:170 (1974).
37.  P. R. Hooper, Proceedings of the 5th Conference on X-ray Analytical Methods, N.V. Philips Gloeilampenfabrieken, Eindhoven, 1966, pp. 76-87.
38.  V. J. Manners, J. V. Craig, and F. H. Scott, *J. Inst. Metals,* *95*:173 (1967).
39.  E. Kunkel, *Anal. Chem.*, *270*:126 (1974).
40.  R. Tertian, Doctoral Thesis, University of Paris, June 7, 1972.
41.  F. Bernstein, *Advan. X-ray Anal.*, *6*:436 (1962).
42.  F. Bernstein, *Advan. X-ray Anal.*, *6*:436 (1962).

# 8

## Qualitative Analysis

### 8.1  INTRODUCTION

While the major thrust of this text is *quantitative* x-ray spectrochemical analysis, it must *not* be assumed that *qualitative* x-ray
spectrochemical analysis is trivial or has no important applications.  In many instances, the analyst is asked to identify the entire suite of elements present in the sample.  Often one is required to go one step further and estimate the overall composition
of a sample without consuming the time needed for a thorough quantitative analysis.  The problems associated with qualitative x-ray
spectrochemical analysis generally differ depending upon the dispersion method utilized, i.e., wavelength dispersion or energy
dispersion.  These two techniques will be discussed separately in
the paragraphs that follow.

### 8.2  QUALITATIVE IDENTIFICATION:  GENERAL

As with any spectral analysis method, a group of spectral lines com-
prises a fingerprint of the unknown element.  If more than one
element is present, as is generally the case, spectral overlap may

Table 8.1   Spectral Line Interferences Found in Wavelength
Dispersion

| X-ray Spectrochemical Analysis | Intensity |
|---|---|
| 1. Coherent and incoherently scattered continuum source | Strong increases as atomic number decreases; has a maximum value at low Bragg angles. |
| 2. Coherently scattered tube characteristic lines | Strong increases as atomic number decreases. |
| 3. Incoherently scattered tube characteristic lines | Moderate to strong intensity increases as atomic number decreases. |
| 4. Tube contamination lines Cu (target substrate), W (filament, deposition), Ni, Ca, Fe (Be window vacuum seal) | Generally increases in intensity with tube age. |
| 5. L spectra (or M spectra) of higher atomic number elements overlapping K spectra of lower atomic element | |
| 6. Higher order lines diffracted by analyzing crystal | See Table 8.2. |
| 7. Nondiagram lines $K\alpha_3$, $K\alpha_4$ satellites | Increases as atomic number decreases. |
| 8. Background caused by crystal fluorescence | |

Table 8.2   Relative Intensity of Higher Order Lines
of BaK$\alpha$ $\lambda$ = 0.387 Å [4]

| Line | First Order | Second Order | Third Order |
|---|---|---|---|
| $K\alpha_1$ | 100 | 80 | 22 |
| $K\beta_{1,3}$ | 52 | 22 | 7 |
| $K\beta_2$ | 14 | 6 | 2 |

occur and the analyst must use care to avoid error.  In addition,
certain interfering lines will always be present (see Table 8.1)
and must be considered.  If K spectral lines are excited, the
analyst should look for both the K$\alpha$ and the K$\beta$ to verify the iden-
tity.  If overlap has masked one or both of these lines, it may be

possible to use higher order lines from the same series (Table 8.2)
or lines from a different series to obtain a positive identification.  L spectra can also be used to identify elements with atomic
numbers greater than 40.  In this instance, the analyst initially
looks for the $L\alpha$ line, then uses other lines including $L\beta$ and $L\gamma$
lines to verify the choice.

M series lines can also be used for initial identification or
verification, but unlike the majority of the K and L series, the
spectral distribution and specific line intensities of M series
lines is variable.  This is because many M lines originate from
partially filled atomic orbitals or in some cases from molecular
orbitals.  Although this problem is by no means unique to the M
line series, it can be particularly severe in this case.

## 8.3  QUALITATIVE ANALYSIS BY WAVELENGTH DISPERSION

As discussed earlier in this text, x-ray spectra are decidedly
simpler than optical emission spectra, but even so, spectral line
interferences do occur.  Table 8.1 lists several types of spectral
line interference commonly encountered in x-ray spectrochemical
analysis using wavelength dispersion.

Figures 8.1 through 8.3 illustrate the inherent simplicity of
x-ray spectra showing the K, L, and M spectra of several common elements.  The K spectrum is the simplest and is used for the qualitative analysis of elements whose atomic numbers are less than 70.
The L and M spectra are used for the analysis of higher atomic number elements (L spectra above Z = 35 and M spectra above Z = 71)
with the L spectra being used most commonly.

Within any spectral series (K, L, or M) certain lines will have
dominant intensities and these are generally chosen as the best
indicators of the presence of a particular element.  The $K\alpha_1$-$K\alpha_2$
doublet is always the dominant line in the K series, although the
$K\beta_1$-$K\beta_3$ doublet can be strong, particularly for the higher atomic
number elements.  Note that the $K\beta$ may be very weak, relative to

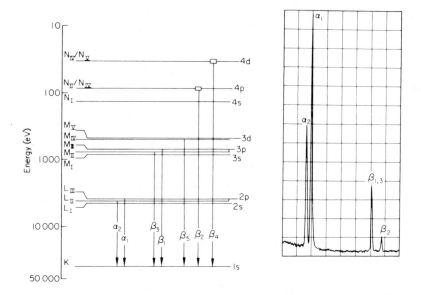

Figure 8.1  The K emission spectrum of tin.  Reprinted from Ref. 4 with permission.

the $K\alpha$, for the low atomic numbers.  Figure 8.4 shows the variation [1,2] in the $K\alpha/K\beta$ line intensity ratio as a function of the atomic number for elements between chlorine (Z = 17) and xenon (Z = 54).  L spectra are more complex and are also subject to considerable variation in relative line intensities.  Table 8.3 shows some typical relative line intensities for elements representing the extremes of the range of commonly used L lines.

## 8.3.1  Spectral Line Intensity

The absolute measured intensity of a spectral line depends upon many factors previously discussed in Chaps. 2 and 3.  Two dominant factors controlling the production of fluorescent x-rays are the fluorescent yield and the excitation overvoltage $(V - V_{\phi})^{1.6}$.  These two factors may be used to obtain reliable, rapid semiquantitative analysis from a qualitative identification spectrum.  This will be discussed in a subsequent paragraph.  Figures 8.5 through 8.7 illustrate the atomic-number dependence of the K fluorescent

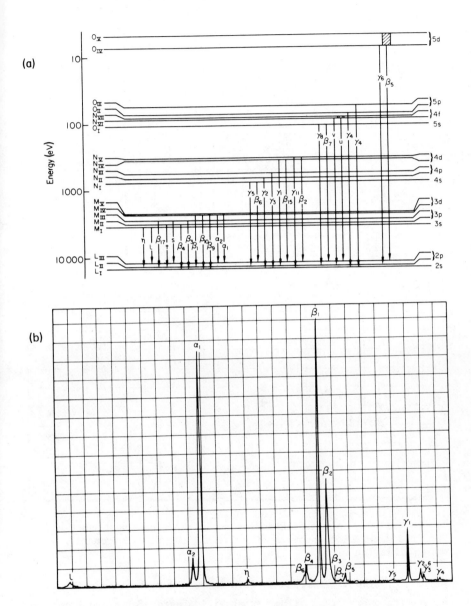

Figure 8.2   (a)   Observed transitions in the gold L spectrum.
(b)   The L emission spectrum of gold.   Reprinted from Ref. 4 with
permission.

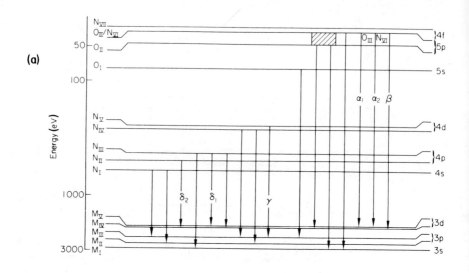

**(a)**

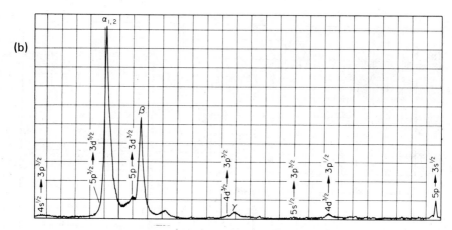

**(b)**

Figure 8.3  (a)  Observed transitions in the tungsten M spectrum.
(b)  The M emission spectrum of tungsten.  Reprinted from Ref. 4
with permission.

yield $[\omega_K]$, the L fluorescent yield $[\omega_L]$, and a combined critical
excitation overvoltage* fluorescent yield factor for K and L spectra.

---

*The value of the exponent in this expression is discussed by Compton
and Allison [5] and ranges from 1.5 for x-ray excitation to 1.67
for electron excitation (see also Sec. 3.2).

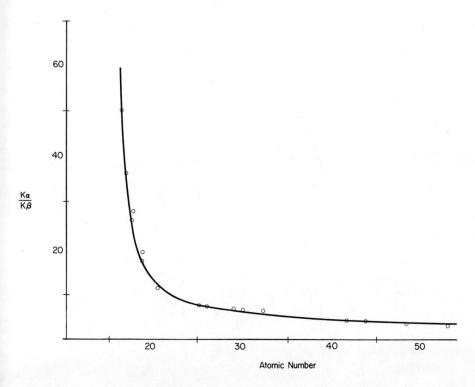

Figure 8.4  Variation in Kα/Kβ ratio with atomic number [1].

Table 8.3  Relative Experimental Line Intensities[a]
for the L Series Lines of Strontium (Atomic Number
38) and Gold (Atomic Number 79).

|              | Strontium | Gold |
|--------------|-----------|------|
| $L\alpha_1$  | 100       | 89   |
| $L\alpha_2$  |           | 12   |
| $L\beta_1$   | 65        | 100  |
| $L\beta_2$   |           | 40   |
| $L\gamma_4$  | 6         | 7    |
| $L\gamma_1$  |           | 19   |

[a]These relative line intensities are dependent upon
the atomic number of the element, as well as detector
response which may vary with photon energy.
Source:  Jenkins [4].

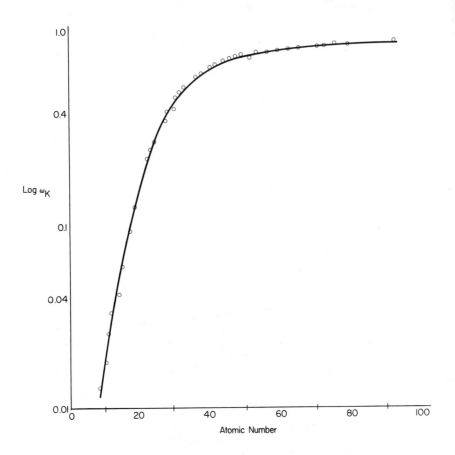

Figure 8.5   The atomic-number dependence of the K fluorescence yield $\omega_K$ [3].

## 8.3.2   Determination of Potential Spectral Contaminants

When using a crystal spectrometer or an energy dispersive instrument, it is necessary to periodically check the spectral contamination level by inserting a chemically pure low-atomic-number organic
sample such as polyethylene (or other similar pure organic compound)
and then scan the entire wavelength range of interest.  The spectrum
scattered in this manner will contain lines from the contamination
sources mentioned in Table 8.1 (items 1-4).  In addition, other contamination sources may appear periodically, such as contamination
on the tube window from powdered or liquid samples, powdered material which may have become lodged in the collimators, and possibly
low intensity lines arising from the sample holder cups.  Since it

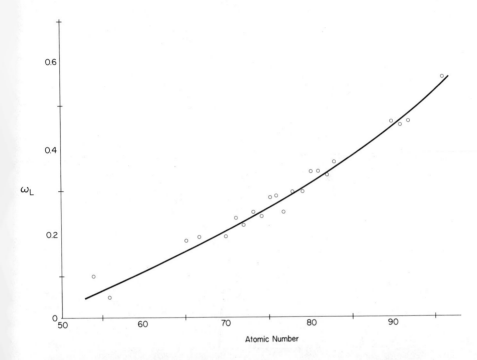

Figure 8.6  The atomic-number dependence of the L fluorescence yield $\omega_{L\alpha}$ [3].

is desirable that qualitative scans record the entire spectrum, such scans should be run without using a manual or automatic pulse height selector, or any other device, such as a primary beam filter, which modifies the spectral output.  Thus, higher order lines will always be present in the recorded spectrum, and these lines may prove useful for identification purposes.

### 8.3.3  Scanning Conditions:  Wavelength Dispersion

It is good experimental practice to run a qualitative scan on any sample submitted for quantitative analysis in order to uncover potential overlap and other interference problems.  In order to scan the entire wavelength (0.2-20 Å), it is necessary to utilize several crystals and two detectors.*  The scanning speed [degrees (2θ) per

_____

*Some manufacturers utilize a tandem detector arrangement in which the fluorescent x-rays must pass through the flow counter before impinging upon the scintillation detector.  In this manner, a higher detection efficiency is achieved over the entire wavelength spectrum.

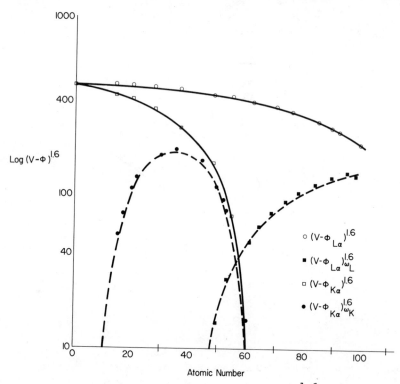

Figure 8.7 Atomic number dependence of $(V - \Phi)^{1.6}$ and the combined factor $\omega(V - \Phi)^{1.6}$ for 50kV excitation using a chromium anode tube.

minute] and time constant must be chosen so as not to introduce spectral line shifting or distortion [6,7]. In practice this is generally not a problem with a modern x-ray spectrometer owing to the rather large detector aperture. A useful rule of thumb states that the product of the scanning speed (degrees per minute) times the time constant (in seconds) should not exceed 2. This will allow qualitative scanning speeds of $4^{\circ}$/min to be used with a time constant of 0.5 s. Although most spectrometers are limited to maximum scanning speeds of $4^{\circ}$/min, it has been shown [6] that a modified spectrometer can be run at a scanning speed as high as $32^{\circ}$/min using a very small time constant (0.04 s) without significant loss of intensity or peak distortion.

Table 8.4  Spectral Interference in Qualitative Analysis
Using the X-ray Energy-Dispersive Analyzer

---

1.  Coherent and incoherent scattered continuum from tube.
2.  Coherently scattered tube or secondary source characteristic
    lines.
3.  Incoherently scattered tube or source characteristic lines.
4.  Contamination lines.
5.  Overlapping of K, L, and M spectra.
6.  Energy resolution varies with photon energy.
7.  Escape peaks.
8.  Incoherent escape distributions (Compton edge)..
9.  Sum peaks.
10.  Low-energy tailing (incomplete charge collection).
11.  Diffraction lines from specimen.

---

To cover a range of 0.5 to 20 $\overset{\text{o}}{\text{A}}$ in the wavelength spectrum, it
is necessary to scan a total of 200 degrees $2\theta$.  At a speed of
$4^{\text{o}}$ $(2\theta)$/min, this will require 50 min of scanning time.


## 8.4  QUALITATIVE ANALYSIS BY ENERGY DISPERSION

Energy dispersive analysis systems can accumulate the spectral data
from the entire energy spectrum in a period of 200 s or less, although in order to obtain high sensitivity over the entire energy
range, it is often necessary to use several excitation modes which
may double or triple the time needed.  In nearly all cases, the
acquisition time is significantly faster than the 40-50 min for the
scanning crystal x-ray spectrometer discussed in Sec. 8.3.3.  Table
8.4 lists spectral line interference problems associated with
qualitative analysis using the energy-dispersive x-ray spectrometer.
    Figures 8.8 through 8.12 are typical energy spectra which
illustrate some of the spectral interference problems associated
with the energy dispersive technique.  Several of the interference
sources listed in Table 8.1 and Table 8.4 are common to both wavelength and energy dispersion, such as items 1 through 5.  In Figs.
8.8 and 8.10, however, one can see several additional interference

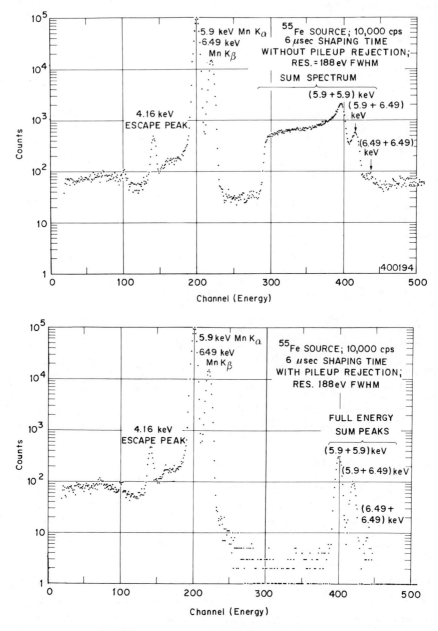

Figure 8.8   (a) $^{55}$Fe emission spectrum showing the prescence of sum spectra.   (b) $^{55}$Fe emission spectrum utilizing Pulse Pileup Rejection.

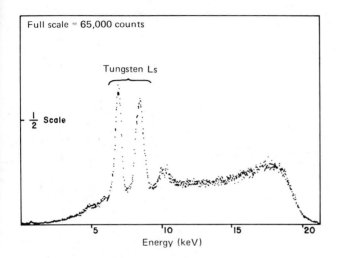

Figure 8.9   Tungsten radiation scattered from a Lucite sample.
Reprinted from Ref. 10 with permission of Kevex Corp., Burlingame,
Calif.

problems which are unique to the energy-dispersive method.   These
are listed in Table 8.4 as items 7 through 10.   Typical of these
problems is that of escape peaks and sum peaks, the origin of which
have been discussed in Chap. 4.   Using pulse-pileup rejection circuitry, only full-energy sum peaks having apparent pulse energies as
shown in Fig. 8.8 will be evident.   These sum peaks can easily be
mistaken for true spectral lines.   An escape peak can be recognized
by looking for its strong *parent* line which lies on the high-energy
side of the escape peak.   The Si(Li)-detector escape peak has an
energy equal to the energy of the parent line less 1.74 keV.   The
probability of Si K radiation escaping from the detector (production
of escape peak) increases as the incident photon energy is decreased.
Thus, low-energy photons having energies slightly more energetic
than the K absorption energy of silicon (1.838 keV) will produce
stronger escape peaks than higher energy incident x-rays (see Sec.
4.3.9).

Coherent and incoherent scattered lines from the excitation
source radiation are shown in Figs. 8.9 and 8.10.   Note that the
coherently scattered line has a lower intensity than the incoherent

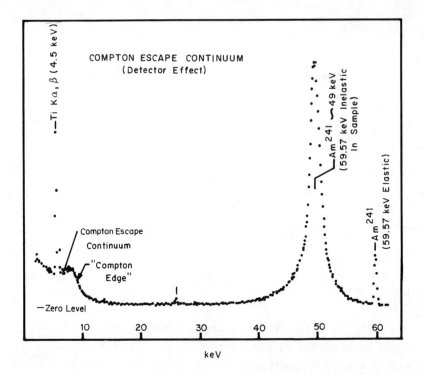

Figure 8.10  Compton escape effects from the detector.  Reprinted
from Ref. 8 with permission of Kevex Corp., Burlingame, Calif.

profile in Fig. 8.10.  This is due to the fact that the sample consists of low-atomic-number elements and the incident x-rays have a
high energy.  Both of these factors cause a large fraction of the
incident photons to suffer energy loss by the Compton process.
This is illustrated in Fig. 8.13 where the excitation source photon
having an energy E strikes an atom in the specimen.  If the specimen atom has a very low atomic number (< 10), the collision will be
largely inelastic resulting in an energy loss.  The scattered photon
will now have an energy $E' < E$, and the magnitude of the energy
change ($\Delta E$) increases with the scattering angle $\psi$ to a maximum value
at $\psi = 180^{\circ}$.  In energy dispersive systems, the scattering angle
can be large, i.e., $> 90^{\circ}$, as shown in Fig. 8.14 which compares
several designs used in energy dispersive analyzers [8].  Angles of

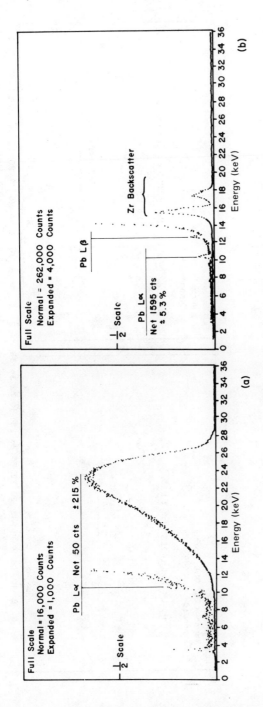

Figure 8.11  Spectrum of 10-ppm lead in aqueous solution taken at a total count rate of 10,000 counts/sec, a counting time of 300 sec, filtration, and various targets:  (a)  direct excitation with a tungsten target x-ray tube operating at 25 kV; (b) secondary target excitation with a zirconium target.  Reprinted from Ref. 10 with permission from Kevex Corp., Burlingame, Calif.

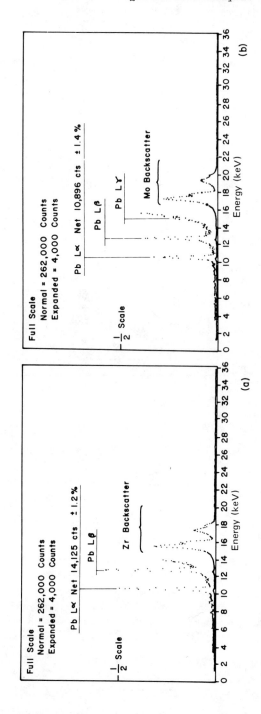

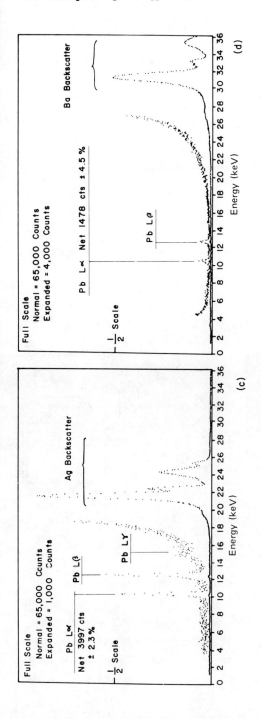

Figure 8.12   Spectrum of 100-ppm lead in aqueous solution using secondary target excitation, a total count rate of 10,000 counts/sec, counting time of 300 sec, filtration, and various targets:   (a) Zr, (b) Mo, (c) Ag, and (d) Ba.   Reprinted from Ref. 10 with permission from Kevex Corp., Burlingame, Calif.

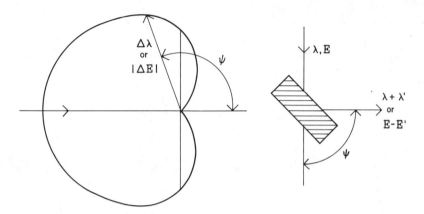

Figure 8.13 The Compton effect (incoherent scattering) showing the effect of the scattering angle on the magnitude of the energy (or wavelength) shift ($\Delta E$ or $\Delta \lambda$).

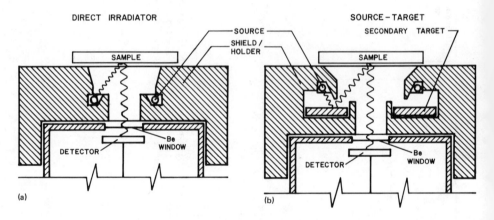

Figure 8.14 Schematic representation of radioisotope excitation system with annular source configuration: (a) direct irradiator; (b) secondary target irradiator. Reprinted from Ref. 8 with permission from Kevex Corp., Burlingame, Calif.

$150^{\circ}$ are not uncommon and large incoherent energy shifts (see Fig. 8.10) may occur. Comparing Figs. 8.9 and 8.10, it can be seen that incoherent scattering by the excitation source is generally not a serious problem unless direct high-energy excitation is used in conjunction with a low-atomic-number sample. The position of a suspected incoherent scattered tube line will always be on the low-

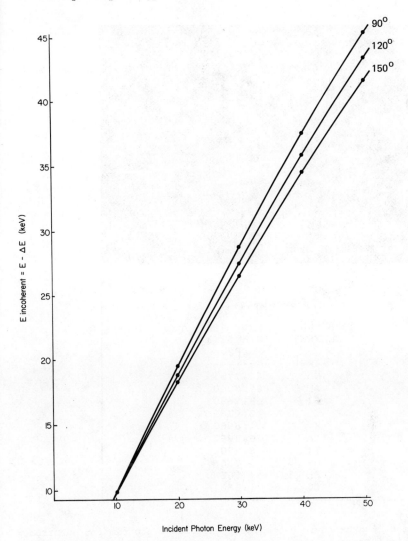

Figure 8.15  Position of a Compton scattered line having the energy
$E - \Delta E$, where E is the energy of the incident photon.  $E - \Delta E$ is
calculated for three values of the scattering angle $\psi$ (90, 120, and
$150^{\circ}$) according to Eq. (8.1).

energy side of the parent line.  If the scattering angle $\psi$ (see Figs.
8.13 and 8.14) is known, then $\Delta E$ may be calculated from Eq. (8.1) or
$E - \Delta E$ may be estimated from Fig. 8.15 which shows $E - \Delta E$ plotted
for values of $90^{\circ}$, $120^{\circ}$, and $150^{\circ}$.

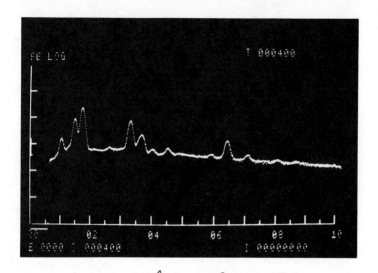

```
****REPORT****
SAMPLE    1
ELEMENT     INTENSITY
               CPS
-------     ----------
  NA         44.7751
  AL        288.8060
  SI        768.7690
  CL          3.9762
  K         234.3980
  CA         30.4993
  TI          8.7902
  MN          3.7366
  FE         41.1654
  CU          1.5881
```

Figure 8.16  KLM ID is a completely automatic program for spectrum analysis.  The only input required is the spectral data.  The program then identifies all elements and their associated energy lines and calculates the intensity above background for each line in the spectrum.  All of this is accomplished without operator intervention or use of standard samples.  Reprinted by courtesy of EG&G ORTEC.

$$\Delta E = \frac{E^2 (1 - \cos \psi)}{511 + (1 - \cos \psi)E} \tag{8.1}$$

where E is the photon energy in kiloelectronvolts.

## 8.4.1  Methods of Qualitative Analysis

In order to qualitatively analyze an energy dispersive spectrum, it is necessary to accurately measure the energies of the various spectral lines.  Subsequently, one must remove from consideration interference lines such as those discussed in Table 8.4.

Energy measurement may be done manually by moving a calibrated dial or alphanumeric display on a cathode-ray screen.  Alternatively, several instrument systems will automatically "peak list" the entire spectrum and print a listing of spectral lines and their background-corrected integrated intensities (see Fig. 8.16).  When interfering lines (Table 8.4) have been removed, the list of energies can be manually searched for spectral line combinations which are characteristic of specific elements.  For example, if a strong line is suspected to be caused by a $K\alpha$ energy, the list is searched for a $K\beta$ line having the appropriate intensity relation to the $K\alpha$ (see Fig. 8.17).  If a $K\beta$ line can not be found, the original $K\alpha$ line assumption is probably in error and another assumption must be made (i.e., $L\alpha$, $L\beta$).  This manual procedure is complicated by line overlapping if several elements are present in the spectrum.  Computer software is available on some systems to project the spectral line locations of suspected elements directly on the cathode-ray-tube display of the contents of the multichannel analyzer.  More sophisticated systems using dedicated computers can automatically analyze the spectrum and assign elemental designation to each line (see Fig. 8.16).

## 8.5  SEMIQUANTITATIVE ANALYSIS

Jenkins [4] has shown that a reasonably accurate semiquantitative analysis can be performed using the background-corrected peak heights (wavelength dispersion) or the background-corrected peak areas (energy dispersion).  These intensities are corrected using the weighted fluorescent yields $\omega'$ and excitation overvoltage factors F, according to Eqs. (8.2) and (8.3), and then normalized.

Table 8.5  Estimated Composition of an 7075 Aluminum Alloy

| Element | Line | $\omega$ | g | $\omega' = \omega g$ | $F^a$ | $I_{measured}$(counts/s) | $I_{corrected}$(counts/s) | %(Calculated) | %(Actual) |
|---------|------|----------|------|----------------------|-------|--------------------------|---------------------------|---------------|-----------|
| Al | K$\alpha$ | 0.026 | 0.94 | 0.024 | 156 | 2370 | 633 | 93.3 | 89.3 |
| Cu | K$\alpha$ | 0.425 | 0.88 | 0.375 | 85 | 1140 | 35.8 | 5.3 | 5.7 |
| Zn | K$\alpha$ | 0.458 | 0.88 | 0.403 | 79 | 310 | 9.7 | 1.4 | 1.6 |

$^a$All elements measured at 25 kV.

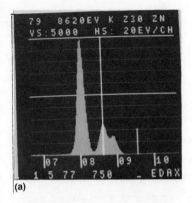
(a)

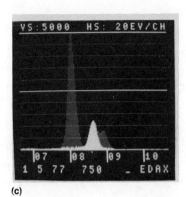
(c)

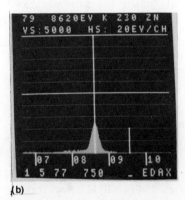
(b)

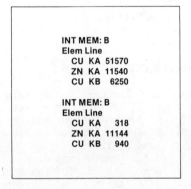

Figure 8.17  Electronically stripped energy-dispersive analysis
spectra revealing the prescence of a weak zinc line on a strong
copper line.  Scan (a) shows the original spectrum consisting of
strong copper emission, the weak zinc line is labelled.  Scan (b)
shows the spectrum stripped of copper.  Scan (c) shows the stripped
and original spectra superimposed upon each other.  Also shown is
the computer output of the stripped spectrum.  Reprinted with permission from Edax International Inc., Prairie View, Illinois.

$$I_{corrected} = \frac{I_{measured}}{\omega' \times F} \tag{8.2}$$

where $\omega'$ is the product of the fluorescent yield and the probability
factor g (see Table 8.5).  In addition,

$$F = (V - \Phi)^{1.6} \tag{8.3}$$

Table 8.5 lists some typical data obtained by application of this method. It will be noted that quite reasonable concentration estimates were obtained even for this rather difficult case of low- and high-atomic-number elements. Magnesium was also present at about 2.6%, but was not measured, causing the estimated value of aluminum to be high. If widely separated elements on the periodic chart are being analyzed, it is necessary to utilize an additional correction to account for detection efficiency of the different energies (wavelengths) being measured. This correction factor can be approximated empirically by measuring the counts per second per percent (cps/%) for a wide range of elements using the same excitation conditions.

REFERENCES

1.  E. J. McGuire, *Phys. Lett.*, *33A*:288 (1970).
2.  J. H. McCrary, *Phys. Rev.*, *A4*:1745 (1971).
3.  W. Bambynek, *Rev. Mod. Phys.*, *44*:716 (1972).
4.  R. Jenkins, in *An Introduction to X-ray Spectrometry*, Heyden, London, 1974.
5.  A. H. Compton and S. K. Allison, in *X-rays in Theory and Experiment*, 2nd Ed., D. Van Nostrand, New York, 1935.
6.  E. F. Kaelble, ed., *Handbook of X-rays*, McGraw-Hill, New York, 1967.
7.  H. P. Klug and L. E. Alexander, in *X-ray Diffraction Procedures*, John Wiley, New York, 1954.
8.  Rolf Woldseth, in *X-ray Energy Spectrometry*, Kevex Corporation, Burlingame, Calif.,
9.  D. A. Gedcke, *X-ray Spectrom.*, *1*:129 (1972).
10. D. E. Porter, *X-ray Spectrom.*, *2*:85 (1973).

# 9

## Basic Problems in Quantitative Analysis

9.1  GENERAL

Quantitative analysis by x-ray emission involves a series of steps, each of which must be carefully controlled if accurate data are to be obtained.  Although accuracies of the order of a few tenths of one percent are possible, and indeed being obtained by many hundreds of x-ray laboratories all over the world, a thorough understanding of specimen heterogeneity and absorption phenoma is necessary if the many potential pitfalls of the technique are to be avoided.  There are almost as many actual methods of x-ray analysis in use as there are laboratories employing x-ray spectrometers, and what follows is a description of basic procedures rather than specific details of individual methods.  The literature in x-ray analysis is vast, and the reader will find a wealth of information on specific applications in the review articles which have appeared over the past 10 years or so [1-7].

Both wavelength and energy dispersive spectrometers are currently employed for quantitative work.  Although the method of excitation, dispersion, and measurement of a given wavelength or energy may differ, once the intensity has been measured, the con-

version of this intensity to concentration is similar in both types
of instrumentation.  Similarly, problems due to heterogeneity,
methods of predicting and correcting for matrix influences, and so
on will be identical.

The starting material in any analytical procedure is the sample
which has been submitted for analysis.  The analyst is rarely involved in the taking of this sample, and one must generally assume
that the given sample is representative of that which is to be
analyzed.  This in itself may be a major assumption, particularly
in those cases where the bulk material, for which a representative
analysis is required, is large in size.  For example, in surveying
an area of land covering many square miles, for a given element or
mineral type, the distribution of the mineral may itself be very
variable and unless many samples are taken, an accurate assessment
of elemental concentration may not be justified.  In general terms,
striving for an accuracy of analysis better than the error due to
sampling is unwarranted since obtaining high accuracy is nearly always time consuming and costly.

There are three major stages in a quantitative determination,
these being first, the preparation of a specimen from the sample
submitted for analysis; second, the excitation of a suitable emission line from each element to be analysed and the measurement of
its intensity; and third, the conversion of the measured intensity
into elemental concentration.  In Chap. 7 (Fig. 7.1), these steps
were indicated along with the more important procedures which are
employed, depending upon the original form of the sample and
special requirements in measuring the x-ray intensity.

The relationship between measured x-ray intensity (I) and element concentration (C) will be of the form

$$C = K.I.M.S.$$

where M represents the interelement effects and S, the specimen
heterogeneity.  This expression shows that four factors may affect
the accuracy in the measurement of C, namely,

1. The factor K: This is a factor which depends upon the design of the spectrometer and the conditions under which the spectrometer is operated. This constant will vary from equipment to equipment, although less so where the equipment is of the same "type." For a given instrument, K is only a constant where all measurements are performed under constant conditions, i.e., same voltage, current, and anode of the x-ray tube; same crystal; same detector; etc. Since K is dependent upon many factors, it is almost impossible to calculate accurately, and in practice its value is determined by calibration. Thus by plotting I.M.S. vs. C for a range of calibration standards, the slope of the curve (I/C) is equal to 1/K, in practice this means assuming a constant M and S and plotting I vs. C, or correcting I for interelement effects, following removal of S by suitable specimen preparation.

2. The intensity I: This is the net intensity of the measured wavelength or energy peak above background. For error levels above the limits set by the mechanical precision or electrical stability of the instrument, counting statistics alone will determine the precision in the measurement of I and hence in the precision in the estimation of C.

3. Interelement effects M: These may include primary and secondary absorption effects, plus enhancement and third-element effects. Since these effects all lead to systematic errors in the measurement of the "true" intensity, the accuracy obtained in the estimation of C will be directly dependent upon how well the quantitative method employed corrects for, or minimizes, interelement effects.

4. Specimen heterogeneity S: This will depend mainly upon the penetration depth of the measured x-rays relative to the average particle size of the specimen. Where both penetration depth and particle size are of the same order, it will also depend upon the elemental distribution within a particle. At the present state of the art, the only reliable way of reducing or eliminating problems arising from specimen inhomogeneity lies in controlling the particle size influence by suitable specimen preparation.

The above shows that careful measurement of K and I will lead to *accurate* results, only where effects of interelement interactions are eliminated, or at least reduced to insignificant proportions. Should these latter effects be ignored, careful measurement of K and I will give only good *precision* and in extreme cases will lead to a very repeatable but incorrect estimate of the concentration.

Specimen preparation has been dealt with previously (see Chap. 7), and it will be assumed from now on that the specimen can be considered homogeneous as far as the penetration of the x-ray beam is concerned. Similarly, the selection of excitation conditions and the basis of measurement of x-ray intensity have also been dealt with in previous sections. The following paragraphs are thus concerned mainly with the conversion of x-ray intensities to elemental concentrations, although further reference will be made to special problems in the measurement of x-ray intensities where this is relevant.

## 9.2   MEASUREMENT OF X-RAY INTENSITY

### 9.2.1   The Net Counting Error

It is important that a sufficient number of counts be taken on each analytical line, and if necessary on the associated background, to ensure that the counting error is not the limiting error in the analysis. It is also important that no systematic errors be introduced in the measurement of the net count intensity, such as those that might accrue due to counting loss. For the wavelength spectrometer, one of the most useful forms of the equations which relates the net counting error $\sigma\%_{net}$ with the peak $(I_p)$ and background $(I_b)$ counting rates is

$$\sigma\%_{net} = \frac{100}{\sqrt{t}} \frac{1}{\sqrt{I_p} - \sqrt{I_b}} \qquad (9.1)$$

In this expression, t is the total counting time made up of the time spent counting at the peak position $t_p$ and the background position $t_b$. One of the conditions for Eq. (9.1) is that the total analysis time be divided up such that

$$\frac{t_p}{t_b} = \sqrt{\frac{I_p}{I_b}} \qquad (9.2)$$

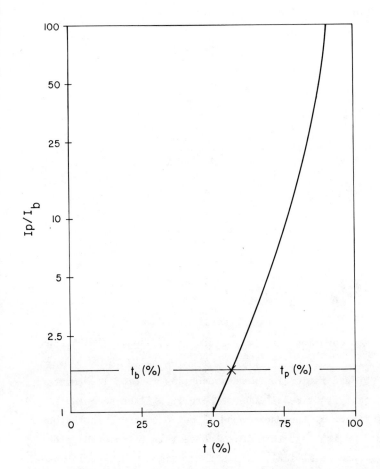

Figure 9.1   Optimum division of peak and background counting times.

Thus, as the peak-to-background ratio $(I_p/I_b)$ increases, a greater percentage of the total analysis time should be spent counting at the peak.   Figure 9.1 shows a graphical representation of the distribution of the total analysis time for different peak-to-background ratios.

It will be seen that as the peak-to-background ratio is increased, a point will eventually be reached where the ignoring of the background will have little influence on the net counting error. As an example, Fig. 9.2 shows a plot of net counting error against

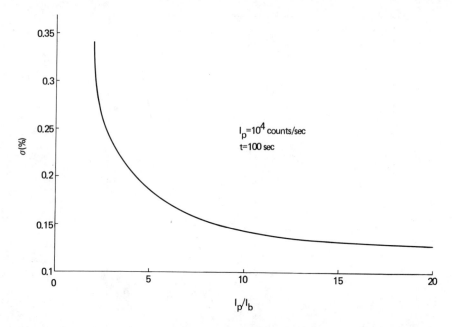

Figure 9.2  Net counting error as a function of peak-to-background ratio.

peak-to-background ratio, for a peak counting rate of $10^4$ counts/s
and a total analysis time of 100 s.  Here it will be seen that
whereas the effect of the background is quite critical in the peak-
to-background region less than 10:1, above this ratio, the effect
of the background is minimal.  It is thus common practice in wavelength-dispersive spectrometers to ignore the background where the
peak-to-background ratio is in excess of 10:1.

In energy-dispersive spectrometers where peak-to-background
ratios are relatively low, background subtraction is nearly always
applied.  In this particular case, however, no additional measurement time is required.  If background is not subtracted but determined as a calibration curve intercept, one additional effect may
have to be considered, this being the drift in the background.
Although the background level may remain essentially constant during short periods of time, over a period of several hours significant fluctuations may be observed.  As an example, Fig. 5.9 shows

the results obtained with an energy-dispersive spectrometer in
which a series of 1000-s live time (about 30 min of real time)
counts were taken over a period of 90 h.  Three curves are shown
representing raw data, long-term drift, and short-term fluctuations.
The long-term drift curve was generated using an 11-point smoothing
routine on the raw data, giving a curve containing drift components
for periods of 5 1/2 h or longer.  The short-term curve was extracted by subtracting the long-term-drift curve from the raw data, and
adding back the mean value.  This gives a curve containing components with periods between 5 1/2 h and 30 min and includes most of
the counting statistical error.  In the example given, the long-
term drift was ± 0.1% and the short-term fluctuations were beyond
counting statistics, less than 0.01%.  A long-term drift of this
order could, of course, introduce a significant systematic error in
a net peak measurement, for low peak-to-background ratios.  For
this reason, it is advisable to carry out a background subtraction.

9.2.2  Estimation of the Background

In those cases where the background is significant, and therefore
has to be measured, problems may occur in actually making the background measurement.  The true background is that counting rate
measured at the analyte wavelength when the analyte concentration is
zero.  Such a measurement is impracticable since the background is
itself dependent upon the matrix composition; thus it is incorrect
to use a value obtained in the *absence* of the analyte to correct
for background when the analyte *is* present.  Hence, the use of a
"blank" determination for the estimation of background in a series
of measurements is dangerous and should only be used in those cases
where the background is matrix independent.  This will be less of a
problem in trace-element analysis.  From this it will be seen that
any measurement of background is at the very best only an estimation since in practice further complications may arise from interfering lines or a background level which is wavelength dependent.

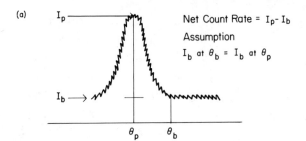

(a)

Net Count Rate = $I_p - I_b$

Assumption

$I_b$ at $\theta_b$ = $I_b$ at $\theta_p$

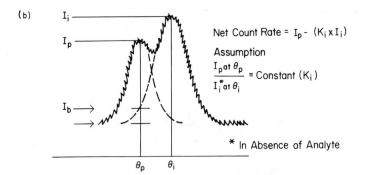

(b)

Net Count Rate = $I_p - (K_i \times I_i)$

Assumption

$\dfrac{I_p \text{ at } \theta_p}{I_i^* \text{ at } \theta_i}$ = Constant $(K_i)$

\* In Absence of Analyte

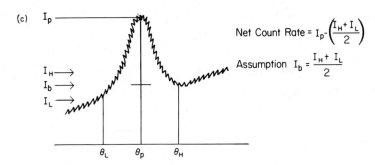

(c)

Net Count Rate = $I_p - \left(\dfrac{I_H + I_L}{2}\right)$

Assumption $I_b = \dfrac{I_H + I_L}{2}$

Figure 9.3  Typical peak-to-background situations in x-ray spectrometry.

Figure 9.3 illustrates three typical situations which might occur in practice when using x-ray spectrometers.  Figure 9.3(a) shows the general case of a peak superimposed upon significant background in a wavelength-dispersive spectrometer.  The same situation occurs with energy-dispersive spectrometers, and in this instance, the angular scale values in Fig. 9.3 should be replaced with energy scales.  A measured wavelength from the analyte element gives a counting rate $I_p$ at the peak position $\theta_p$.  The background is estimated by making a second measurement at a different angle $\theta_b$, typically $1^o(2\theta)$ higher than the analyte angle.  Assuming that $I_b$ at $\theta_p$ is the same as $I_b$ at $\theta_b$, then the net counting rate from the analyte element is $I_p - I_b$.  This is by far the most commonly used method of assessing the background and is generally satisfactory in cases of well-resolved peaks superimposed upon background which is not angular dependent.

It should not be forgotten, however, that where background is being estimated, even as little as $1^o(2\theta)$ away from the peak, the wavelength of the background may differ considerably from that of the peak.  This difference can be as much as 10% for short wavelengths measured at low angles, e.g., Ba K$\alpha$ diffracted with the LiF(200) crystal, and as little as 0.5% for longer wavelengths measured at high angles, e.g., Ca K$\alpha$ with LiF(200).  Also, backgrounds are always better measured to the high-Bragg-angle side of the analyte peak since satellite lines, such as K$\alpha_3$ and K$\alpha_4$, can give significant broadening to the peaks toward the short-wavelength side, particularly in the case of the low-atomic-number elements.

Figure 9.3(c) illustrates the case of a peak superimposed upon a background which is very angular (i.e., wavelength) dependent. This type of situation is relatively common in the measurement of wavelengths of less than an angstrom or so in low average atomic number matrices.  The best method of estimating the background in this case is to make two background readings, $I_H$ and $I_L$, to the high $\theta_H$ and low $\theta_L$ angle side, respectively, of the peak.  The arithmetic mean of the two readings is then taken as the background value.

A further complication is encountered where the analyte line may be partially overlapped by a second line with a peak at $\theta_i$ [Fig. 9.3(b)].  The best solution in this case is to estimate the contribution of i at $\theta_p$, either by calibration or by means of a blank, and to determine the intensity ratio of i at $\theta_p$ and $\theta_i$.  By assuming

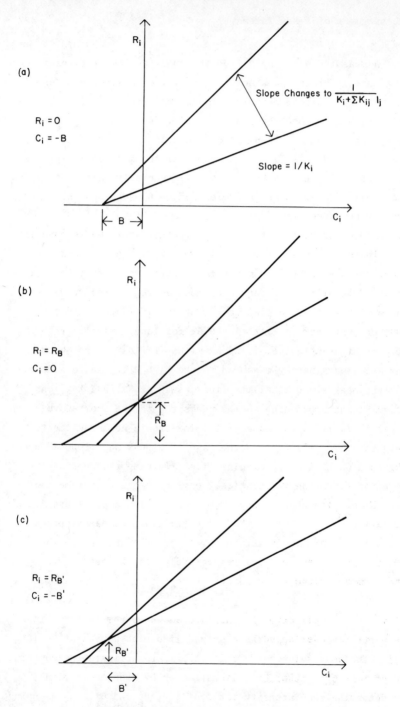

Figure 9.4 Effect of residual background on the calibration curve.

that this ratio $K_i$ is constant, the background at $\theta_p$ can be taken as $K_i I_i$. In the case of energy-dispersive spectrometers where peak areas are normally used, a simple mathematical solution based on area interference can be employed. However, in practice it is more common to employ a computer-applied least-squares fitting technique (see Chap. 6).

### 9.2.3 Effect of the Background on the Calibration Curve

In those cases where the background is not subtracted from a peak intensity (generally always the case with wavelength-dispersive spectrometers and rarely the case with energy dispersive spectrometers) the effect of residual background on the shape of the calibration curve must be considered. Figure 9.4 illustrates what might be observed in practice. Curve (a) shows a typical situation where the slope of count ratio $R_i$ vs. concentration $C_i$ varies considerably due to a wide range of matrix influences. The slope of the curve is $1/K_i$ and this is corrected for matrix effects using an expression of the form $1/(K_i + \Sigma K_{ij} I_j)$, where $K_{ij}$ is a constant representing the effect of element j on element i. $I_j$ is the count rate from element j. In this curve, all slopes pass through the point $R_i = 0$, $C_i = -B$, where B is a background term having the same dimensions as C. Note that this curve gives a range of values of $R_i$, where $C_i = 0$, i.e., when the concentration of the analyte element is zero, the intensity measured at the analyte line (the blank) would be matrix dependent.

The second curve (b) is similar except that the curves all pass through the point $R_i = R_B$, $C_i = 0$. In other words, when the analyte concentration is zero, the background is constant and has the value $R_B$ with the same dimensions as $R_i$. The third curve (c) has all curves passing through $R_i = R_{B'}$, $C_i = -B'$. In this instance, when $C_i$ equals zero, the background intensity is somewhat matrix dependent and a more complex expression of concentration and intensity would be required to force all the curves through the required pivot point. An expression which has been used [8] to correct for this effect is

$$C_i = R_i \ (K_i + \sum_i K_{ij}I_j) + (B_i + \sum_i B_{ik}I_k).$$ (9.3)

In practice, all three of these situations may be encountered, especially in the case of wavelength-dispersive spectrometers, where, as has been previously stated, backgrounds are rarely sub-tracted unless the peak-to-background ratio is less than 10:1. Matrix-dependent background intensity is common in those cases where high background arising from scatter of the primary source is pre-sent. This will generally mean measurement of wavelengths < 0.7 $\overset{o}{A}$ in low-atomic-number matrices where bremsstrahlung source radiation is employed. The second case will occur whenever the background intensity at the measured wavelength or energy is independent of the matrix. In wavelength dispersive systems, this will typically occur in the measurement of longer wavelengths (> 2 $\overset{o}{A}$) where pulse height selection has been employed to remove background harmonics. The third case is an intermediate situation between the first two and is the one most frequently encountered.

## 9.2.4 Effect of Counting Loss

In Sec. 4.6 mention was made of the causes of counting losses, due mainly to deadtime in the detection and counting circuitry. It is important that these counting losses be carefully controlled since in all instances they will lead to systematic errors, which may manifest themselves either as low counts in absolute measurements or as low or high count ratios where the count ratio technique is em-ployed. Figure 9.5(a) demonstrates the effect of counting loss in the absolute counting technique. In this instance, there is a linear relationship between the true count rate and the elemental concentration. As far as the measured count rate is concerned, however, the counting loss causes the measured count rate $I_m$ always to be less than the true count rate $I_t$. Note that in this case, where a true count-rate calibration curve is assumed, the calculated elemental concentration is always low. This effect could of course be corrected for using the deadtime correction equation:

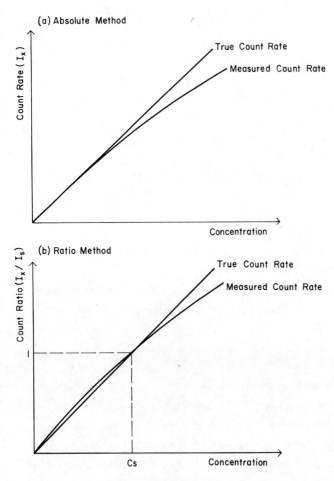

Figure 9.5   Effect of counting loss on the shape of the calibration curve.

$$I_t = \frac{I_m}{1 - I_m t_d} \qquad\qquad (9.4)$$

where $t_d$ is the deadtime.  In practice, the counting loss may be corrected for either by using a mathematical expression of the type shown above or, alternatively, by using counting-loss correction circuitry incorporated in the counting chain.  It should also be appreciated that the counting error $\sigma$ on an observed number of counts $N_m$ is less than would be predicted from the usual expression $\sigma = \sqrt{N}$ as a result of the derandomizing effect of the deadtime. Due to the fact that the *true* number of counts is greater than that actually collected, the correct counting error $\sigma_t$ in the calculated true counts $N_t$ is given by

$$\sigma_t = \frac{\sqrt{N_m}}{1 - I_m t_d} \tag{9.5}$$

where $N_m$ represents the observed number of counts.

Where the ratio method of counting is used [Fig. 9.5(b)], the same measured count rate curve is obtained, but in this case the curve is forced through the point $I_x/I_s = 1$, $C = C_s$, where $C_s$ is the elemental concentration in the reference standard. As will be seen in Fig. 9.5(b), if a linear relationship is assumed, a somewhat higher count ratio than expected (hence a high concentration) will be obtained below a count ratio of unity and vice versa.


## 9.3  ELEMENTAL INTERACTIONS IN HOMOGENEOUS SPECIMENS

Once an accurate measurement of the net intensity of a selected wavelength has been made, one is in a position to establish a relationship between this x-ray intensity and the chemical composition of the sample. As discussed in Secs. 2.9 to 2.11, allowance may have to be made in this relationship for interelement effects including primary and secondary absorption, plus enhancement, including third element effects. As an example, if the concentration $C_i^x$ of an element i is being measured in a specimen of unknown composition and a reference standard containing element i at a known composition $C_i^s$ is available for calibration purposes, and provided both specimen and reference are homogeneous over the penetration depth of the measured wavelength $\lambda_i$, the following relationship will hold

$$C_i^x = C_i^s \left[ \frac{I_i^x}{I_i^s} \right] \left[ \frac{\text{total absorption for } \lambda_i \text{ by } x}{\text{total absorption for } \lambda_i \text{ by } s} \right]$$

$$\times \left[ \frac{\text{total enhancement of } \lambda_i \text{ by } x}{\text{total enhancement of } \lambda_i \text{ by } s} \right] \qquad (9.6)$$

$I_i^x$ and $I_i^s$ represent the measured intensities of $\lambda_i$ in specimen and reference, respectively. A linear relationship between concentration and intensity or intensity ratio will only occur in those cases where, first, there is no significant absorption difference for $\lambda_i$ by specimen and reference, and second, where the total enhancement of $\lambda_i$ by other elements present in specimen and reference is the same. In practice, absorption effects are *always* present, whereas enhancement effects are not necessarily present.

One of the great advantages of the x-ray technique is that enhancement and absorption effects are to a large extent predictable, and assessment of potential matrix problems can readily be made by reference to standard mass absorption coefficient and wavelength tables. For example, if one considers a series of binary alloys of Fe/Cr, Fe/Co, and Fe/Ni the slopes of the calibration curves for chromium, cobalt, and nickel can be predicted as illustrated in Fig. 9.6. The series of figures at the top shows the position of the characteristic lines of the elements in question along with the absorption curves. The corresponding lower figures show the shapes of the respective calibration curves.

*Iron/Chromium*

In this instance, Cr Kα is the measured wavelength and it will be seen that both iron and chromium have very low absorptions for this wavelength. It will also be noted, however, that the Fe Kα line lies just to the short-wavelength side of the chromium absorption edge and is thus able to enhance Cr Kα. The calibration curve for chromium shows a positive deviation from linearity (i.e., decreasing

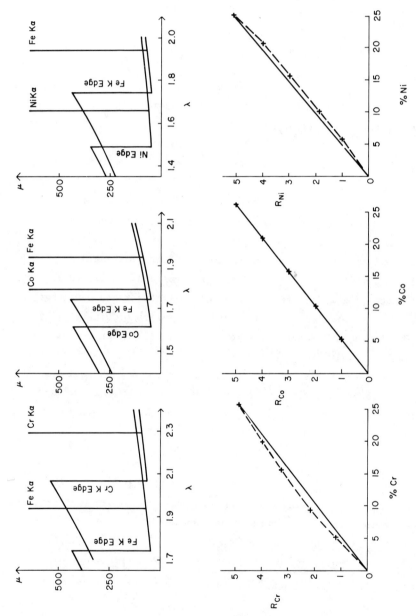

Figure 9.6 Absorption and enhancement effects in the Fe/Cr, Fe/Co, and Fe/Ni systems.

slope with increasing analyte concentration) because as more
chromium is added, there is less iron present to enhance the
chromium.

*Iron/Cobalt*

In this instance, the measured wavelength is Co Kα, and again both
iron and cobalt have low absorption for Co Kα.  Unlike the previous
case, the Fe Kα line is now to the long-wavelength side of the absorption edge of the analyte element and hence cannot enhance Co Kα.
Thus the calibration curve for Co Kα is reasonably straight over
the whole concentration range.

*Iron/Nickel*

Here the analyte element is nickel and again enhancement is not
possible.  This time, however, there is a strong absorption of
Ni Kα by iron and this is about eight times greater than the nickel
self-absorption.  The calibration curve for nickel shows negative
deviation from linearity (i.e., increasing slope with increasing
analyte concentration) because as the concentration of nickel is
increased, the lower absorbing nickel replaces the higher absorbing
iron.

Not only can one predict the presence of an interelement effect, one can also go a long way to predicting the magnitude of such
an effect.  For example, if a steel sample containing chromium,
iron, and nickel were being analyzed over the range 16 to 20%
chromium, with a fixed concentration of nickel at 8% and iron as the
remainder, use of absorption tables allow the data to be presented
in the form shown in Fig. 9.7.  Only characteristic lines to the
short-wavelength side of the absorption edge of the analyte element
can excite the characteristic lines of that element, but in this
case the characteristic lines of both iron and nickel fulfill this
condition, hence both will enhance Cr Kα, iron more so than nickel
because it is closest to the chromium absorption edge.  Further to
this, since nickel can also enhance iron, nickel will also display
a third-element effect on chromium via iron.  As far as secondary

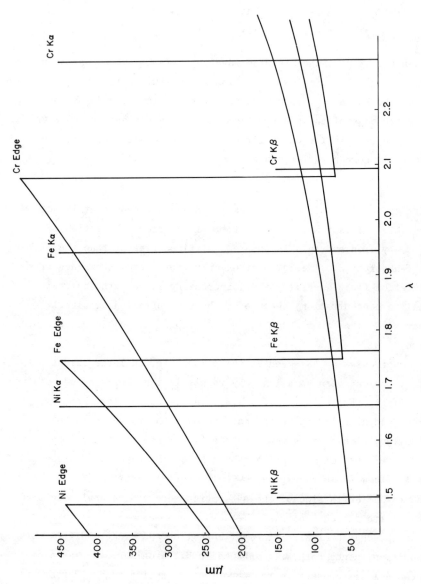

Figure 9.7  Absorption and enhancement effects in the Ni/Fe/Cr system.

absorption is concerned, it is apparent that Cr $K\alpha$ is weakly absorbed by both iron ($\mu$ = 115) and nickel ($\mu$ = 146) relative to its
own self-absorption ($\mu$ = 90). The total variation in secondary absorption over the 16-20% chromium content variation can be calculated using Eq. (2.11), i.e.,

$$\mu(16\% \text{ Cr}) = (90 \times 0.16) + (115 \times 0.76) + (146 \times 0.08) = 113.5$$
$$\mu(20\% \text{ Cr}) = (90 \times 0.20) + (115 \times 0.72) + (146 \times 0.08) = 112.5$$

which is less than 1% change in secondary absorption. In this instance, by far the dominant effect would be the enhancement of
chromium.

The converse would be true if the analyte element were nickel,
say over the concentration range 5 to 15%, with the chromium concentration fixed at 18% and with iron the remainder. In this instance,
neither iron nor chromium can enhance nickel; so the only matrix
effect will be that due to absorption. Again, the total variation
in secondary absorption can be calculated, viz,

$$\mu(5\% \text{ Ni})\ = (316 \times 0.18) + (397 \times 0.77) + (61 \times 0.05) = 365.6$$
$$\mu(15\% \text{ Ni}) = (316 \times 0.18) + (397 \times 0.67) + (61 \times 0.15) = 332.0$$

in this case about a 9.2% change in secondary absorption.

The above exercise is useful in rapidly predicting the approximate magnitude of an absorption effect in a given matrix. Although
it is slightly more tedious to perform a similar calculation to include both secondary and primary absorption, it is frequently worth
the extra effort since calculations using secondary absorption
effects alone tend to overestimate the change in total absorption.
The complete absorption calculation is made easier if one is able
to use the concept of effective wavelength (see Sec. 10.7.3). As
an example, one could recalculate the previously considered example
of nickel varying in concentration between 6 and 10%, this in time
however, using the total absorption terms $\alpha$ rather than the mass
absorption coefficients $\mu$. The total absorption $\alpha_j W_j$ (see Chap. 2)
is given by

$$\sum \alpha_j W_j = \sum W_j [\mu_j(\lambda_e) + A\mu_j(\lambda_i)] \tag{9.7}$$

where j represents each matrix element present at a weight fraction
$W_j$, $\lambda_e$ the effective excitation wavelength, $\lambda_i$ the measured wavelength. A is the geometric constant, which arises because of the
difference in incident and takeoff angles of the spectrometer. The
effective excitation wavelength for Ni K$\alpha$ is two thirds of the
absorption-edge value of nickel, i.e., 2/3 × 1.488 = 0.992 $\overset{o}{A}$ and
tables yield the following values:

$\mu Cr(\lambda_e) = 80$     $\mu Cr(Ni\ K\alpha) = 316$     $\alpha_{Cr/Ni} = 601$

$\mu Fe(\lambda_e) = 99$     $\mu Fe(Ni\ K\alpha) = 397$     $\alpha_{Fe/Ni} = 754$

$\mu Ni(\lambda_e) = 122$     $\mu Ni(Ni\ K\alpha) = 61$     $\alpha_{Ni/Ni} = 223$

using a value of A equal to 1.65. Note from Eq. (2.34) that the
constant A = $(\csc \psi_2)/(\csc \psi_1)$.

Inserting the $\alpha$ values in place of the $\mu$ values used in the
previous calculation, the range in total absorption is now 694.6
(6% Ni) to 673.4 (10% Ni) or a 7.6% change in absorption compared
with 9.2% using the secondary absorption only.

It will be appreciated that these predictions of possible
matrix effects have been done without any actual measurements at
all being made. In practice, one might confirm the prediction by
running a series of analyzed calibration standards on the spectrometer and visually inspecting the graphic relationship between x-ray
intensity and chemical composition.


9.4  PREPARATION OF THE CALIBRATION CURVE

9.4.1  Types of Calibration Curves

The most common method of calibrating the x-ray spectrometer is to
use a calibration curve in which a series of standard samples are
used to prepare a plot of counting rate against elemental concentra-

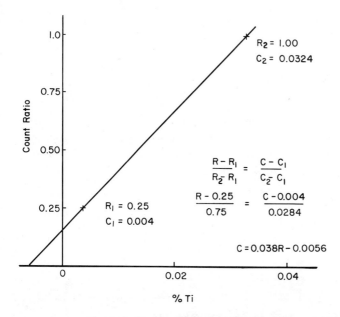

Figure 9.8   Derivation of the equation of a calibration line.

tion.   A "reasonable fit" of the data points to a straight line is usually taken as an indication of the absence of significant matrix effects and the curve is then used for analysis.   The equation of the line may be derived using standard methods as illustrated in Fig. 9.8.   Two points, $(X_1, Y_1)$ and $(X_2, Y_2)$, on the curve are selected and these data inserted in the equation

$$\frac{X - X_1}{X_2 - X_1} = \frac{Y - Y_1}{Y_2 - Y_1}$$

which is then simplified to the form $Y = mX + b$.   In this instance, the Y axis represents count ratio and the X axis, concentration. The derived equation can then be used for the calculation of other concentrations from measured count ratios.   Where some form of computational facility is available, it may be useful to employ regression analysis to fit the data to the best regression line (see also Sec. 5.5.1).

## 9.4.2  Regression Analysis

If the following straight-line relationship exists between the concentration C of an element giving intensity I and background B,

$$C = mI + B$$

and $\delta$ is taken as the difference between the actual value and the corresponding "true" value on the calibration curve.  For each of the n data points the value of $\delta$ is given by

$$\delta_i = mI_i + B - C_i \tag{9.8}$$

Equation (9.8) is raised to the second power to obtain the best least-squares fit:

$$\mathscr{S} = \sum_i \delta_i^2 = \sum_i (mI_i + B - C_i)^2 \tag{9.9}$$

The parameters m and B must be calculated for a minimum value of $\mathscr{S}$, and this is achieved by equating the first derivative of $\mathscr{S}$ to zero; thus

$$\frac{d\mathscr{S}}{dm} = \sum_i 2(mI_i + B - C_i)I_i = 0$$

and

$$\frac{d\mathscr{S}}{dB} = \sum_i 2(mI_i + B - C_i) = 0$$

which can be reduced to

$$m \sum_i I_i^2 + B \sum_i I_i - \sum_i C_i I_i = 0$$

and

$$m \sum_i I_i + nB - \sum_i C_i = 0$$

from which equations for m and B can be derived.

In the extension of this least-squares procedure to cover interelement correction terms, other partial differentials must be derived for each term required, i.e. of the form

ELEMENT 01   NI

| | CHEMICAL | CALCULATED | DIFFERENCE |
|---|---|---|---|
| 01 | 12.500 | 12.865 | -.365 |
| 02 | 9.600 | 9.402 | .198 |
| 03 | 10.800 | 10.691 | .109 |
| 04 | 14.000 | 14.275 | -.275 |
| 05 | 15.200 | 14.331 | .869 |
| 06 | 12.500 | 12.399 | .101 |
| 07 | 6.260 | 5.988 | .272 |
| 08 | 12.450 | 12.057 | .393 |
| 09 | 9.490 | 9.245 | .245 |
| 10 | 20.600 | 21.756 | -1.156 |
| 11 | 9.470 | 9.200 | .270 |
| 12 | 9.480 | 9.285 | .195 |
| 13 | 9.520 | 9.145 | .375 |
| 14 | 9.330 | 9.144 | .186 |
| 15 | .370 | .602 | -.232 |
| 16 | .370 | .606 | -.236 |
| 17 | .400 | .630 | -.230 |
| 18 | .400 | .631 | -.231 |
| 19 | .560 | .781 | -.221 |
| 20 | .560 | .768 | -.208 |
| 21 | 2.160 | 2.189 | -.029 |
| 22 | 2.160 | 2.191 | -.031 |
| SIGMA | | .40281 | |

| M(I) | B(I) |
|---|---|
| 12.56490 | .25612 |

$$\overset{*}{C}_{Ni} = I_{Ni} \cdot 12 \cdot 565 + 0 \cdot 256$$

Figure 9.9  Regression analysis output fitting nickel data to a straight line.

$$\frac{d\mathscr{S}}{dK_{ij}} = 0$$

As an example [8], the data given in Fig. 9.9 were obtained from a series of 22 nickel, iron, chromium standards with concentrations of nickel varying between 0.37 and 20.6%.  The listed data are in the form directly outputted from a regression program [8]. The column of figures listed under "chemical" are the chemical values of the standards entered in pairs with the measured x-ray intensities.  The constants indicated under M(I) and B(I) are the slope and background values obtained from the regression analysis yielding the equation for $C_{Ni}$.  The data listed under "calculated" are the calculated concentrations of nickel obtained using the given equations.  The last column of figures are the differences between chemical and x-ray values, and the value following "sigma"

```
*  ER
*  XC    1    NI  2
*  MC    1    2
*  DS    1    5
*  MP
```

ELEMENT  01   NI

|      | CHEMICAL | CALCULATED | DIFFERENCE |
|------|----------|------------|------------|
| 01*  | 12.50    | 12.98      | -.48       |
| 02   | 9.60     | 9.63       | -.03       |
| 03   | 10.80    | 10.86      | -.06       |
| 04   | 14.00    | 14.20      | -.20       |
| 05*  | 15.20    | 14.24      | .96        |
| 06   | 12.50    | 12.48      | .02        |
| 07   | 6.26     | 6.20       | .06        |
| 08   | 12.45    | 12.45      | .00        |
| 09   | 9.49     | 9.43       | .06        |
| 10   | 20.60    | 20.54      | .06        |
| 11   | 9.47     | 9.41       | .06        |
| 12   | 9.48     | 9.55       | -.07       |
| 13   | 9.52     | 9.38       | .14        |
| 14   | 9.33     | 9.34       | -.01       |
| 15   | .37      | .40        | -.03       |
| 16   | .37      | .40        | -.03       |
| 17   | .40      | .42        | -.02       |
| 18   | .40      | .42        | -.02       |
| 19   | .56      | .57        | -.01       |
| 20   | .56      | .56        | -.00       |
| 21   | 2.16     | 2.13       | .03        |
| 22   | 2.16     | 2.14       | .02        |

SIGMA              .07919

```
        M(I)              B(I)
     14.85360           .00872

        M(I,J)           B(I,J)
      -.93618
      -.99453
```

$$* \quad C_{ni} = I_{ni}\left[14.854 - 0.936 I_{ni} - 0.995 I_{cr}\right] + 0.009$$

Figure 9.10  Regression analysis output for nickel with corrections for nickel and chromium.

is the standard deviation of these differences.

The fit in the given example is not good and this is due to the influence of chromium and iron on the nickel.  In the given example, it is apparent from the spread of the data about the calibration line, as well as from the calculated standard deviation, that there is a significant matrix effect on the measured nickel intensity. The regression analysis was therefore repeated employing interelement correction, and the results of this are shown in Fig. 9.10. The instructions at the top of the figures are operator commands used in setting up the analysis.  As an example, DS means delete

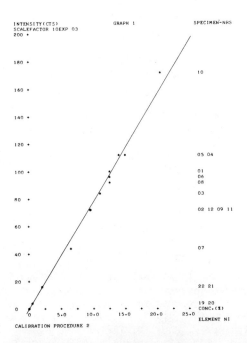

INTENSITY(CTS)                        GRAPH 1                    SPECIMEN-NBS
SCALEFACTOR 10EXP 03

CALIBRATION PROCEDURE 2

Figure 9.11   Regression analysis output data using the printer
terminal as a plotter.

standard--in this instance standards 1 and 5.   In fact, these data
are not used in the regression analysis, but the x-ray data is calculated and reported; however, an asterisk (*) is printed by the
sample number to remind the operator that these standards have not
been used.   As before, four columns of figures follow giving
sample number, chemical concentration, x-ray concentration, and
difference.   Since corrections were requested for Ni (1) and Cr (2),
the values of these correction factors are also printed out under
M(IJ).   The actual working equation is indicated on the bottom of
the sheet.

Comparison of the new standard deviation (0.0792) with the
original (0.4028) shows a fivefold improvement.   For the convenience
of the operator, a plot routine allows the printing of the results
in graphical form on the teleprinter, and as an example, Fig. 9.11

shows the output of the Ni data just given.  Intensities are given
on the ordinate with the appropriate scale factor (in this instance,
10 EXP 03, i.e., $10^3$), and concentrations are given on the abcissa.
In fact, a 50 × 50 network is used with a scale indication given at
intervals of 5 units.  The scale can be chosen by the operator by
instructions given through the teleprinter.  Each datum is given as
a star (*) and the corresponding specimen number is given at the
right-hand side of the page.  The actual calibration line is drawn
in by hand.

## REFERENCES

1. *X-Ray Fluorescence Spectrometry Abstracts, Vols. 1-6,* Science
   & Technology Agency Ltd., London, 1970-1975.
2. *Advances in X-Ray Analysis, Vols. 1-18,* Plenum, New York, 1957-
   1974.
3. *Analytical Chemistry Reviews,* published April of each even year.
4. R. Jenkins,  X-ray fluorescence analysis, *M.T.P. International
   Review of Science, Physical Chemistry Series One, 13:*95, Medical
   Technical Publishing Company, Oxford, 1973.
5. M. J. Owers and H. I. Shalgosky, *J. Phys. E, 1974:*593.
6. J. R. Rhodes, in *Energy Dispersion X-Ray Analysis,* ASTM Special
   Technical Publication STP485, ASTM, Philadelphia, 1970, p. 243.
7. D. A. Gedcke, *X-ray Spectrom., 1:*129 (1972).
8. R. Jenkins, J. de Klerck, and S. van Gelder, Philips Scientific
   and Analytical Equipment Bulletin No. FS31, Philips, Eindhoven,
   1964.

# 10

# Methods and Models for Quantitative Analysis

## 10.1 GENERAL

It has been shown in previous chapters that for "homogeneous speci-
ments," characteristic x-ray line intensity is subject to random
and systematic errors determined mainly by instrumental limitations
and artifacts, plus matrix effects arising from some or all of the
elements making up the analyzed specimens. It was also stated that
specimen heterogeneity problems are best reduced by adequate specimen preparation rather than in attempting to express the x-ray
intensity/concentration relationship in terms of particle size distribution and particle statistics. Whereas the previous chapter
was concerned mainly with the causes and prediction of random and
systematic errors in quantitative analysis, this chapter is concerned with *methods* of quantitative analysis, which to a large extent means the compensation for matrix interferences in homogeneous
specimens. It is thus assumed that in all cases the specimen has
been rendered homogeneous. It is also assumed that the magnitude of
random and systematic errors arising from the instrument are known
and are controllable within the range of accuracy required by the
quantitative method in question. Finally, it is assumed that random

errors due to counting statistics will not represent the limiting
factor in the ultimate accuracy of a given determination.


## 10.2   PROVISION AND CHOICE OF STANDARD REFERENCE MATERIALS

One of the greatest problems which the analyst is likely to encounter is the provision of standard reference materials, and this
is generally the major factor in the choice of a given quantitative
procedure.  It is, however, important to appreciate that the function of the standard reference material is really twofold.  First,
it is required to determine the sensitivity (i.e., the instrument
response per unit change in concentration) of the spectrometer for a
given selected wavelength or energy, i.e., element.  This use is
really practical convenience since although it is, in principle,
possible to calculate the x-ray photon yield from an element under a
given set of experimental conditions, in practice, it is far easier
to use a single reference standard to actually measure the counts
per second per percent given by the spectrometer.  Provided nothing
changes between the measurement of the standard and a given unknown
specimen, it can then be assumed that the only difference in sensitivity of the spectrometer for the two measurements is that due to
matrix effects.  In this context, the standard reference material is
referred to as an *instrument standard* since its sole function is to
determine instrument sensitivity.  Indeed, many commercially available spectrometers allow the automatic ratioing of x-ray line intensities for unknown specimen and instrument standard.  In principle at least, only one instrument standard is required for a given
range of elements in a given matrix type.

The second function of the standard reference material is to
provide the means of establishing the relationship between x-ray
intensity and chemical composition, either for the evaluation of
potential matrix effects, or for the provision of a working curve.
In this instance, the function of the standard reference material is
to act as a "calibration standard" and many such standards may be

required to adequately cover the elemental concentration ranges required.  Since, in practice, the required standards may not be
available, many quantitative methods have been devised to minimize
the need for standard reference materials.


## 10.3  TYPES OF QUANTITATIVE METHODS FOR THE ANALYSIS OF HOMOGENEOUS MATERIALS

There is a wide diversity of quantitative methods available to the
x-ray spectroscopist, and it is convenient to break these down into
four major sections.  Table 10.1 indicates these major sections,
and further lists individual procedures within a given section.
Also shown in the table are the matrix effects for which the technique provides adequate correction, plus an indication of the number
of *calibration* reference standards required and the range of applicability of the technique.  A cross in parentheses indicates partial
overcoming of the indicated effect.  It will be appreciated that
such a table represents a broad generalization and should only be
considered as a rough guide of the potential applicability of a
given procedure.  Again it should be noted that only the dilution
techniques remove problems due specifically to heterogeneity, and
in all other cases a sample is assumed to be homogeneous.

For many years, the greater portion of quantitative x-ray
spectrometry was carried out using "in-type" analysis in which
calibration standards are used to set up calibration curves and
these curves then used to estimate chemical composition from measured x-ray characteristic line intensities from unknown samples.
This method was essentially borrowed from the ultraviolet emission
spectroscopist; the latter technique was well established at the
time of the development of x-ray fluorescence methods.  The in-type
method is still widely employed today, even with its inherent
limitations and need for many calibration standards, and its great
popularity probably stems from its ease of application and minimum
requirement for computational facilities.  More recently, a better

Table 10.1  Types of Matrix Correction Procedure

| Treatment of Matrix Effect | Corrects for | | | Calibration Reference Standards Needed | Concentration Covered |
|---|---|---|---|---|---|
| | Absorption | Enhancement | Heterogeneity | | |
| Ignoring | | | | | |
| 1. In type | -- | -- | -- | Many | Narrow |
| 2. Thin film | -- | -- | -- | Few | Moderate |
| Minimizing | | | | | |
| 1. Dilution | (x) | (x) | (x) | Few | Moderate |
| 2. Double dilution | x | x | x | Few | Wide |
| Compensating for using | | | | | |
| 1. Scattered source lines | x | -- | -- | None[a] | Narrow |
| 2. Mass absorption coefficients | x | -- | -- | None[a] | Moderate |
| Correction for based on | | | | | |
| 1. Fundamental parameters | x | x | -- | None[a] | Wide |
| 2. Calculated $\alpha$-coefficients | x | x | -- | Some | Wide |
| 3. Effective wavelength | x | -- | -- | None[a] | Moderate |
| 4. Measurement of $\mu$ | x | x | -- | None[a] | Moderate |
| 5. $\Delta C$ from regression | x | x | -- | Many | Wide |
| 6. $\Delta I$ from regression | x | x | -- | Many | Moderate |

[a]Standards may be required for the establishment of instrument sensitivity and tube spectrum.

widespread understanding of x-ray physics, plus the advent of the
cheap, high-speed minicomputer, is resulting in a much broader
application of the more versatile mathematical correction procedures.

At this stage of development of x-ray spectrometry, there is a
fair degree of agreement as to the form of the basic mathematical
relationship between the intensity of a characteristic x-ray wavelength and chemical composition. Early work by Sherman [1] described the basic form of this relationship and Sherman's work has
been more recently extended by Shiraiwa and Fujino [2-3] who found
that a factor of 1/2 had been left out of the Sherman equation.
The currently employed form of the "fundamental relationship" has
been given in Table 2.2 and such a relationship allows for the excitation of the analyte element by continuous radiation from a broad
spectrum of wavelength, or energies, with correction for primary and
secondary absorption. Additional expressions allow correction for
enhancement and third element effects [2,3]. Although it is, in
principle, possible to utilize the fundamental equation for the
solution of matrix effects and the provision of quantitative analysis without calibration standards, there are four major problems to
overcome before this ultimate goal can be achieved. First, the
algorithm which must be employed is obviously unwieldy and requires
a fair degree of computation for the provision of a solution, in
addition to requiring much computer storage space for the fundamental constants. Something of the order of 60,000 words of core space
would be required in a reasonably sophisticated computer for the
program in Fortran. Second, the accuracy of many of the fundamental
constants is currently no better than ±5%, and although these data
are being steadily improved [4], there is still a fair way to go
before satisfactory accuracy (approximately ±1%) can be obtained for
the whole wavelength range. A third problem is that the integral
form of the continuous exciting radiation remains somewhat unwieldy.
Effects of anode self-absorption, window absorption, and contamination make calculation of the x-ray tube spectrum from first principles difficult and unreliable, and a separate experiment is

generally used to measure the spectral distribution rather than to
calculate it [5].  Fourth, the models proposed to date ignore
multiple scattering.  These effects become important at high energies in low-atomic-number matrices where coherent and incoherent
scattering are significant relative to photoelectric absorption.
As a result of these four difficulties, the use of so-called fundamental methods for quantitative  analysis remains somewhat limited
and semifundamental, and purely empirical methods still enjoy the
greater popularity.

The various mathematical correction procedures form the bulk of
this particular chapter and are classified under type-four methods.
Also covered in the following sections are the more commonly employed alternative methods of quantitative analysis listed in
Table 10.1.


## 10.4  TYPE-ONE METHODS:  IGNORING MATRIX EFFECTS

### 10.4.1  In-Type Analysis

Equation (9.6) shows that a linear relationship will exist between
characteristic x-ray line intensity and elemental concentration,
provided that there is no significant change in total absorption
or enhancement over the range of the calibration line.  Since the
total absorption range is itself dependent upon the concentration
range, it is obvious that by limiting the calibration range *of all
interfering elements,* one can limit the absorption range.  In other
words, by matching the matrix of the calibration standard with that
of the unknown specimen, a direct correlation can be made between
measured x-ray intensity and elemental composition.  This is the
basis of so-called in-type or type-matching analysis, which is by
far the most commonly employed method of quantitative analysis.
In principle, this technique offers a powerful and almost univer-
sally applicable means of elemental analysis; but in practice,
there are three potentially major problems that should always be
considered.

Table 10.2   Absorption Data for Copper, Zinc, Lead, and Tin Alloys

| Analyte Element | Cu | Zn | Sn | Pb |
|---|---|---|---|---|
| Analyte wavelength $\lambda_i$ (Å) | 1.54 | 1.44 | 0.49 | 1.18 |
| Effective wavelength $\lambda$ (Å) | 0.92 | 0.86 | 0.28 | 0.66 |
| Total absorption[a] $\alpha_j$ for each matrix element j | | | | |
| j = Cu | 186 | 151 | 31.4 | 358 |
| j = Zn | 206 | 166 | 34.3 | 393 |
| j = Sn | 529 | 434 | 34.3 | 240 |
| j = Pb | 590 | 482 | 87.4 | 274 |

[a] $\alpha_j = \mu_j(\lambda) + A \cdot \mu_j(\lambda_i)$ with A taken as 1.65.

First, it must be realized that as far as absorption effects are concerned, it is the absorption range that limits the working curve. As an example, for the determination of Cu in the system Cu, Zn, Pb, Sn, the total absorption values are listed in Table 10.2. Assuming a reference standard of composition 60% Cu, 30% Zn, 5% Sn, and 5% Pb, the total absorption for Cu $K\alpha$ would be 229, made up of 112 from Cu and 117 from the rest of the matrix. Provided that the relative concentrations of Zn, Pb, and Sn remained the same, i.e., giving an $\alpha$ (remainder) equal to

$$(0.75 \times 206) + (0.125 \times 529) + (0.125 \times 590) = 294$$

the general expression for the total sample absorption $A_{Cu}$ for Cu $K\alpha$ would be

$$\begin{aligned} A_{Cu} &= 186W + 294(1 - W) \\ &= 294 - 108W \end{aligned}$$

where W is the weight fraction of copper.   Alternatively,

$$A_{Cu} = 294 - 1.08 \, C_{Cu} \qquad (10.1)$$

where $C_{Cu}$ is the weight percentage of copper.   A similar exercise for lead in the same matrix yields

$$A_{Pb} = 345 - 0.71 \, C_{Pb} \qquad (10.2)$$

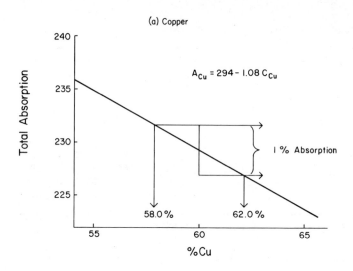

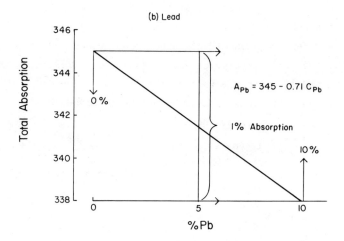

Figure 10.1   Effect of absorption range on the accuracy of a
measurement of concentration.

Figure 10.1 shows graphical representations of Eqs. (10.1) and
(10.2) for copper and lead, respectively.  It is possible to use
these curves to predict acceptable calibration ranges for given
absorption variations.  For example, an absorption variation of 1%,
yielding an accuracy of 1%, would allow the range 58 to 62% copper

Table 10.3   Calibration Standards for Tin

---

a.  Poor calibration standards for tin (concentrations in percent)

| Cu | Zn | Sn | Pb | (Sn Kα) |
|----|----|----|----|---------|
| 60 | Remainder | 0 | 0 | 32.6 |
| 60 | Remainder | 2 | 2 | 33.6 |
| 60 | Remainder | 4 | 4 | 34.7 |
| 60 | Remainder | 6 | 6 | 35.7 |
| 60 | Remainder | 8 | 8 | 36.8 |
| 60 | Remainder | 10 | 10 | 37.9 |

b.  Ideal calibration standards for tin (concentrations in percent)

| Cu | Zn | Sn | Pb | (Sn Kα) |
|----|----|----|----|---------|
| 70 | Remainder | 0 | 8 | 36.5 |
| 60 | Remainder | 2 | 4 | 34.7 |
| 70 | Remainder | 4 | 6 | 35.5 |
| 60 | Remainder | 6 | 0 | 32.6 |
| 65 | Remainder | 8 | 10 | 37.7 |
| 65 | Remainder | 10 | 2 | 33.5 |

---

to be covered.  Similarly, if an accuracy of ±0.2% were sought, the corresponding ±0.2% absorption would allow a range between 59.6 and 60.5% copper.  Note, however, that the same ±1% in absorption for lead would allow a calibration range between 0 and 10% of lead.

A second potential problem source in in-type analysis lies in the selection of the calibration standards, since it is necessary that the calibration standards reflect not only the analyte range but also the range of all potential absorbers (and enhancers). Consider again the case of the Cu, Zn, Sn, Pb alloys discussed previously.  Let us assume that it is necessary to construct a calibration curve for tin over the concentration range 0 to 10%.  One set of commercially available standards have the approximate compositions listed in Table 10.3, part a.  At first sight, these standards might appear ideal, and indeed count data for the intensity of Sn Kα plot on a smooth curve (see Fig. 10.2).  Curve B would be predicted from the α values.  A smooth curve of decreasing slope might also seem reasonable at first sight since tin is a rather heavy element relative to the dominant matrix elements copper and zinc, and this appears to be a typical case of self-absorption.

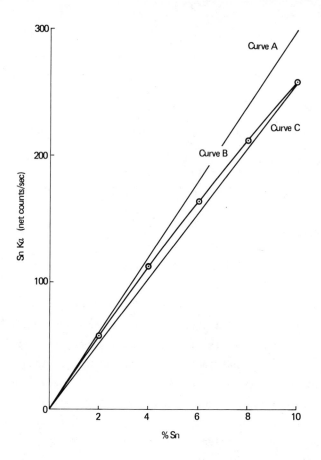

Figure 10.2   Calibration curves for Sn Kα.

However, close inspection of the individual absorption data shown
in Table 10.2 reveals that whereas zinc and tin have identical absorption values for Sn Kα with the value for copper very similar,
the absorption value for lead is 2 1/2 times greater.  The curvature
in the tin calibration graph is in fact not due to tin at all, but
to lead, the concentration of which increases in the calibration
standards at the same rate as tin.

    Due mainly to the fact that there is a strong counteracting
effect of primary absorption in tin, the curve for tin in Cu/Zn is
an almost straight line as shown in curve A, Fig. 10.2.  Curve A

thus represents the line for tin in Cu/Zn at 0% lead.   Curve C
represents a similar line for tin in Cu/Zn to which 10% Pb has been
added.   Hence, in a matrix of Cu/Zn with lead varying from 0 to 10%,
all tin data would actually fall within the wide spread between the
curves A and C.   The influence of this on the estimation of tin concentration from curve B would be disastrous in those cases where
the lead concentration also varied.   A better selection of calibration standards would have quickly revealed this problem and such a
selection is shown in Table 10.3, part b.

The third potential problem that may be encountered involves
the actual calibration curve obtained.   The check that is generally
employed to test whether in-type analysis can be used is to plot
characteristic x-ray intensity as a function of elemental concentration.   A straight line with an acceptable deviation of data points
is generally taken as a guarantee that the method is satisfactory.
What often happens, however, is that the curve is not straight, and
the deviation of data points is not acceptable.   The analyst is now
faced with a real problem since the deviation may be due to one or
more of the following causes:

1.   Insufficient precision in the collection of count data
2.   Insufficient accuracy in the provided analyses of the
     standards
3.   Absorption or enhancement problems in the sample matrices
4.   Heterogeneity problems

An analytical procedure should never be rejected unless all four of
these error sources has been properly investigated.

## 10.4.2   Thin-Film Analysis

From Eq. (2.29) and Table 2.2 the intensity $I_i(\lambda_i)$ for a wavelength
$\lambda_i$ excited by a single wavelength $\lambda_o$ from a specimen of finite
thickness T can be written as

$$I_i(\lambda_i) = \frac{G_i Q_{if}(\lambda_o) I_o(\lambda_o)}{\alpha_T(\lambda_i)} \left(1 - e^{-\alpha_T(\lambda_i)\rho T}\right) \tag{10.3}$$

In this equation, $G_i$ is a geometric constant including solid-angle
terms, $Q_{if}(\lambda_o)$ the excitation probability, and $\alpha_T(\lambda_i)$ the total

absorption for $\lambda_i$ as given in Eq. (2.34). T is the specimen thickness and $\rho$ the density. If the weight fraction $W_i$ is separated from the excitation probability term and all constant terms collected into one constant $K_i$, Eq. (10.3) becomes

$$I_i(\lambda_i) = \frac{K_i W_i}{\alpha_T(\lambda_i)} \ (1 - \exp[-\alpha_T(\lambda_i)\rho T]) \tag{10.4}$$

It can be shown that as $\alpha_T(\lambda_i)\rho T$ becomes small, then the term $\exp[-\alpha_T(\lambda_i)\rho T]$ tends to $1 - \alpha_T(\lambda_i)\rho T$. Substituting in Eq. (10.4) gives

$$I_i(\lambda_i) = K_i \ W_i \rho T \tag{10.5}$$

This equation is independent of matrix absorption and enhancement terms and can be conveniently employed for thin-film analysis since the relationship between $I_i(\lambda_i)$ and $W_i$ is dependent only upon the mass thickness $\rho T$ of the specimen. The method can be directly applied to existing thin films such as paints [6], or alternatively, thin films can be deliberately contrived by evaporating specimen solutions onto plastic films [7]. Excellent calibration curves can be obtained provided that the limitations of Eq. (10.5) are appreciated.

Chung and his co-workers [8] have pointed out that the percent error in the approximation $\exp(-\alpha\rho T) = 1 - \alpha\rho T$ is of the form

$$\%Error = [1 - (1 - \alpha\rho T) \exp \alpha\rho T] \times 100 \tag{10.6}$$

Typical values for this error are given in Table 10.4 along with combinations of $\alpha$, $\rho$, and T, giving an introduced error of 1%. It should be noted that for $\alpha$ values falling within the typical range of 100 to 500 $cm^2/g$ with a density of 2-5 $g \cdot cm^3$; thickness values lie in the range of 1 to 10 $\mu m$.

Table 10.4  Errors Introduced in the Approximation
$\exp(-\alpha\rho T) = 1 - \alpha\rho T$ (upper), and Combinations of $\alpha\rho T$
Giving an Introduced Error of 1%, i.e., $\alpha\rho T = 0.135$
(lower)

| $\alpha\rho T$ | % Error |
|---|---|
| 0.01 | 0.005 |
| 0.1 | 0.535 |
| 0.135 | 1.00 |
| 0.2 | 2.29 |
| 0.5 | 17.6 |
| 0.75 | 47.1 |
| 1.00 | 100 |

| $\alpha(cm^2/g)$ | $\rho(g \cdot cm^3)$ | $T(\mu m)$ |
|---|---|---|
| 100 | 1 | 13.5 |
| 200 | 5 | 1.35 |
| 10 | 4 | 33.8 |
| 250 | 2 | 2.7 |
| 500 | 8 | 0.34 |

## 10.5  TYPE-TWO METHODS:  MINIMIZATION OF MATRIX EFFECTS

### 10.5.1  Dilution Methods

The point has been made in previous sections that the major cause
of interelement interactions is that due to variations in the total
composition of the specimens being analyzed.  It has also been shown
that in-type analysis is based on narrowing down the calibration
range so that the absorption range is also decreased.  It is similarly possible to reduce the absorption range by dilution of the
specimen.  As an example, if one were analyzing a specimen of potassium carbonate for the presence of calcium carbonate, a large absorption effect of the potassium on the Ca K$\alpha$ line would be observed.
The total absorptions of $CaCO_3$ and $K_2CO_3$ for Ca K$\alpha$ radiation are 383
and 1381, respectively.  Over a calibration range of 0 to 10% $CaCO_3$,

i.e., 0 to 4% calcium, the change in total absorption would be just over 7%. By diluting a specimen of absorption S with an absorber of total absorption A, the total absorption D of the diluted specimen would be

$$D = WS + (1 - W)A \qquad\qquad\qquad (10.7)$$

where W is the weight fraction of the original specimen in the diluted specimen. Thus by diluting the specimens with sodium tetraborate with an A value of 238, a 9:1 absorber to sample dilution ratio would reduce the absorption range down to 2.8%. Similarly, a 49:1 absorber to sample dilution ratio would reduce the absorption range to 0.77%. Diluting the samples in this way would of course reduce the number of calcium atoms per unit mass of diluted specimen, in fact, from 4 to 0.08% in the example cited. Fortunately, however, the counting rate from Ca Kα would not drop off in the same proportion because as the absorption of the sample decreases, the penetration of the Ca Kα line increases and more specimen volume contributes to the measured signal. Again, in the example quoted, the penetration of Ca Kα is 3.1 μm, 11.7 μm, and 15.3 μm, respectively. In practice, it is found that the characteristic line intensity drops off as the square root of the dilution ratio.

Dilution methods are often used for the dual purpose of producing a homogeneous specimen as well as reducing matrix effects. In the given case, the sample could have been fused with borax to give a glass disk or even dissolved in dilute hydrochloric acid to give a calcium-containing solution. Either of these methods would have obviated possible heterogeneity problems.

## 10.5.2  Double-Dilution Methods

In the simple case where only primary and secondary absorption effects are significant and third element and enhancement effects can be ignored, there is a simple relationship between the weight fraction and the measured characteristic line intensity:

$$I_i^x = \frac{K_i W_i^x}{\alpha_i^x} \tag{10.8}$$

where $K_i$ is a constant under a given set of experimental conditions. Where a standard contains a known weight fraction $W_i^s$ of the same element i, has a total absorption $\alpha_i^s$ and gives a characteristic line intensity $I_i^s$ a similar expression holds:

$$I_i^s = \frac{K_i W_i^s}{\alpha_i^s} \tag{10.9}$$

Dividing Eq. (10.8) by Eq. (10.9) eliminates $K_i$ and gives

$$\frac{I_i^x}{I_i^s} = \frac{W_i^x \alpha_i^s}{W_i^s \alpha_i^x} \tag{10.10}$$

Equation (10.10) contains three ratio terms, an intensity ratio, a weight fraction ratio, and an absorption ratio. A single experiment giving values of $I_i^x$ would not allow solution for $W_i^x$ even though $W_i^s$ were known since the absorption ratio term is unknown. However, it is possible to vary the relationship between I and W by changing $\alpha$, for example, by diluting the specimen with an absorber using a known dilution ratio. A pair of such dilutions would yield two simultaneous equations in $\alpha$ and W in which $\alpha$ could be eliminated. This is the basis of the double-dilution technique proposed by Tertian [9,10] and offers a powerful technique for the analysis of mixtures for single elements where no standards are available. One useful expression for the application of the double-dilution method is

$$\frac{W_i^x}{W_i^s} = \frac{[(I_i^x)_1 \times (I_i^x)_2]/[(I_i^x)_2 - (I_i^x)_1]}{[(I_i^s)_1 \times (I_i^s)_2]/[(I_i^s)_2 - (I_i^s)_1]} \tag{10.11}$$

in which $W_i^x$ and $W_i^s$ are the weight fractions of element i in the unknown sample and standard, respectively, $(I_i^x)_1$ and $(I_i^x)_2$, are the

Table 10.5  Data Obtained on a Mixture of $ZnO^a$ and $Fe_2O_3{}^b$ Using the Double-Dilution Method

| Run No. | Mixture | Counts/s (Zn Kα) | Counts/s (Fe Kα) |
|---------|---------|------------------|------------------|
| 1 | 0.35 g unknown<br>6.65 g borax | AX1 = 9012 | BX1 = 11084 |
| 2 | 0.70 g unknown<br>6.30 g borax | AX2 = 13527 | BX2 = 20384 |
| 3 | 0.35 ZnO<br>6.65 g borax | AS1 = 13368 | -- |
| 4 | 0.70 g ZnO<br>6.30 g borax | AS2 = 20773 | -- |
| 5 | 0.35 g Fe₂O₃<br>6.65 g borax | -- | BS1 = 35897 |
| 6 | 0.70 g Fe₂O₃<br>6.30 g borax | -- | BS2 = 62063 |

$^a$For ZnO, WAX = [(AX1 × AX2)/(AX2 - AX1)]/[(AS1 × AS2) /(AS2 - AS1)] × 100 = 72.0% ZnO.

$^b$For $Fe_2O_3$, WBX = [(BX1 × BX2)/(BX2 - BX1)]/[(BS1 × BS2)/(BS2 - BS1)] × 100 = 28.5% $Fe_2O_3$.

count rates obtained from element i in the unknown sample at dilution 1 and dilution 2, and $(I_i^s)_1$ and $(I_i^s)_2$ are two count rates similarly obtained on the standard.  It is common practice to try to use the same two sets of absolute weights of unknown/diluent and standard/diluent since this simplifies the arithmetic.  If exactly the same weights are not used, the left-hand side of Eq. (10.11) must be adjusted accordingly.

Table 10.5 lists a set of data obtained by a group of relatively inexperienced students at an x-ray school who were using the double-dilution method for the first time.  The unknown mixture was 72% ZnO and 28% $Fe_2O_3$.  One group of students were asked to determine the zinc using pure ZnO as a reference, and the other to do iron using pure $Fe_2O_3$ as a reference.  Sodium tetraborate was the diluent, and the samples were made into beads by fusion.  Extremely good results were obtained even though this elemental combination shows a very marked interelement interaction.

Two pitfalls must always be avoided in this otherwise generally applicable technique.  First, the dilution ratios must be chosen carefully since, as will be seen from Eq. (10.11), count-rate-difference terms occur in the relation.  If the absorption difference is small, then the intensity difference will be small, and the counting error in the intensity difference will be large.  On the other hand, if the dilution ratio difference is too large, the x-ray intensity from the now very dilute lower dilution will itself be so low that the count error may be significant.

A second problem to avoid is that of deadtime.  Where equipment without automatic deadtime correction is employed, a large error may accrue since measurement pairs are all being done at low and high count rates, respectively.  Thus, the high count rate (i.e., the lower dilution) will have a greater count loss than the low count rate (i.e., the higher dilution ratio).  It is thus essential to correct all of the count data for deadtime.

## 10.6  TYPE-THREE METHODS:  COMPENSATION FOR MATRIX EFFECTS

It is sometimes possible to utilize x-ray information from the analyzed specimen, over and above characteristic line intensities, to correct for possible matrix interferences.  Methods for doing this fall into two broad categories, namely use of scattered lines and use of internal standards.  In the first of these techniques, use is made of the fact that the primary radiation not only excites characteristic lines from the specimen but is also scattered.  Since the degree and type of scatter is matrix dependent, a rough and ready characterization of the matrix can be made by study of the scattering characteristics of the specimen.  Such a method of matrix "compensation" is particularly useful in that it requires no special sample treatment or large numbers of calibration standards for its application, and further, once a correction model has been established, only one or two additional counting measurements are required to yield the scattering data.

Where internal standards are used, an additional element may be added which is affected in the same way by the specimen matrix as the analyte element, thus yielding an intensity ratio of analyte wavelength intensity to internal standard wavelength intensity, which is matrix independent. Although this method may sound attractive at first sight, in practice it is unwieldy and time consuming since a separate internal standard may have to be chosen for each analyzed element. In addition to this, the ideal internal standard element may not be available in a convenient form and even if it is, may itself cause further matrix complications. There is, however, one special case of the use of internal standards which is utilized extensively, and this is the case where the analyte element itself is used as the internal standard, using a technique called *spiking*. This technique is especially useful for the preparation of hard-to-come-by secondary standards such as trace heavy elements in whole rock analysis.

## 10.6.1 Use of Scattered Source Lines

The technique of using scattered radiation as an internal standard was one of the first matrix correction procedures to be employed in quantitative analysis [11]. Although it is one of the oldest techniques, it is probably the least understood, which probably explains why it has not found more widespread application. The basis of the model is that since the fluorescence intensity $I_f$ is inversely proportional to the total matrix absorption $\alpha_T$ and since the mass absorption coefficient is approximately proportional to $Z^4$, it follows that $I_f$ is proportional to $Z^{-4}$. Similarly, since the sum $I_s$ of the coherent and incoherent scatter is also inversely proportional to $\alpha_T$ and since coherent scatter is roughly proportional to $Z^2$, it follows that if the scattering is mainly coherent, $I_s/I_f$ is approximately proportional to $Z^2$. This ratio is thus far less matrix dependent than is $I_f$ alone. Where incoherent scatter is significant, the ratio of $I_s/I_f$ is proportional to somewhere between the first and second power of Z.

The original Anderman model [11] did not incorporate measurement at a specific wavelength, and since this obviously plays an
important role in the scattering process, the application of matrix
compensation using scattered line intensities has proceeded along
rather empirical lines.  Most of the scattering theory is based on
an extension of the isolated atom model (see also Sec. 2.4) and as
such falls far short of providing an adequate mathematical form.
Molecular scattering effects in the form of diffraction also come
into play, as do instrumental artifacts such as the higher order
scattering mentioned in Sec. 2.12.  More recent attempts [12,13] to
rationalize the previous explanations have produced more complex
correction algorithms and have to date, still failed to provide
correction models yielding a generally acceptable accuracy.  The
scattered line method thus must still be regarded as a somewhat
empirical technique but nevertheless extremely useful under specific
circumstances.

A good example of the use of scattered source lines to correct
for matrix problems is the work of Burkhalter [14] in the analysis
of silver ores.  In this particular application, it was necessary to
determine silver in the concentration range of 8 to 80 ppm in a wide
range of matrices which were predominantly silicaeous but with varying amounts of Fe, Zn, Ba, and Pb.  In order to evaluate the potential effects of the matrix elements on the Ag K$\alpha$ intensity, a series
of silver standards were prepared, each containing 0.1% Ag, but with
the addition of 5% of each of the major interfering elements.
Table 10.6 lists the composition of the standards along with the observed counting rates on the Ag K$\alpha$.  It will be clearly seen that
these counting rates vary by more than a factor of 2 for standards
of identical silver content.  Also shown in the table is the intensity of the incoherently scattered line from the excitation source,
which in this case happened to be an $I^{125}$ radioisotope emitting Te K
radiation.  Again the intensities of the Compton scattered line are
very variable, but the ratio of the Ag K$\alpha$ to Compton scatter intensities is reasonably constant, in fact varying by ± 6%.  By employing

Table 10.6   Intensity Data Taken on Silver Standards

| Sample Matrix | $I_{Ag\ K\alpha}$ (counts/s) | $I_{Compton}$ (counts/s) | $I_{Ag\ K\alpha}/I_{Compton}$ |
|---|---|---|---|
| $SiO_2$ | 200.0 | 1890 | 0.106 |
| $SiO_2$ + 5% Fe | 150.0 | 1420 | 0.106 |
| $SiO_2$ + 5% Zr | 90.1 | 454 | 0.106 |
| $SiO_2$ + 5% Ba | 139.0 | 1440 | 0.096 |
| $SiO_2$ + 5% Pb | 83.8 | 754 | 0.111 |

a quantitative method in which he considered the ratio of the Ag Kα line to the Compton scattered line, Burkhalter was able to determine silver in variable matrix ores to within ± 20%.

10.6.2   Use of Measured Mass Absorption Coefficients

It will be seen from Eq. (9.6) that for homogeneous specimens, in those cases where enhancement effects are negligible, the counting rate obtained for a given concentration of an element in a certain matrix is inversely proportional to the total absorption of that matrix.  It has also been indicated that the secondary absorption term is usually the predominant effect in the total absorption of a specimen; thus it would seem feasible that if some means existed of measuring or estimating the mass absorption coefficient of a specimen for a given wavelength or energy, an absorption correction could be applied.  If the mass absorption coefficients of the specimen for the characteristic radiations are unknown (as will generally be the case), they can be measured by inserting a known thickness of the specimen between the specimen and detector.  The mass absorption coefficient μ is calculated from the Beer-Lambert law

$$I = I_o \exp(-\mu\rho x) \qquad\qquad\qquad (10.12)$$

where I is the counting rate measured with the absorption specimen in place, and $I_o$ is the measured rate after removal.  The mass thickness ρx is computed by dividing the weight of the absorption specimen by its measured area.  The intensity measurements I and $I_o$

are made on the analyte line with the regular specimen in the normal specimen position.  In those cases where the specimen gives
only a small number of characteristic x-ray photons, e.g., in the
case of trace analysis, an alternative specimen giving the required
characteristic line at a reasonable intensity can be used in the
normal specimen position.  This method works well only for small
values of $\mu \rho x$ where a reasonably high counting rate I can be obtained.  This means high-energy lines (short wavelengths), thin
absorption specimens, and low-atomic-number matrices.  In practice,
the intensities I and $I_o$ are measured as the number of counts acquired in a fixed time t, corrected for deadtime losses.  That is,
the number of counts

$$N_o = I_o t$$
and   N = It

are measured.  The mass absorption coefficient is computed from

$$\mu = \frac{\ln N_o - \ln N}{\rho x} \qquad (10.13)$$

A simple error analysis shows that the percent error in the mass
absorption coefficient will be

$$\frac{\sigma_\mu}{\mu} \times 100\% = 100\% \left[ \left(\frac{\sigma_{\rho x}}{\rho x}\right)^2 + \frac{1}{N_o (\ln N_o - \ln N)^2} \right.$$

$$\left. + \frac{1}{N(\ln N_o - \ln N)^2} \right]^{1/2} \qquad (10.14)$$

where $\sigma_\mu$ is the predicted standard deviation in $\mu$, $\sigma_{\rho x}$ is the predicted standard deviation in the determination of $\rho x$, and the
second and third terms in the square brackets are the predicted
standard deviations due to the measurement of $N_o$ and N, respectively.
It is not too difficult to achieve a 1% accuracy in the measurement
of $\rho x$.  However, if the error contributions of the second and third
terms are each to be less than 1%, it requires

Technique                               Experiment

A
Transparent Sample

B
Dillution
Thick Sample

Sample

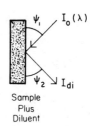

Sample
Plus
Diluent

C
Element Behind
Thin Sample

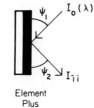

Element

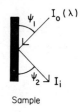

Element
Plus
Sample

Sample

Figure 10.3   Measurement of mass absorption coefficients.

$$N > \frac{10^4}{[\ln(N_0/N)]^2} = \frac{10^4}{(\mu\rho x)^2}$$

For small values of $\mu\rho x$, N is not much different from $N_0$, and a large number of counts are required to measure the difference with precision. For example, for $\mu\rho x = 0.1$ more than $10^6$ counts must be accumulated for N. The corresponding $N_0/N = 1.11$ indicates that the counting rate conditions can be simultaneously optimized for both N and $N_0$.

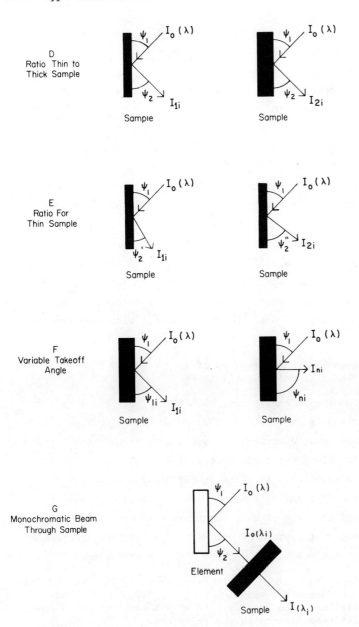

D
Ratio Thin to
Thick Sample

E
Ratio For
Thin Sample

F
Variable Takeoff
Angle

G
Monochromatic Beam
Through Sample

For large values of $\mu\rho x$ the required counting time to determine N become excessive.  For example, for $\mu\rho x$ = 15, the required counts for N is 44 and the ratio $N_o/N$ = 3.3 × $10^6$.  If the spectro-

meter is set up to count the higher rate $N_o/t$ at $10^4$ counts/s, it
will require 4 h to acquire 44 counts for N. A small fraction of
this time is sufficient to measure the rate $N_o/t$ with sufficient
precision to compute $N_o$. From this calculation it is obvious why
the method is suitable only for low values of $\mu\rho x$. This is one of
the major reasons why the accuracy of tabulated mass absorption
coefficients for high-atomic-number elements at long wavelengths is
poor.

     The use of measured secondary absorption coefficients has been
used with considerable success [16] particularly in the area of
whole rock analysis [17].

     Although it is possible, in principle at least, to measure the
primary absorption in a similar way by the positioning of a specimen between source and specimen, the close coupling in most commercially available spectrometers, plus the fact that the primary beam
is generally divergent, make the use of such a procedure impracticable.

10.6.3  Methods for the Correction of Absorption by Use
        of Mass Absorption Coefficients

Measurement of the secondary mass absorption coefficient by the
method described in Sec. 10.6.2 is by no means the only way, and
several authors have described the various techniques which are
available [18,19]. Sparks and Ogle [19] have summarized these
methods diagrammatically as shown in Fig. 10.3. Figure 10.4 indicates the errors in the concentration $\Delta C/C$ due to errors in the
intensity $\Delta I/I$ for each of the techniques. In each of the seven
methods a beam of radiation of intensity $I_o(\lambda)$ strikes the specimen
at an average angle $\psi_1$. The methods are then as follows:

     A.  Transparent sample: This is essentially the same as the
         method described previously in Sec. 10.4.2 for thin-film
         analysis. Note from Fig. 10.4 that this method is the
         least sensitive to intensity errors.
     B.  Dilution of a thick sample: This may be the single- or
         double-dilution technique described in Secs. 10.5.1 and
         10.5.2.

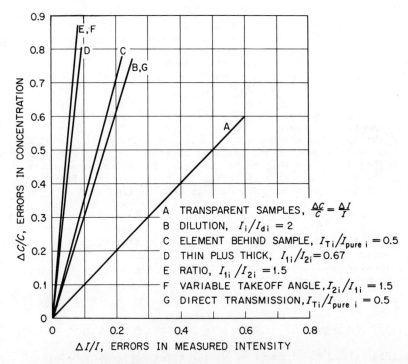

Figure 10.4   $\Delta I/I$ errors in measured intensity.

C.  Element behind thin sample:  In this instance, the specimen is mounted as a thin film on a substrate of a pure
element i.  The concentration $C_i$ of the element in the
specimen is then given by

$$C_i = \frac{I_i Q_i (\sin \psi) \ln(I_{T,i}/I_{pure\ i})}{\rho_s T(I_{T,i}/I_{pure\ i} - 1)} \qquad (10.15)$$

This method has been employed with some success by Giauque
and Jacklevic [20].

D.  Ratio of thin sample to thick sample:  In this case, the
intensities of characteristic lines from thin $I_{1i}$ and thick
$I_{2i}$ samples are employed.  The concentration of the analyte
element is given by

$$C_i = \frac{I_{2i} Q_i (\sin \psi_1)[-\ln(1 - I_{1i}/I_{2i})]}{\rho_s T} \qquad (10.16)$$

E.  Ratio for thin sample:  In this particular case, a thin
sample is again used, but the intensity of element i is

measured at two different takeoff angles $\psi_2'(I_{1i})$ and
$\psi_2''(I_{2i})$. The concentration of element i is then given by

$$C_i = \frac{I_{2i}Q_i(\sin\psi_1)\ln(I_{1i}/I_{2i})}{\rho_s T(I_{1i}/I_{2i} - 2)} \qquad (10.17)$$

This method is only easily applicable to systems where
the takeoff angle is readily variable.

F.  Variable takeoff angle:  This is, in principle at least,
    a powerful method and has been described in detail by
    Ebel [21].  In this technique, a series of measurements of
    characteristic line intensity $I_{ni}$ are taken at different
    takeoff angles $\psi_n$.  The data is then extrapolated to
    $\psi_n = 0$, giving a simple solution of the basic intensity
    equation.  Mass absorption coefficients are measurable by
    this method to an accuracy of about 5%, but the method has
    the major disadvantage that special instrumentation is
    required.

G.  Monochromatic beam through sample:  This is the technique
    previously described in Sec. 10.6.2.


10.7  TYPE-FOUR METHODS:  MATHEMATICAL CORRECTION FOR MATRIX EFFECTS

10.7.1  Historical Development of Mathematical Correction Models

With the regrowth of interest in x-ray spectrometry and specifically
x-ray fluorescence spectrometry in the early 1950s, interest
rapidly grew in the provision of absolute methods of relating characteristic x-ray wavelength intensity and chemical composition.
Gillam and Heal [22] were the first to provide an adequate algorithm
which included secondary fluorescence and their work was followed
three years later in 1955 by Sherman's classic work [1] on the
fundamental relationship between x-ray intensity and composition.

The earlier absolute methods were, however, greeted with little
enthusiasm by the practicing x-ray spectroscopists of that era.
The small cheap digital computer had yet to burst upon the scientific world and computers at that time were large, complex and expensive, such that only the more sophisticated analytical laboratories
could afford the luxury of a facility capable of handling the required algorithms.  Alternative means of quantitative analysis were

sought and although the tried and proven standard addition techniques of the early 1930s provided excellent data in powder and solution analysis, they were equally unsuccessful in bulk sample analysis when addition of the necessary standards was physically impossible.  Typical of this latter category were the fields of ferrous
and nonferrous metallurgy, into which x-ray fluorescence was beginning to creep as an alternative to the ultraviolet (UV) emission
spectrometer.  Some of the highly alloyed formulations were proving
difficult for the UV technique to handle and the x-ray method, although at that time, much slower and considerably less sensitive,
seemed to offer both greater accuracy and precision.  Much of the
UV work up to that time had been done by type standardization and a
good number of well-analyzed primary and secondary standards were
available.

As a result, interest began to develop in empirical relationships between characteristic x-ray wavelength emission and chemical
composition.  Sherman himself had discussed this [23] in 1954, the
year before the publication of his theoretical equations, and his
scheme involved the use of regression analysis to solve equations,
the parameters of which were to be calculated with data from samples
of known composition.  The following equations are typical of this
approach which is generally attributed to Beattie and Brissey [24].

$$W_a = R_a(W_a + K_{ba}W_b + K_{ca}W_c)$$
$$W_b = R_b(W_b + K_{cb}W_c + K_{ab}W_a)$$
$$W_c = R_c(W_c + K_{ac}W_a + K_{bc}W_b)$$

The equations refer to a ternary system of elements a, b, and c,
where W refers to the weight fraction of the element and R a count
rate or count ratio to the pure element standard.  The values K are
constants representing the effect of element a on element b, etc.
These equations are unwieldy to the extent that a complete analysis
is required since the concentration terms occur on both sides of the
equations.  Furthermore, since n + 1 equations are available for n

unknowns (since $\Sigma W_i = 1$, i.e., 100%) there are complications in solving the equations due to overdefinition of the matrix. Only n equations can be used in solving for n unknowns.

A far more practical approach was that of Lucas-Tooth and Price [25] who, in 1961, suggested a linear expression relating the percentage of the nth element in the mth specimen to its intensity $I_{nm}$:

$$W_{nm} = \alpha_n + I_{nm}(K_o + \sum_x K_{nm} I_{xm})  \qquad (10.18)$$

$K_o$ and $\alpha_n$ approximate the slope and background constants set by the sensitivity of the spectrometer for the wavelength in question. The constants $K_{nm}$ represent the influence of element m on n as did the equivalent constants in the Beattie-Brissey relationship. $I_{xm}$ represents the intensity of a line from an interfering element x. As the equations obtained for a given elemental system are linear, they are easily soluble and a slide rule was all that was required once the constants had been determined.

One of the problems with the Lucas-Tooth/Price model was the effects of instrumental drifts on the intensity $I_{nm}$. Lucas-Tooth and Pyne later modified the intensity equation introducing the concept of apparent percentage [26], i.e., assume $(W_{nm})$apparent $= \mu \cdot I_{nm} \simeq (W_{nm})$chemical, then

$$W_{nm} = \alpha_n + W_{nm}^{app}(K_o + \sum_x K_{nx} W_{xm}^{app})  \qquad (10.19)$$

This was done to overcome problems due to the magnitude of the raw-intensity term I and to attempt to isolate instrumental influences and drifts by means of the factor $\mu$. These models are still current and are generally referred to as intensity correction models.

At about the same time as the introduction of the intensity correction models, Lachance and Traill [27] noted that by substitution with the (n + 1)th equation in the Beattie-Brissey model, a much more workable model could be obtained. For a binary mixture ab,

$$W_a = R_a(W_a + K_{ba} W_b)$$

Since

$$W_a + W_b = 1$$

substitution gives

$$W_a = R_a(1 + \alpha_{ab}W_b)$$

where

$$\alpha_{ab} = K_{ba} - 1$$

The form of the equation after substitution is such that complete matrix solution is not necessary.  Also, the interelement correction factor $\alpha_{ab}$ does have some theoretical significance, and it is fairly easy to show [28] that where absorption effects predominate and the effective wavelength approximation is accurate, the constant $\alpha$ is equal to

$$\alpha_{ab} = \frac{\mu_b(\lambda) + A\mu_b(\lambda_a)}{\mu_a(\lambda) + A\mu_a(\lambda_a)} - 1 \qquad (10.20)$$

In this equation, $\mu_b(\lambda)$ and $\mu_a(\lambda)$ are the respective mass absorption coefficients of elements b and a for the effective wavelength $\lambda$ for the excitation of $\lambda_a$.  A is a geometric constant equal to between 1 and 2 (see Sec. 2.11).

By about the mid-1960s, automatic x-ray spectrometers were becoming relatively commonplace, minicomputers were also beginning to make their appearance on the scene.  Although these early models were reasonably cheap, core memory was still expensive and the difference between 4 and 8K of direct access core was almost a factor of 2 in the price.  Nevertheless, a 4K computer with suitable I/0 devices and interfaces still only represented around 30% of the cost of a semiautomatic, wavelength-dispersive sequential spectrometer and equipment manufacturers were quick to realize that in addition to adding some reasonably inexpensive computational facility to the spectrometer, the minicomputer also offered a cheap alternative for the increasingly expensive, hardware-control logic circuitry required for the automatic selection of spectrometer parameters such as wavelength and count data acquisition. As a result, the computer controlled spectrometer made its appearance around 1965 and nearly all of the early models were equipped with a meager 4K of computer core. Since about 2/3 of this capacity was required for the control of the spectrometer, precious little computation was in fact available and only relatively primitive intensity correction models could be provided. Nevertheless, with the advent of the computer-controlled spectrometer, a new era in mathematical correction procedures was entered. Shortcomings in some of the earlier models rapidly became apparent, particularly in those cases where wide calibration ranges were attempted. Other difficulties were encountered in attempting to apply correction factors obtained on one spectrometer to a different, but nominally identical machine. This latter difficulty was particularly troublesome and obviously had to be solved if universal correction constants were to become a reality.

Added to this, more questions were being raised about the applicability of the Lachance-Traill-type models to those cases where significant secondary fluorescence, or enhancement, was present. Current practice, at that time, was to treat enhancement as "negative absorption," which simply changed the sign of the correction factor. Figure 10.5 illustrates some of the basic shapes of intensity/concentration curves. The upper curve illustrates the case of absorption on a binary, or pseudo-binary, mixture ij. The curve is hyperbolic and may show positive or negative deviation depending upon the relative absorption coefficients $\mu_i$ and $\mu_j$. Rasberry and Heinrich [29] pointed out, however, that the curve for enhancement, also shown in Fig. 10.5, cannot be adequately described by a hyperbolic function, and in 1970, they proposed an entirely new model in which predominant absorption, or predominant secondary fluorescence, was treated separately.

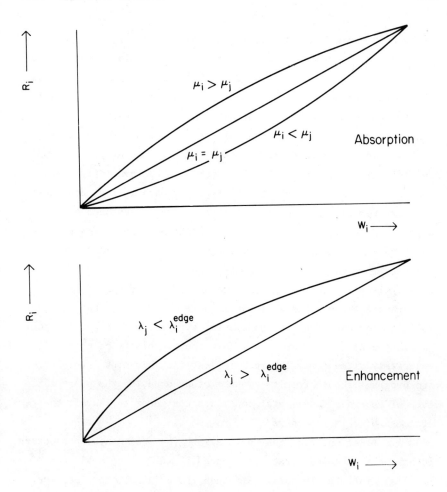

Figure 10.5  Basic shapes of intensity/concentration curves.

10.7.2  The Fundamental Parameters Approach

The problem in adequately describing the spectral distribution of
the primary radiation inhibited the development of absolute inten-
sity/concentration algorithms until 1968, when Birk's group at the
U.S. Naval Research Laboratory published details of their "fundamental parameters" approach [30].  This method differed from the

previously published absolute methods principally in the use of
measured primary spectra [5] rather than in calculated data.  The
value of such an absolute method is inestimable since this virtually
eliminates the need for standards.  Errors in the applications of
the method do accrue because of difficulties in measuring the x-ray
tube spectrum and current uncertainties in certain fundamental data
such as mass absorption coefficients and fluorescent yields, but
sufficiently accurate data will undoubtedly be eventually forth-
coming.  Although the fundamental-type approach is most desirable,
the computer facility required for this type of approach has been
beyond that provided by the conventional minicomputer, mainly because of the requirement for extensive reference tables.  Recent
advances in minicomputer technology, however, now provide a 1 to
2-μs cycle time, 16K, 16-bit minicomputer, with floppy disks for
less than $10,000.  One would anticipate, therefore, that the basic
facility for applying the fundamental-type correction models should,
within the next few years, be within the capability of most
moderately sized x-ray laboratories and flexible high-level-langu-
age programs should result.  On the other hand, at the present time,
lack of *accurate* fundamental constants continues to inhibit the general use of this procedure.

A further question that arises with the fundamental parameters
technique is the inaccuracy introduced in the use of published
spectral distribution data.  Although the original experiment performed by Gilfrich and Birks [5] was well conceived to give the required data; such an experiment is beyond the capabilities of the
average x-ray laboratory.

In the application of the fundamental parameters technique,
algorithms of the general form given in Table 2.2 are employed.  The
usual procedure involves the prior measurement of the x-ray tube
spectrum and replacement of the integral in the basic equation by an
expression of the form

$$\int_{\lambda_{min}}^{\lambda_{abs}} \frac{Q_{if}(\lambda_o)I_o(\lambda_o)\ d\lambda_o}{\mu(\lambda_o)\ csc\ \psi_1 + \mu(\lambda_i)\ csc\ \psi_2}$$

$$= \sum_k \frac{Q_{if}(\lambda_o)D_k(\lambda_o)I_k(\lambda_o)\ \Delta\lambda_o}{\mu(\lambda_o)\ csc\ \psi_1 + \mu(\lambda_i)\ csc\ \psi_2} \qquad (10.21)$$

In this expression $I_k(\lambda_o)$ is the measured intensity for a finite wavelength interval $\Delta\lambda_o$. The additional term $D_k(\lambda_o)$ is required to select that portion of the spectrum effective in the excitation of $\lambda_i$. The term is set to equal unity for $\lambda_o < \lambda_{i\ abs}$ and to zero for $\lambda > \lambda_{i\ abs}$. Where divergent or convergent beam geometries are employed, it may also be necessary to integrate over the range of incident and/or takeoff angles. Table 10.7 lists the values for $\lambda_o$ and $I(\lambda_o)\ \Delta\lambda_o$ taken from the original work of Birks and his co-workers. It seems likely that the energy dispersive spectrometer may soon offer a solution to the problem of the rapid and accurate determination of tube spectra and recent measurements [31] with carefully calibrated Si(Li) detectors have yielded interesting results along these lines.

## 10.7.3  Use of the Effective Wavelength Concept

In 1971, Stephenson proposed an interesting alternative to the difficulty in precisely evaluating the complete integral of the primary spectral distribution, by use of his program CORSET [32]. In this program, Stephenson went back to the effective wavelength concept and replaced the primary spectral source distribution in the absolute equation by such an effective wavelength. Stephenson proposed that the CORSET program could serve as a useful interim alternative until spectral source distributions could be routinely established for individual spectrometers.

The concept of the effective wavelength is an old one and is best illustrated by reference to Fig. 10.6. Here an element i is

Table 10.7   Values of the Relative Integrated Intensity for Tungsten (OEG50) and Chromium (OEG50) X-ray Tubes

Tungsten target (OEG50) 45 kV (C.P.); $\Delta\lambda = 0.02$ A

**Continuum**

| $\lambda^a$(Å) | $I_\lambda\Delta\lambda$ | $\lambda^a$(Å) | $I_\lambda\Delta\lambda$ | $\lambda^a$(Å) | $I_\lambda\Delta\lambda$ | $\lambda^a$(Å) | $I_\lambda\Delta\lambda$ | $\lambda^a$(Å) | $I_\lambda\Delta\lambda$ |
|---|---|---|---|---|---|---|---|---|---|
| 0.29 | 15.5 | 0.75 | 53.9 | 1.21[b] | 33.4 | 1.69 | 21.1 | 2.15 | 8.2 |
| 0.31 | 36.6 | 0.77 | 51.8 | 1.23 | 36.2 | 1.71 | 20.1 | 2.17 | 7.8 |
| 0.33 | 56.8 | 0.79 | 49.9 | 1.25 | 35.8 | 1.73 | 19.2 | 2.19 | 7.5 |
| 0.35 | 76.6 | 0.81 | 48.2 | 1.27 | 35.4 | 1.75 | 18.3 | 2.21 | 7.3 |
| 0.37 | 96.2 | 0.83 | 46.6 | 1.29 | 35.0 | 1.77 | 17.5 | 2.23 | 7.0 |
| 0.39 | 111.1 | 0.85 | 45.2 | 1.31 | 34.6 | 1.79 | 16.8 | 2.25 | 6.7 |
| 0.41 | 116.4 | 0.87 | 44.0 | 1.33 | 34.1 | 1.81 | 16.1 | 2.27 | 6.4 |
| 0.43 | 114.6 | 0.89 | 42.9 | 1.35 | 33.5 | 1.83 | 15.5 | 2.29 | 6.1 |
| 0.45 | 109.0 | 0.91 | 42.0 | 1.37 | 33.0 | 1.85 | 14.9 | 2.31 | 5.9 |
| 0.47 | 104.5 | 0.93 | 41.2 | 1.39 | 32.5 | 1.87 | 14.3 | 2.33 | 5.7 |
| 0.49 | 99.1 | 0.95 | 40.4 | 1.41 | 32.0 | 1.89 | 13.7 | 2.35 | 5.4 |
| 0.51 | 93.9 | 0.97 | 39.6 | 1.43 | 31.5 | 1.91 | 13.1 | 2.37 | 5.1 |
| 0.53 | 89.0 | 0.99 | 38.8 | 1.45 | 30.9 | 1.93 | 12.6 | 2.39 | 4.9 |
| 0.55 | 84.4 | 1.01 | 38.1 | 1.47 | 30.3 | 1.95 | 12.2 | 2.41 | 4.7 |
| 0.57 | 80.3 | 1.03 | 37.4 | 1.49 | 29.7 | 1.97 | 11.7 | 2.43 | 4.5 |
| 0.59 | 76.6 | 1.05 | 36.7 | 1.51 | 29.0 | 1.99 | 11.2 | 2.45 | 4.2 |
| 0.61 | 73.3 | 1.07 | 36.1 | 1.53 | 28.3 | 2.01 | 10.8 | 2.47 | 3.9 |
| 0.63 | 70.1 | 1.09 | 35.6 | 1.55 | 27.6 | 2.03 | 10.4 | 2.49 | 3.7 |
| 0.65 | 67.0 | 1.11 | 35.1 | 1.57 | 26.8 | 2.05 | 10.0 | 2.51 | 3.5 |
| 0.67 | 64.1 | 1.13 | 34.5 | 1.59 | 26.0 | 2.07 | 9.7 | 2.53 | 3.3 |
| 0.69 | 61.2 | 1.15 | 33.9 | 1.61 | 25.2 | 2.09 | 9.4 | 2.55 | 3.1 |
| 0.71 | 58.6 | 1.17 | 33.4 | 1.63 | 24.3 | 2.11 | 9.0 | 2.57 | 2.9 |
| 0.73 | 56.2 | 1.19 | 33.0 | 1.65 | 23.2 | 2.13 | 8.6 | 2.59 | 2.7 |
|  |  |  |  | 1.67 | 22.1 |  |  |  |  |

**Characteristic Lines[c]**

| $L\gamma_3$ 1.06200 2.6 | $L\gamma_2$ 1.06806 10.4 | $L\gamma_1$ 1.09855 27.8 |
|---|---|---|
| $L\beta_2$ 1.24460 180 | $L\beta_1$ 1.281809 407 | $L\alpha$ 1.47639 592 |
|  | $L\ell$ 1.6782 13.8 |  |

being excited by a band of continuous radiation, bounded by the minimum wavelength of the continuum $\lambda_{min}$ and the absorption edge $\lambda_{i\ abs}$ of element i.  As one moves to shorter wavelengths, away from $\lambda_{i\ abs}$, the x-ray photons become less and less effective in usefully

| $\lambda^a$(Å) | $I_\lambda\Delta\lambda$ | $\lambda^a$(Å) | $I_\lambda\Delta\lambda$ | $\lambda^a$(Å) | $I_\lambda\Delta\lambda$ | $\lambda^a$(Å) | $I_\lambda\Delta\lambda$ | $\lambda^a$(Å) | $I_\lambda\Delta\lambda$ |
|---|---|---|---|---|---|---|---|---|---|
| 0.29 | 3.00 | 0.71 | 14.0 | 1.13 | 10.0 | 1.55 | 5.75 | 1.97 | 3.39 |
| 0.31 | 6.60 | 0.73 | 14.0 | 1.15 | 9.78 | 1.57 | 5.62 | 1.99 | 3.29 |
| 0.33 | 8.80 | 0.75 | 14.0 | 1.17 | 9.31 | 1.59 | 5.50 | 2.01 | 3.19 |
| 0.35 | 9.94 | 0.77 | 14.0 | 1.19 | 9.25 | 1.61 | 5.38 | 2.03 | 3.09 |
| 0.37 | 11.0 | 0.79 | 14.0 | 1.21 | 9.00 | 1.63 | 5.25 | 2.05 | 2.98 |
| 0.39 | 12.0 | 0.81 | 13.9 | 1.23 | 8.76 | 1.65 | 5.12 | 2.07[b] | 5.92 |
| 0.41 | 12.9 | 0.83 | 13.8 | 1.25 | 8.52 | 1.67 | 5.00 | 2.09 | 8.77 |
| 0.43 | 13.5 | 0.85 | 13.7 | 1.27 | 8.29 | 1.69 | 4.88 | 2.11 | 8.54 |
| 0.45 | 13.9 | 0.87 | 13.6 | 1.29 | 8.07 | 1.71 | 4.77 | 2.13 | 8.30 |
| 0.47 | 13.8 | 0.89 | 13.4 | 1.31 | 7.84 | 1.73 | 4.66 | 2.15 | 8.07 |
| 0.49 | 13.8 | 0.91 | 13.2 | 1.33 | 7.62 | 1.75 | 4.55 | 2.17 | 7.84 |
| 0.51 | 13.8 | 0.93 | 13.0 | 1.35 | 7.42 | 1.77 | 4.44 | 2.19 | 7.61 |
| 0.53 | 13.8 | 0.95 | 12.7 | 1.37 | 7.22 | 1.79 | 4.33 | 2.21 | 7.39 |
| 0.55 | 13.8 | 0.97 | 12.4 | 1.39 | 7.02 | 1.81 | 4.22 | 2.23 | 7.17 |
| 0.57 | 13.8 | 0.99 | 12.1 | 1.41 | 6.83 | 1.83 | 4.11 | 2.25 | 6.96 |
| 0.59 | 13.8 | 1.01 | 11.8 | 1.43 | 6.65 | 1.85 | 4.01 | 2.27 | 6.74 |
| 0.61 | 13.8 | 1.03 | 11.4 | 1.45 | 6.48 | 1.87 | 3.90 | 2.29 | 6.53 |
| 0.63 | 13.8 | 1.05 | 11.1 | 1.47 | 6.32 | 1.89 | 3.79 | 2.31 | 6.32 |
| 0.65 | 13.8 | 1.07 | 10.9 | 1.49 | 6.16 | 1.91 | 3.69 | 2.33 | 6.13 |
| 0.67 | 13.9 | 1.09 | 10.6 | 1.51 | 6.02 | 1.93 | 3.59 | 2.35 | 5.94 |
| 0.69 | 13.9 | 1.11 | 10.3 | 1.53 | 5.89 | 1.95 | 3.49 | 2.37 | 5.74 |

Chromium target (OEG50) 45 kV (C.P.); $\Delta\lambda = 0.02$ Å

Continuum

[a] $\lambda$ for continuum is the middle of the $\Delta\lambda$ interval.

[b] $L_{III}$ edge occurs at 1.216 Å ($I_\lambda \Delta\lambda$ is 26.1 from 1.200 to 1.216 and 7.3 from 1.216 to 1.220 Å).

[c] $\lambda$ for lines is from Bearden. $\Delta\lambda$ for lines is natural line breadth (from Blohkin, *Physics of X-rays*, AEC Translation 4502).

exciting $\lambda_i$. The effective wavelength is considered to be one single wavelength between $\lambda_{min}$ and $\lambda_{i\ abs}$, which has the same effectiveness in the excitation of $\lambda_i$ as the whole continuum. In practice, where the contribution from characteristic source lines is insignificant, this turns out to be roughly 2/3 of $\lambda_{i\ abs}$. In 1962, Kalman and Heller [36] pointed out that the mean value theorem would predict such a wavelength, but more recent discussion has

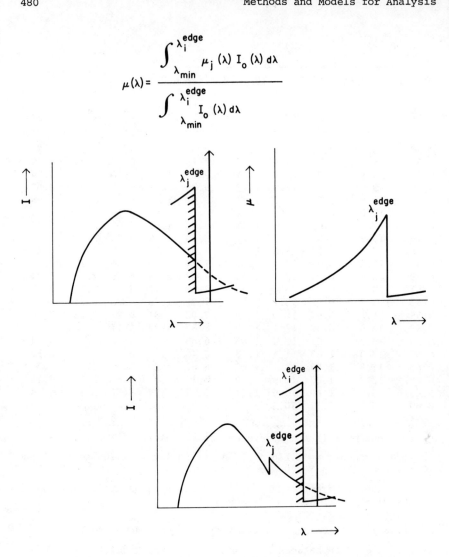

$$\mu(\lambda) = \frac{\displaystyle\int_{\lambda_{min}}^{\lambda_i^{edge}} \mu_j(\lambda)\, I_o(\lambda)\, d\lambda}{\displaystyle\int_{\lambda_{min}}^{\lambda_i^{edge}} I_o(\lambda)\, d\lambda}$$

Figure 10.6  Effective wavelengths and the absorption function.

centered around how this theorem should be applied.  Mencik [21] has recently suggested the effective wavelength be incorporated in the form shown in the upper part of Fig. 10.6, since this would better take into account the detailed distribution of the values of the mass absorption coefficients of all of the matrix elements, as a function of the wavelength of the primary radiation.  As an

Linear model:

$$W_i/R_i = K_i$$

Lachance/Traill:

$$W_i/R_i = K_i \left[ 1 + \sum_j \alpha_{ij} W_j \right]$$

Rasberry/Heinrich:

$$W_i/R_i = K_i \left[ 1 + \sum_j \alpha_{ij} W_j + \sum_k \beta_{ik} W_k / (1+W_i) \right]$$

Claisse/Quintin:

$$W_i/R_i = K_i \left[ 1 + \sum_j \alpha_{ij} W_j + \sum_j \gamma_{ij} W_j^2 \right]$$

Figure 10.7  Forms of the alpha correction models.

example, in Fig. 10.6, element j with its absorption edge $\lambda_{j \ abs}$ to the short-wavelength side of $\lambda_{i \ abs}$ obviously modifies the spectral distribution of the absorption function for element i.

The effective wavelength concept is useful since it avoids the inclusion of an awkward integral in any absolute method of relating x-ray intensity and chemical composition.  It is also of great use in estimating approximate total absorption values using an expression of the form shown in Eq. (2.34).

10.7.4  Alpha Correction Models

Among the earlier spectroscopists to realize the limitations of the Lachance-Traill model were Claisse and Quintin.  In 1967, these workers pointed out [33] that the Lachance-Traill model really could not allow for the polychromatic nature of the incident radiation and suggested the use of an expanded form of the original model.  Figure 10.7 shows the forms of the three main alpha correction models currently in use, and Fig. 10.8 shows an overview of all of the main types of correction procedures.  Note that in Fig. 10.7 all of the models are concentration correction models since all of the

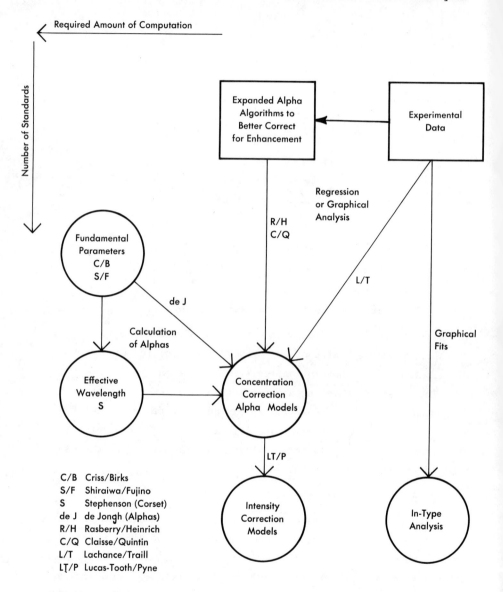

Figure 10.8  Mathematical correction models for quantitative x-ray spectrometry.

correction terms are made up of products of alpha correction constants and weight fractions.  The Rasberry-Heinrich model, alluded to earlier, is shown.  Here the α coefficients are used when the

significant effect of element j on element i is absorption, and the
β coefficients are used when the predominant effect of element k on
element i is enhancement.   The Claisse-Quintin model is also shown
here, albeit in a somewhat shortened form.   This contains additional
higher order terms which allow for the polychromatic nature of the
x-ray beam.   One of the more important conclusions of the study by
Claisse and Quintin was that the coefficients could be calculated
where the spectral distribution of the primary beam was known.   Such
an approach has been employed by de Jongh [34] who, by use of his
program "ALPHAS" has used the Shiraiwa-Fujino fundamental equation
to calculate the alpha correction factors in the Lachance-Traill
equation.   This approach is particularly useful in that it allows
the correction factors to be applied with the small on-line computers being used to control the spectrometer, where the factors
themselves are calculated with a large off-line computer.   The form
of the equations employed is as follows:

$$C_1 = (D_1 + E_1 R_1)(100 + \alpha_{11}C_1 + \alpha_{12}C_2 + \cdots)$$

$$C_2 = (D_2 + E_2 R_2)(100 + \alpha_{21}C_1 + \alpha_{22}C_2 + \cdots)$$

$$\cdots\cdots\cdots\cdots\cdots\cdots\cdots\cdots\cdots\cdots\cdots\cdots$$

$$C_n = (D_n + E_n R_n)(100 + \alpha_{n1}C_1 + \alpha_{n2}C_2 + \cdots)$$

A calibration reference matrix is used to define the line
$C = (D + ER) \times (100 + \Sigma\alpha_{ij}\ \Delta\overline{C}_j)$ in which C is concentration, E the
reciprocal of the slope, R the intensity ratio of unknown element to
calibration reference, and D the intercept on the concentration axis.
$\Delta\overline{C}_j$ represents the concentrations of all elements j in mid-concentra-
tion range.   The D + ER term is independent of matrix effects and
the constants E and D can be reevaluated from time to time and if
necessary updated, for example, if an x-ray tube has to be replaced.
For the reasons already stated, the least accurately known factor in
the intensity/concentration equation is the alpha factor, but the
overall effect of the total correction term relative to the D + ER
term can be reduced by careful choice of the calibration range.

Table 10.8  Experimental and Theoretical Alpha Coefficients

|                        | Mg      | Al      | Si      | Ca      | Fe      |
|------------------------|---------|---------|---------|---------|---------|
| Experimental           | -0.006  | -0.014  | +0.007  | +0.079  | +0.061  |
| Mg Theory              | -0.013  | -0.010  | -0.004  | -0.004  | +0.088  |
| Experimental           | +0.097  | -0.012  | -0.015  | -0.005  | +0.079  |
| Al Theory              | +0.097  | -0.005  | -0.003  | -0.001  | +0.090  |
| Experimental           | +0.081  | +0.090  | -0.012  | +0.006  | +0.086  |
| Si Theory              | +0.097  | +0.110  | +0.004  | +0.003  | +0.092  |
| Experimental           | +0.086  | +0.093  | +0.127  | +0.155  | +0.081  |
| Ca Theory              | +0.095  | +0.109  | +0.129  | +0.144  | +0.098  |
| Experimental           | +0.028  | +0.038  | +0.044  | +0.168  | +0.086  |
| Fe Theory              | +0.029  | +0.033  | +0.039  | +0.153  | +0.032  |

For example,

$$Ca = (D + ECa\ K\alpha)(100 + 0.095\ Mg + 0.109\ Al + \cdots + 0.098\ Ca)$$

Based on the average composition:

| | | | |
|---|---|---|---|
| $MgO$ | 20% | | |
| $Al_2O_3$ | 20% | | 1 g sample |
| $SiO_2$ | 20% | | |
| $CaO$ | 20% | Fused | 2 g $La_2O_3$ |
| $Fe_2O_3$ | 20% | | 9 g $Li_2B_4O_7$ |

---

Since the reference standard contains all matrix elements, the correction term should contain concentration difference terms $\Delta W_j$ rather than absolute concentration terms $W_j$. The transform from $\Delta W_j$ to $W_j$ follows:

$$\frac{W_i}{R_i} = k(1 + \sum \alpha_{ij}\ \Delta W_j)$$

$$= k(1 - \sum \alpha_{ij}\ \overline{W}_j + \sum \alpha_{ij}\ W_j)$$

$$\frac{W_i}{R_i} = k'(1 + \sum \alpha'_{ij}\ W_j)$$

in which

$$\alpha'_{ij} = \frac{\alpha_{ij}}{1 - \sum \alpha_{ij}\overline{W}_j} \qquad k' = k(1 - \sum_i \alpha_{ij}\overline{W}_j)$$

Table 10.9  Analysis of Basic Slag (all data given in percent)

|  | Nominal Concentration | Counting Error (2σ) | Measured Concentrations | |
|---|---|---|---|---|
|  |  |  | Theoretical α Values | Experimental α Values |
| MgO | 4.63 | 0.10 | 4.62 | 4.61 |
| $Al_2O_3$ | 0.77 | 0.02 | 0.71 | 0.65 |
| $SiO_2$ | 11.2 | 0.09 | 11.0 | 11.3 |
| CaO | 43.2 | 0.16 | 41.8 | 41.9 |
| $Fe_2O_3$ | 22.7 | 0.14 | 22.3 | 22.4 |

where $\overline{W}_j$ represents the concentration of the interfering element j
in the reference matrix.  It will be seen that the new equation is
of the same form as the original and that $\alpha'_{ij}$ is related to $\alpha_{ij}$.
Note also that where values of $\alpha_{ij}$ are to be calculated, this must
be done for an assumed average composition of a hypothetical matrix
defined by $\sum_j \overline{W}_j$, this being the reference matrix.

10.7.5  Determination of Interelement (Alpha) Coefficients

Whichever of the alpha correction models are employed, two methods
are available for the estimation of the alpha coefficients, namely
calculation from first principles or regression analysis using experimental data from many standards.  As an example, Table 10.8
lists such experimental (regression) and theoretical (calculated)
Alpha coefficients established for the 5 component system MgO, $Al_2O_3$,
$SiO_2$, CaO, $Fe_2O_3$ diluted in a borax glass.  Fair agreement is found
in general, with differences of around 10% for the more significant
correction terms.  Table 10.9 shows a typical analysis of a basic
slag which has been performed [35] using these theoretical and experimental alphas.  It will be seen that agreement is excellent even
though the slag composition is very different from the chosen average composition of 20% for the major oxides.

    The absolute accuracy in the calculated value of $C_i$ will depend
both upon the accuracy of the α correction constants employed and
the concentration differences of the interfering elements in the

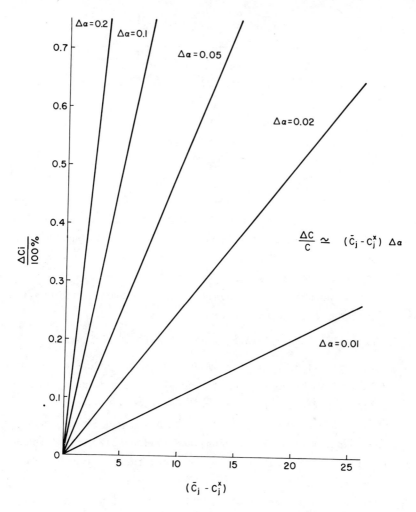

Figure 10.9  Required accuracy of the α constants.

analyzed sample and the calibration reference matrix.  For example,
if the error in α is Δα and the concentration difference term for an
interfering element j between unknown and reference matrix is
$\overline{C}_j - C_j^x$, then the resulting error $\Delta C_i/C_i$ in the analyzed elements
will be given by

$$\frac{\Delta C_i}{C_i} \simeq \Delta\alpha(\overline{C}_j - C_j^x)/100 \qquad \text{for } \alpha_{ij}(\overline{C}_j - C_j^x) \ll 100$$

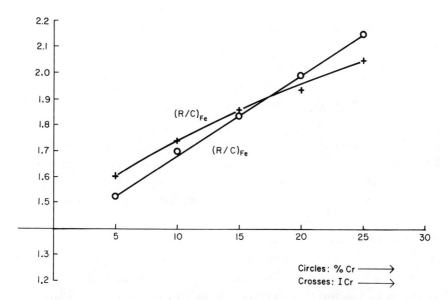

Figure 10.10 Graphical estimation of the $\alpha$ factor based on concentration (circles) and intensity (crosses).

Figure 10.9 illustrates some plots of $\overline{C}_j - C_j^x$ against $(\Delta C_i / C_i) \times 100\%$ for $C_i = 100\%$ using different values of $\Delta \alpha$.

An acceptable error in $C_i$ would lie between 0.05 and 0.2%, i.e., 100 ± 0.05% to 100 ± 0.2%. The curves illustrate that the *absolute* accuracy in $\alpha$ would have to be better than ± 0.01 where the concentration range of the interfering element is greater than 10%. Conversely, if the concentration range of the interfering element was only a few percent, an absolute accuracy in of ± 0.02 would be acceptable. In general terms, it can be stated that the required absolute accuracy in $\alpha$ lies between 0.01 and 0.05.

A further great advantage of the alpha-type models is that it is easy to determine the $\alpha$ factor by graphical means. As an example, Fig. 10.10 (circles) illustrates such a determination in a binary alloy series of iron and chromium. Rearrangement of the Lachance-Traill equation into the form of a straight-line relationship, followed by plotting $(W/R)_{Fe}$ against $W_{Cr}$, allows the determination of the factor $\alpha_{Cr/Fe}$ since this is the slope of the curve.

A useful practical point to bear in mind is that the process of
estimating the α factor in this way allows the checking of the
validity of the model.  For example, in this particular case, a good
straight-line correlation was obtained over a concentration range
of 0 to 30% of chromium, indicating that the model is applicable to
these matrix circumstances over this concentration range.

     A similar exercise could be performed with the Lucas-Tooth/Pyne
model [Fig. 10.10 (crossed points)].  Here it will be seen that when
W/R for iron is plotted against, in this instance, the intensity of
Cr Kα a less satisfactory fit is obtained and although the model
might give an acceptable accuracy over a somewhat restricted calibration range, this range is never as wide as with the Lachance-
Traill model.  The reason for this is to be found in the fact that
the two approaches differ only in the form of the correction term.
In the Lachance-Traill model, this term is the sum of the products
of the correction constants and the *concentrations* of the interfering elements.  Since the intensity of the line from the absorbing
element may *itself* not give a linear relationship with concentration, the concentration correction models invariably work over a
much wider calibration range.  On the other hand, the concentration
correction models always involve the solution of simultaneous equations and much more computation is required than with the intensity
correction models.

## 10.7.6  Problems in the Application of Regression Analysis

The greatest danger in the application of regression methods is the
tendency to disregard theory altogether and to blindly employ corrections for matrix effects that common sense would predict could
not possibly exist.  The use of single- or multiple-regression computer programs for the determination of the values of α correction
factors is particularly dangerous since one is usually dealing not
with true mathematical identities but rather with x-ray intensities
and chemical analysis data, which are subject to significant random
and systematic errors.  Although it is true that large, high-

capacity digital computers allow the use of built-in check procedures, the tendency in spectrometry today is toward the use of the
small interfaced computer which is used jointly for spectrometer
control, data collection, and conversion. Since, to date, the total
storage capacity utilized is rarely more than 16K words with a word
length of 12 to 16 bits, large and complicated regression programs
are not often available.

In the determination of $\alpha$ correction constants, the results of
a badly applied multiple-regression analysis may not be immediately
apparent since the check frequently employed is to re-insert all
correction factors in the mathematical model, relating elemental
concentration and measured intensity, and to recalculate elemental
concentrations from intensities measured on the calibration standards. However, the excellence of the correlation between "chemical"
and "x-ray" concentrations found for the calibration standards
is far more likely to depend upon the number of regression constants
taken than on the actual regression procedure itself. Often the
analysis of a new standard not used for the calibration will yield
a large error indicating the true error in the calibration.

The following example may serve to illustrate the type of problem which can occur. The analytical problem involved the measurement of arsenic in a tin bronze which contained significant quantities of lead. The Pb L$\alpha$ line ($\lambda$ = 1.176 Å) interferes with the
As K$\alpha$ line (1.177 Å); so a line overlap correction has to be made.
A correction can, in principle, be made by use of a line overlap
correction for lead, with possible additional corrections for matrix
absorption. Two sets of intensity measurements were made on each
standard, the first of these $I_{As}$ was made at 1.176 Å and corresponds
to the summation of the Pb L$\alpha$ and As K$\alpha$ intensities. The second,
$I_{Pb}$ was made at the Pb L$\beta_1$ line ($\lambda$ = 0.982 Å) which is free from
interference. Table 10.10 lists the measured intensities and arsenic concentrations in the nine standards available. All intensities are given in the form of intensity ratios.

Table 10.10   Arsenic and Lead in Tin Bronze

| No. | $I_{As}$ | $I_{Pb}$ | $I_{Cr}$ | Chem % (As) | x-ray % (As) Regression A | x-ray % (As) Regression B |
|-----|----------|----------|----------|-------------|---------------------------|---------------------------|
| 1 | 0.79804 | 0.76385 | 0.84203 | 0.0411 | 0.0401 | 0.0378 |
| 2 | 1.07493 | 1.01538 | 0.96635 | 0.0351 | 0.0411 | 0.0367 |
| 3 | 0.79305 | 0.92657 | 0.93214 | 0.0363 | 0.0294 | 0.0336 |
| 4 | 2.76811 | 2.38996 | 1.08975 | 0.0278 | 0.0263 | 0.0338 |
| 5 | 1.95562 | 1.75941 | 1.00219 | 0.0311 | 0.0308 | 0.0345 |
| 6 | 1.71218 | 1.14430 | 0.95999 | 0.0351 | 0.0349 | 0.0352 |
| 7 | 0.66131 | 0.72168 | 0.93723 | 0.0386 | 0.0390 | 0.0363 |
| 8 | 4.35260 | 3.61930 | 1.01606 | 0.0373 | 0.0381 | 0.0326 |
| 9 | 0.77350 | 0.78095 | 1.01190 | 0.0351 | 0.0393 | 0.0369 |
|   |         |         |         |        | $\sigma = 0.0025$ | $\sigma = 0.0035$ |

Equation for regression A:

$$C_{As} = I_{As}\,(-0.00781 + 0.00195 \times I_{Pb}) + 0.07617 - 0.03515 \times I_{Cr}$$

max slope = -0.0008          max background = +0.0466
min slope = -0.0064          min background = +0.0379

Equation for regression B:

$$C_{As} = I_{As}\,(0.018923) + 0.0440 - 0.02537 \times I_{Pb}$$

max slope = min slope = 0.018923

max background = +0.0238
min background = -0.0497

---

*Regression Analysis (Inclusion of Wrong Interelement Correction)(A)*

This is an example of an *incorrect* approach which appears at first
sight to give acceptable results.  The approach in this case was
incorrect due to a bad choice of interelement correction.  The correction procedure was of the form (see Ref. 8, Chap. 9)

$$C_{As} = I_{As}\,[K_1 + K_2\,I_{Pb}] + [B + K_3\,I_{Cr}] \qquad (10.22)$$

Total slope      Total background

Here it will be seen that the slope of the calibration curve
is determined by the constants $K_1$ and $K_2$ and a slope correction
factor is being applied by the $K_2 I_{Pb}$ term.  The total background is
determined by the B and $K_3$ constants with a background correction
factor applied in the form $K_3 I_{Cr}$.  Here, $I_{Cr}$ is the intensity of

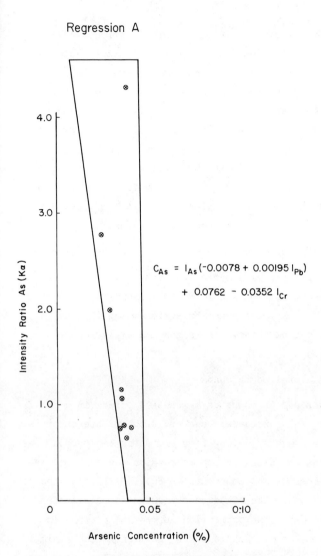

Figure 10.11 Correlation between arsenic characteristic line intensity and concentration showing incorrectly determined working equation and limits of this equation.

the scattered line from the chromium target x-ray tube, which is being used as a measure of background variations.

Table 10.10 lists the data obtained by regression analysis and gives values for the constants in Eq. (10.22) as well as the

calculated concentrations of arsenic using that equation. At first
sight, the standard deviation of 0.0025% looks very acceptable and
the correlation between x-ray and chemical data shown in Table
10.10 also looks reasonable. On closer inspection, however, it is
found that the method gives ludicrous results. By putting the maximum and minimum values of $I_{Pb}$ and $I_{Cr}$ into Eq. (10.22), it is possible to calculate the maximum and minimum values of slope and
background (see Table 10.10). Figure 10.11 plots the limits of the
correction curves and also indicates the uncorrected $I_{As}$ data. It
will be clearly seen that the curves always slope the wrong way,
i.e., increasing concentration gives decreasing intensity.

Regression Analysis (Use of Poor Range of Calibration Standards) (B)

In the previous example, the major error was that completely the
wrong approach was used. Since both lead and arsenic concentrations
affect the value of $I_{As}$, a background correction using $I_{Pb}$ should
be employed. In the second method, the algorithm employed was

$$C_{As} = I_{As}(K_1) + (B + K_2 I_{Pb}) \qquad\qquad (10.23)$$
$$\quad\text{Slope}\quad\ \text{Total background}$$

where the total background is adjusted by the $K_2 I_{Pb}$ term in which $K_2$
would be negative; in this instance, however, problems are encountered due to the poor range of calibration standards employed.

Table 10.10 lists the data for the regression analysis and, as
before, gives the values of the constants in Eq. (10.23) plus the
calculated concentrations. The standard deviation in this instance
was 0.0035%. The limits of the correction curves are plotted in
Fig. 10.12 using the value of the slope and the maximum and minimum
background values. The shape of the correction curve looks far more
reasonable in this instance, since indeed one would theoretically
predict a family of parallel curves. Not so reasonable, however, is
the correlation between x-ray and chemical values shown in Fig.
10.13. The scatter of points in this case is certainly not distributed evenly along the 1:1 relationship, and in fact all x-ray data
are within ± 0.0026 of the mean of 0.352%. In other words, this

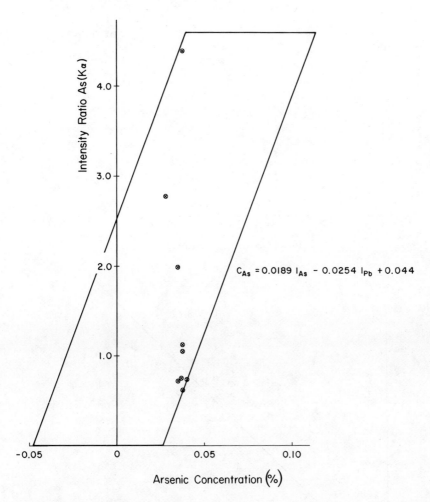

Figure 10.12  Correlation between arsenic characteristic line intensity and concentration showing correctly determined working equation and its limits.

method yields approximately the same result in every case, irrespec-
tive of the arsenic concentration.

In both of the examples cited above, the incorrect data (although a mathematically correct regression) has resulted from a poor

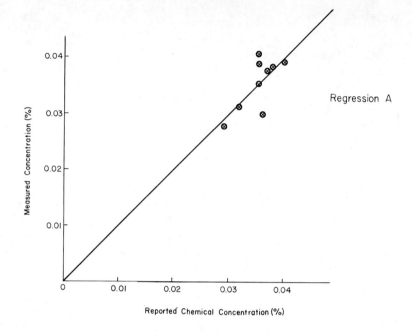

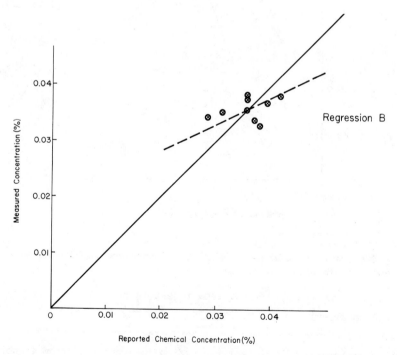

Figure 10.13  Correlation between measured and reported concentrations.

494

range of standard samples.  Of the nine standards available, all
lie within 0.0345 ± 0.006%.  What is really required to cover the 0
to 0.05% arsenic concentration level are eight standards with
0.040, and 0.050%, or as close to that distribution as possible.
and 0.040%, or as close to that distribution as possible.

Careful planning of concentrations in the standards and careful analysis of errors in determined regression parameters is manda-
tory for achieving a reliable analysis via empirical corrections.


## 10.8   CHECK OF THE EFFECTIVENESS OF A CORRECTION METHOD

The only true test of the success of a selected method of quantitative analysis is its effectiveness in removing systematic and random errors yielding an acceptable degree of analytical accuracy.
When comparing standard (e.g., chemical) and x-ray data, the observed error $\sigma_o$ will be made up of both errors in the standards
$\sigma_{chem}$ and then in the x-ray quantitative $\sigma_{x-ray}$, i.e.,

$$\sigma_0^2 = \sigma_{chem}^2 + \sigma_{x-ray}^2$$

*Only* where $\sigma_{chem}^2$ can be assumed to be negligible relative to
$\sigma_{x-ray}^2$ can $\sigma_o$ be assumed to approximate $\sigma_{x-ray}$.  In practice, such a
circumstance is generally unrealistic and a far more likely situa-
tion is that one is assuming to obtain a value of $\sigma_{x-ray}$ approximately equal to $\sigma_{chem}$.  This in turn means that a more reasonable
value of $\sigma_o$ is $\sigma_{x-ray} \times \sqrt{2}$.  In other words, when comparing x-ray
and standard data, the observed deviation is likely to be too large
by about $\sqrt{2}$.

Correlation of x-ray values with chemical concentrations have
been found to follow the empirical relation

$$S = F \sqrt{C} \qquad\qquad (10.24)$$

where S is an index of reliability, C the percentage level, and F a
constant.  Although S has no theoretical statistical significance,

it is found in practice that it resembles closely the standard
deviation $\sigma$.  This can be seen from the following.

If all other errors have been reduced to insignificant proportions, the limiting error is that due to counting statistics.  In
this case

$$\sigma_N = \sqrt{N} = \sqrt{It}$$

where N is the number of counts collected in time t at counting
rate I.  Since I = N/t provided that the error in t is negligible,

$$\sigma_I = \frac{\sigma_N}{t} = \frac{\sqrt{It}}{t} = \sqrt{\frac{I}{t}}$$

If the concentration C is proportional to I, i.e.,

$$I = CK$$

the error in C, i.e., $\sigma_C$, will be proportional to the error in I,
i.e., $\sigma_I$.  Thus

$$\sigma_C = \frac{\sigma_I}{K} = \frac{1}{K}\sqrt{\frac{I}{t}} = \frac{1}{K}\sqrt{\frac{CK}{t}} = F\sqrt{C}$$

where $F = 1/\sqrt{Kt}$.

Typical values of F for "good" quantitative x-ray spectrometry
vary between 0.005 and 0.05, e.g., for a value of F = 0.01, the
standard deviation would be 0.01 at the 1% concentration level and
0.03 at the 9% concentration level.  It should be noted that F is
dependent upon the constant K which is the sensitivity of the spectrometer for the given element, and only where adequate sensitivity
is maintained will the relationship in Eq. (10.24) be maintained.
The factor F is also time dependent provided that random errors
other than that due to counting statistics are not significant.

REFERENCES

1.  J. Sherman, Spectrochim. Acta, 7:283 (1955).
2.  T. Shiraiwa and N. Fujino, Jap. J. Appl. Phys., 5:886 (1966).
3.  T. Shiraiwa and N. Fujino, X-ray Spectrom., 3:64 (1974).
4.  Bambynek, Rev. Mod. Phys., 44:716 (1972).
5.  J. V. Gilfrich and L. S. Birks, Anal. Chem., 40:1077 (1968).

6. J. D. McGinness, R. W. Scott, and J. S. Mortensen, *Anal. Chem.*, *41*:1858 (1969).

7. T. N. Rhodin, *Anal. Chem.*, *27*:1857 (1955).

8. F. H. Chung, A. J. Lentz, and R. W. Scott, *X-ray Spectrom.*, *3*:172 (1974).

9. R. Tertian, *Fluorescence x théorie et pratique de l'analyse des solutions liquides ou solides par une méthode de double measure*, Thesis, University of Paris VI, June 7, 1972.

10. R. Tertian, *Spectrochim. Acta*, *27B*:159 (1972).

11. G. Anderman and J. W. Kemp, *Anal. Chem.*, *30*:1306 (1958).

12. K. P. Champion, J. C. Taylor, and R. N. Whittem, *Anal. Chem.*, *38*:109 (1966).

13. D. L. Taylor and G. Anderman, *Anal. Chem.*, *43*:712 (1971).

14. P. G. Burkhalter, *Anal. Chem.*, *43*:10 (1971).

15. Leake, *Chem. Geol.*, *5*:7 (1969).

16. M. L. Salmon and J. P. Blackledge, *Norelco Reporter*, *3*:68 (1956).

17. K. Norrish and R. M. Taylor, *Clay Minerals Bull.*, *5*:98 (1962).

18. A. Lubecki, *J. Radioanal. Chem.*, *2*:3 (1969).

19. C. J. Sparks and J. C. Ogle, Proceedings of the First Annual N.S.F. Contaminants Conference, Oak Ridge, August 1973, p. 421.

20. R. D. Giauque and J. M. Jacklevic, *Advan. X-ray Anal.*, *15*:169 (1971).

21. H. Ebel, *Advan. X-ray Anal.*, *13*:68 (1969).

22. E. Gillam and H. T. Heal, *Brit. J. Appl. Phys.*, *3*:353 (1952).

23. J. Sherman, *A.S.T.M. Spec. Tech. Publ.*, *157*:27 (1956).

24. H. J. Beattie and R. M. Brissey, *Anal. Chem.*, *26*:950 (1954).

25. H. J. Lucas-Tooth and B. J. Price, *Metallurgia*, *54*:149 (1961).

26. H. J. Lucas-Tooth and C. Pyne, *Advan. X-ray Anal.*, *7*:523 (1964).

27. G. R. Lachance and R. J. Traill, *Can. Spectrosc.*, *11*:43 (1966).

28. R. Jenkins in *An Introduction to X-Ray Spectrometry*, Heyden, London, 1974, p. 132.

29. S. D. Rasberry and K. F. J. Heinrich, *Anal. Chem.*, *46*:81 (1974).

30. J. W. Criss and L. S. Birks, *Anal. Chem.*, *40*:1080 (1968).

31. T. C. Loomis and H. D. Kieth, *X-ray Spectrom.*, *5*:104 (1976).

32. D. A. Stephenson, *Anal. Chem.*, *43*:310 (1971).

33. F. Claisse and M. Quintin, *Can. Spectrosc.*, *12*:129 (1967).

34. W. K. de Jongh, *X-ray Spectrom.*, *2*:151 (1973).

35. R. Jenkins, *Advan. X-ray Anal.*, *18*:372 (1974).

36. Z. M. Kalman and L. Heller, *Anal. Chem.*, *34*:946 (1962).

37. Z. Mencik, *X-ray Spectrom.*, *4*:108 (1975).

# 11

## Trace Analysis

### 11.1 TRACE CONCENTRATIONS AND SPECIMENS OF LIMITED QUANTITY

When the elements of interest in a specimen occur at extremely low concentrations, or when the available amount of material in a specimen is very small, it is desirable to apply the special techniques of trace analysis. In most fluorescence spectrometers the analyzed area on the surface of the specimen is designed to be approximately 5 cm$^2$. If the specimen is larger in area than this sensitive region and is much thicker than the thickness yielding 99% of the emitted x-ray intensity, then it can be considered to be of infinite dimensions. Many of the quantitative models and analysis techniques have been developed for this type of specimen. Occasionally, it is necessary to analyze a specimen which is thin compared to the 99% yield thickness or a specimen which cannot cover the total sensitive area of the fluorescence spectrometer. This type of specimen is considered to be of limited quantity. More frequently, specimens of unlimited quantity are available, but several or all of the elements of interest are present at trace concentration levels.

The term *trace concentration* will be used in this chapter for concentrations sufficiently low to suffer significant loss of

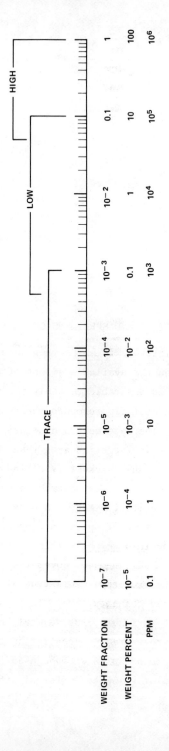

Figure 11.1  Concentration categories.  Reprinted by courtesy of EG&G ORTEC.

analytical precision due to low peak-to-background ratios or low
counting rates.  Consequently, concentration levels considered to
be trace will vary depending on sample composition and the sensitivity of the individual fluorescence spectrometer.  Figure 11.1 illustrates the trace-, low-, and high-concentration ranges for specimens of unlimited quantity.  In this instance, low concentrations
are considered to be concentrations where good analytical precision
can be obtained, but the presence of the element does not significantly affect the intensity measured on other elements.  The boundaries between what could be considered trace, low, or high are somewhat indefinite.  This fact is reflected in an overlap of the ranges
in Fig. 11.1.  Although the definition of the trace-, low-, and
high-concentration ranges is arbitrary, this choice of categories
proves to be useful in selecting appropriate quantitative techniques
in fluorescence analysis.

Both trace concentration and limited quantity specimens are
characterized by low counting rates and poor peak-to-background
ratios on the analyte line.  For this reason, most of the discussion
in Secs. 11.2 to 11.4 and 11.6 is applicable to both types.  There
are some additional methods that are appropriate for specimens of
limited quantity.  These are discussed in Sec. 11.5.  Frequently,
trace concentrations permit simplification of the quantitative
models.  These changes are discussed in Sec. 11.7.


## 11.2  COUNTING STATISTICS NEAR THE LOWER LIMIT OF DETECTABILITY

For trace analysis, counting rates are low and peak-to-background
ratios are generally poor.  This means that the dominant error in
trace analysis is normally due to counting statistics.  Since peak-
to-background ratios are low, it becomes important to measure and
subtract the background contribution accurately.

In a few ideal situations, it is possible to get by without
directly measuring the background on the calibration standards or
the unknowns.  This is illustrated in Fig. 11.2.  In this case, the

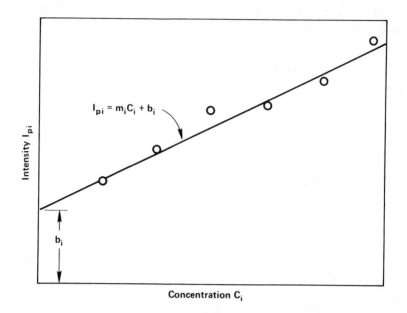

Figure 11.2  A linear calibration curve for a trace element in a
matrix with unchanging composition.  Reprinted by courtesy of EG&G
ORTEC.

matrix composition was known not to vary significantly from specimen
to specimen, and the concentration of the analyte element "i" is
low enough that a linear relationship between intensity and concentration can be assumed.  Peak intensities were measured on several
well-characterized standards with a wavelength spectrometer and
plotted against known concentration.  The y-axis intercept of the
straight line obtained by a least-squares fit to the data provides
an indirect estimate of the background $b_i$.  This calibration curve
can be used to compute the concentration of the analyte in "unknown"
specimens having identical matrix composition without measuring the
background on each specimen.  Note that the background $b_i$ can include contributions from the following sources.

1.  The x-ray tube continuum scattered by the specimen
2.  Characteristic lines in the x-ray tube spectrum scattered
    by the specimen (anode lines and contaminant lines)
3.  Characteristic lines fluoresced in the specimen chamber or
    in the x-ray spectrometer itself and recorded by the
    spectrometer.

4.   Background generated in the specimen or x-ray spectrometer due to characteristic lines of other elements fluoresced in the specimen (i.e., weak inelastic scattering mechanisms in the specimen, the Si(Li) detector background shelf, higher order diffraction

Consequently, to use the indirect method for background removal, the background sources 1 to 4 must be guaranteed to be constant from specimen to specimen.  The constancy of the background is difficult to guarantee in practical analytical situations.  Hence, a more reliable method is to make a background determination on each specimen for the peaks of interest.  Note, however, that a direct background measurement still may not eliminate contributions 2 and 3 above; so that a small y intercept may still result as in Fig. 11.2.

## 11.2.1   The Wavelength Dispersive Spectrometer

Figure 11.3 illustrates the measurement of intensity on the analyte peak and on the corresponding background for a wavelength dispersive spectrometer.  If $I_p$ is the counting rate measured at the peak position over a time $t_p$ and $I_b$ is the intensity measured at the background position over a time $t_b$, then the net intensity in the peak above background is

$$I_{p-b} = I_p - I_b \qquad\qquad (11.1)$$

and the percent standard error in $I_{p-b}$ is given by Eq. (4.200) in Sec. 4.9.2 as

$$(\sigma\%)_{net} = \frac{100\%}{\sqrt{t_p + t_b}(\sqrt{I_p} - \sqrt{I_b})} \qquad\qquad (11.2)$$

Recall that Eq. (11.2) applies for the optimum choice of counting time, where

$$\frac{t_p}{t_b} = \left(\frac{I_p}{I_b}\right)^{1/2} \qquad\qquad (11.3)$$

Also, errors from sources other than counting statistics are assumed to be negligible.  For trace analysis, $(I_p/I_b) \to 1$, and equal counting times are generally chosen for the peak and background

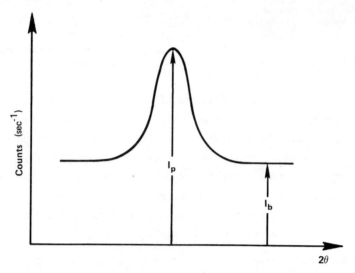

Figure 11.3  Peak and background measurements for the wavelength-dispersive spectrometer.  Reprinted by courtesy of EG&G ORTEC.

intensities.  If the background under the peak is not flat but has a gentle slope as a function of $2\theta$, two background determinations may be made at equal angular increments above and below the peak and these two values averaged.  In this case, a counting interval of $t_b/2$ is spent at each background position.

Statistical fluctuations in the measured background contribute not only to the error in measured concentration, but also limit the minimum level of concentration which can be detected with confidence.  When the net counts in the peak $I_{p-b}t_p$ approach the magnitude of the statistical error in the background counts $\sqrt{I_b t_b}$, it becomes difficult to detect the presence of the peak.  This effect is also apparent in Eq. (11.2).  As $I_p \rightarrow I_b$, the error in $I_{p-b}$ approaches infinity.  The question arises as to what minimum concentration is detectable with a reasonable level of confidence.  This concentration level is called the *minimum detectable limit,* or the *lower limit of detection.*

The level of confidence demanded is somewhat arbitrary.  However, it is frequently chosen to be 95%.  To understand what the 95%

confidence means, and to develop the proper formula for the minimum detectable limit $C_{MDL}$, it is necessary to define a measurement process. For a wavelength dispersive spectrometer, the detection test is carried out in the following way. First, the spectrometer is set to record the background to one side of the expected peak position. The background is counted for a time $t_b$ to get the measured background counts $N_b$. Note that the value $N_b$ is subject to statistical variations. Provided x-ray generator drift is negligible, $N_b$ is a sampling of a Poisson distribution with a mean $\overline{B}$ and a standard deviation $\sigma = \sqrt{\overline{B}}$. Next, the spectrometer is set to the peak position, and a number of counts $N_p$ is recorded in an equal time $t_p = t_b$. To test for the presence of a peak, the difference $\delta$ is examined.

$$\delta = N_p - N_b \tag{11.4}$$

The difference $\delta$ must be larger than the uncertainty due to the background in order to claim the presence of a peak. To set a decision level on $\delta$, it is necessary to examine the probability distribution of $\delta$ if no peak is present. In this case, both $N_p$ and $N_b$ are samples of the same distribution with mean $\overline{B}$ and standard deviation $\sigma = \sqrt{\overline{B}}$. That is, $N_p$ and $N_b$ are both samples of the same background level. For large B, it can be shown that $\delta$ is normally distributed with a mean $\overline{\delta} = 0$ and a standard deviation $\sigma_\delta = \sqrt{2\overline{B}}$. By reference to tables of the error function, it is found that there is a 10% probability of measuring a $\delta$ either larger than $+1.645\sigma_\delta$ or less than $-1.645\sigma_\delta$. In the case of the minimum detection limit, only the upper limit is of interest. Hence, there is a 5% probability of measuring a value of $\delta$ larger than $+1.645\sigma_\delta$. If the decision level for detection is set at

$$\delta_{MDL} = 1.645\sigma_\delta = +1.645\sqrt{2\overline{B}} \tag{11.5}$$

and no peak is present, then there is only a 5% probability of falsely reporting a peak present due to statistical fluctuations in the background. This is the 95% confidence decision level, and the decision level for $\delta$ defines the lower limit of detection in terms of observed counts.[*]

$$\begin{aligned}(N_{p-b})_{MDL} &= (N_p - N_b)_{MDL} \\ &= \delta_{MDL} \\ &= 1.645 \sqrt{2\overline{B}} \\ &\simeq 1.645 \sqrt{2N_b}\end{aligned} \qquad (11.6)$$

Note that the mean value of the background $\overline{B}$ has been approximated by the measured background $N_b$.

Equation (11.6) can be written in terms of intensities by noting that

$$I_b = \frac{N_b}{t_b} \qquad\qquad I_{p-b} = \frac{N_{p-b}}{t_b}$$

so that

$$(I_{p-b})_{MDL} = 1.645 \sqrt{\frac{2I_b}{t_b}} \qquad (11.7)$$

If concentration is related to intensity by

$$I_p = mC + I_b \qquad (11.8)$$

then Eq. (11.7) can be converted to concentrations using the slope constant m.

$$\begin{aligned}C_{MDL} &= \frac{(I_{p-b})_{MDL}}{m} \\ &= \frac{1.645}{m} \left(\frac{2I_b}{t_b}\right)^{1/2}\end{aligned} \qquad (11.9)$$

---

[*] Currie defines this as a decision level and defines the detection limit at a higher level of concentration where the probability of reporting the peak present is 95% when the peak is in fact present. Currie's treatment is more rigorous. See L. A. Currie, in *X-Ray Fluorescence Analysis of Environmental Samples* (T. G. Dzubay, ed.), Ann Arbor Science Publishers Inc., Michigan, 1977, pp. 289-306.

Since equal time is spent counting the background and the peak. Eq. (11.9) can be expressed in terms of the total counting time $t = t_p + t_b = 2t_b$ as

$$C_{MDL} = \frac{1.645}{m} \left(\frac{2I_b}{t/2}\right)^{1/2}$$

$$= \frac{1.645(2)}{m} \left(\frac{I_b}{t}\right)^{1/2}$$

$$= \frac{3.29}{m} \left(\frac{I_b}{t}\right)^{1/2} \tag{11.10}$$

At a concentration decision level equal to $C_{MDL}$, there is only a 5% probability of reporting the element present when it is not.

It can be shown from Eq. (11.2) that the percent standard deviation in the measured concentration at $C_{MDL}$ is $\geq 60.8\%$, and approaches 60.8% as $I_b t_b$ approaches infinity. Furthermore, it can be shown that

$$(\sigma\%)_{net} \rightarrow \frac{100\ C_{MDL}}{1.645\ C} \tag{11.11}$$

as

$$I_b t_b \rightarrow \infty \tag{11.12}$$

if equal peak and background counting times are used, and $C/C_{MDL}$ is the ratio of actual concentration to the predicted minimum detectable limit. From Eq. (11.11) it is clear that concentrations must be at least six times $C_{MDL}$ if percent standard deviations of 10% are desired in the measured concentration.

The computed $C_{MDL}$ is used more often as a figure of merit describing instrument performance than as an analytical tool. In this respect a word of caution is in order. A value quoted for $C_{MDL}$ is meaningless unless all of the three pieces of information listed below accompany the figure.

1.  The total counting time (real time, not livetime)
2.  The formula used to compute $C_{MDL}$
3.  The specimen composition for which $C_{MDL}$ was measured

It is necessary to state the total counting time since $C_{MDL}$ is inversely proportional to the square root of the counting time.  It is
possible to improve the minimum detectable limit by a factor of 2
by increasing the counting time by a factor of 4.  There is always
some counting time beyond which it is impractical to run because
the capability to analyze many specimens is severely compromised,
or because instrument drift becomes a limitation.  For this reason
detection limits are usually quoted for counting times in the range
of 10 to 600 s.  It is important to quote the formula used to compute $C_{MDL}$ since it is possible to choose slightly different definitions of $C_{MDL}$, usually due to a different choice of confidence
limits.  The minimum detectable limit depends on the instrument
sensitivity expressed by the slope constant m.  Since m is greatly
dependent upon specimen composition, the $C_{MDL}$ value can only apply
to a specific matrix.  Where measurement errors other than counting
statistics contribute to the detection limit, their effect should
also be specified.

Equation (11.10) allows the computation of $C_{MDL}$ from a calibration curve such as the one plotted in Fig. 11.2.  The slope of the
curve is m and the y intercept is taken to be $I_b$.  There is another
method for computing $C_{MDL}$ which employs only a single standard of
known concentration $C_{std}$.  Normally, the composition of the standard is chosen to be identical to that for which $C_{MDL}$ value is desired, except that the concentration of the analyte is chosen to be
10 to 100 times the expected $C_{MDL}$.  The peak and background counting
rates are measured on this standard and use is made of the fact that

$$m = \frac{I_p - I_b}{C_{STD}} = \frac{I_{p-b}}{C_{STD}} \qquad (11.13)$$

Substituting Eq. (11.13) in Eq. (11.10) gives

$$C_{MDL} = \frac{3.29\ C_{STD}}{[(I_{p-b}/I_b)I_{p-b}t]^{1/2}} \qquad (11.14)$$

where the intensities are those measured on the standard.  Equation (11.14) has the advantage of requiring fewer measurements, but it places greater responsibility on the analyst for ensuring that the standard is truly representative of the behavior near the detectable limit.  Interfering peaks, such as described in the list of background sources as items 2 and 3 (Sec. 11.2), can lie under the analyte peak or at the wavelength used for background determination. Consequently, a significant distortion of the true $C_{MDL}$ value can occur when using the single standard method.  Even with the calibration curve method one can be misled unless a measurement is made to ensure that the y intercept $b_i$ is entirely accounted for by the background $I_b$ measured at an adjacent wavelength.  Equation (11.14) is interesting because it shows that the detection limit is a function of the net peak-to-background ratio, the net peak intensity, and the total counting time.  Increase any one of these by a factor of 4 and the $C_{MDL}$ will decrease by a factor of 2.

11.2.2  The Energy-Dispersive Spectrometer

Most of what has been said in Sec. 11.2.1, for the wavelength-dispersive spectrometer also applies to the energy-dispersive spectrometer and will not be repeated.  The major differences between the two systems arise from the use of peak areas instead of peak heights, and the fact that peak and background data are accumulated simultaneously rather than sequentially.  In the equations developed in Sec. 11.2.1 no mention of deadtime losses was made. Such derivations are still reasonably accurate if the deadtime losses are kept low (less than 10%), and all data are corrected for deadtime losses.  For wavelength-dispersive spectrometers this condition is easily met at trace concentration levels.  In energy dispersive systems deadtime losses can be quite high, and the existence of appreciable deadtime significantly alters the standard deviation

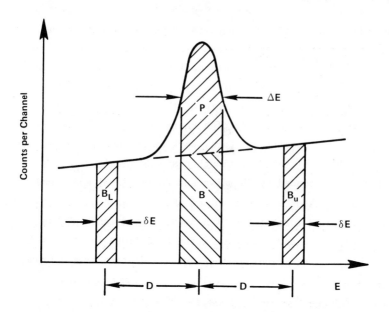

Figure 11.4  Peak and background measurements for the energy-
dispersive spectrometer.  Reprinted by courtesy of EG&G ORTEC.

of the N counts recorded in an elapsed real time, t.  To compensate
for deadtime losses, most energy spectrometers utilize a livetime
clock wherein elapsed time is counted only when the system is not
dead.  Provided that the deadtime loss correction is accurate, the
preset livetime method yields a standard deviation

$$\sigma_N = \sqrt{N} \qquad\qquad\qquad (11.15)$$

where N is the recorded number of counts in the livetime $t_\ell$.  In
the following, it is assumed that Eq. (11.15) holds.

   Figure 11.4 illustrates the measurement of peak intensity for
the energy spectrometer on a spectrum accumulated for a preset
livetime $t_\ell$.  To improve the statistical precision, the counts under
the peak are integrated over an energy interval $\Delta E$.  This interval
is normally centered on the peak and corresponds to a number of
channels $\eta_p$.  For trace element work, $\Delta E$ is just slightly wider than
the FWHM of the peak.  The integral of the counts under the peak
is

$$N_t = P + B \tag{11.16}$$

where P is the net area of the peak above background, and B is the contribution due to background.  Some means of estimating the background B must be chosen so that the net peak area can be determined as

$$P = N_t - B \tag{11.17}$$

The expected standard deviation for P in this case is

$$\sigma_P = \sqrt{\sigma_{N_t}^2 + \sigma_B^2}$$

$$= \sqrt{N_t + \sigma_B^2} \tag{11.18}$$

and the relative standard deviation for P is

$$\frac{\sigma_P}{P} = \frac{\sqrt{N_t + \sigma_B^2}}{N_t - B}$$

$$= \frac{\sqrt{N_t + \sigma_B^2}}{P} \tag{11.19}$$

The contribution to the statistical error due to the background estimate, $\sigma_B$ will depend on the method used to estimate B.  Several techniques are available and should be considered at low concentrations where the background can contribute significantly to the statistical error.

Certainly $\sigma_P$ can be minimized if $\sigma_B^2 = 0$.  One can make $\sigma_B^2$ insignificant with respect to $N_t$ by using a very large number of channels to estimate B.  If the background is linear through the peak and for a number of channels $\eta_B/2$ symmetrically located on either side of the peak, then these $\eta_B$ channels can be integrated to estimate the background under the peak.  The number of counts integrated in the $\eta_B$ channels will be related to the estimated background under the peak by

$$N_B = \frac{B}{\eta_P} \eta_B \tag{11.20}$$

With this measurement, Eq. (11.17) becomes

$$P = N_t - \frac{\eta_P}{\eta_B} N_B \qquad\qquad\qquad (11.21)$$

and

$$\sigma_B = \frac{\partial B}{\partial N_B} \sigma_{N_B} = \frac{\eta_P}{\eta_B} \sqrt{N_B} \qquad\qquad (11.22)$$

By choosing $\eta_B \gg \eta_P$, the term $\sigma_B^2$ in Eq. (11.18) can be made insignificant compared to $N_t$ to yield

$$\frac{\sigma_P}{P} \approx \frac{\sqrt{N_t}}{P} = \frac{\sqrt{P + B}}{P} \qquad\qquad (11.23)$$

Often the background cannot be adequately represented by a straight
line. For this situation some other suitable mathematical function can be least-squares fitted to the background over a wide
region to reduce $\sigma_B$, although the calculation of $\sigma_B$ will become
somewhat more complicated.

More frequently, very few channels of background are available
on either side of the peak. A typical solution is to utilize $\eta_P/2$
channels of background on either side of the peak where $\eta_P$ is an
even number. If it is assumed that these background regions are
symmetrically located on either side of the peak, then Eqs. (11.19),
(11.21), and (11.22) give, for $\eta_B = \eta_P$,

$$\frac{\sigma_P}{P} = \frac{\sqrt{N_t + B}}{N_t - B} = \frac{\sqrt{P + 2B}}{P} \qquad\qquad (11.24)$$

Equation (11.24) is a frequently quoted formula for the relative
standard deviation of the net peak intensity. It should be appreciated that it applies to a specific method of background estimation
and does not necessarily represent the best that can be achieved.
It is, however, a convenient method since most multichannel analyzers make it easy to integrate the peak and background channels.
Since the total number of channels integrated in the peak and the
background are the same, a simple subtraction produces the net peak
counts. Figure 11.4 illustrates this method if $\Delta E = 2\delta E$ is chosen.

Obtaining the best statistical precision is important near the lower limit of detection. Hence, it is important to use a peak integration interval $\Delta E$ which minimizes the statistical error. By assuming a gaussian peak shape and using either Eq. (11.23) or Eq. (11.24), it can be demonstrated that there is an optimum width, $\Delta E$, which minimizes the relative standard deviation. The optimum integration interval is a function of the ratio of the peak height above background to the height of the background. Figure 11.5 presents the results of such a calculation for Eq. (11.23). The figure of merit which is plotted is given by

$$ F_m = \left(\frac{\sigma_P}{P}\right)^{-1} \tag{11.25} $$

(See also Sec. 4.9.3 and Fig. 4.65.)

A high value of $F_m$ corresponds to a small statistical error. Each curve has been normalized to unity at its maximum. The line-to-background ratio (L/B) is the ratio of the net peak height above background to the background height. Figure 11.5 clearly shows that $\Delta E$ should be set at 1.2 times the FWHM of the characteristic peak when the minimum detection limit is approached. At high line-to-background ratios, the width of the peak integration region should be approximately twice the FWHM. These general rules can also be applied to Eq. (11.24) since the choice of the optimum integration region width is fairly insensitive to the choice between Eqs. (11.23) and (11.24). More precisely, the curves for Eq. (11.24) can be obtained from Fig. 11.5 by changing the line-to-background labels to L/B = 20, 2, and 0.2.

With energy dispersive spectrometers it is convenient to carry out all data reduction on the spectrum accumulated for the preset livetime $t_\ell$. Intensities are derived by dividing the net peak counts by the livetime. Thus, the net intensity in the peak is given by

$$ I_{p-b} = \frac{P}{t_\ell} = \frac{N_t - B}{t_\ell} \tag{11.26} $$

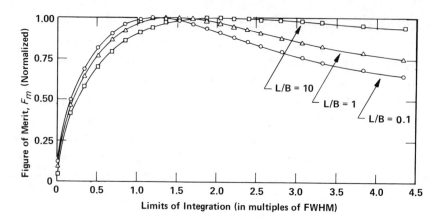

Figure 11.5  Figure of merit $F_m = (\sigma_p/P)^{-1}$ as a function of the integration limits and the line-to-background ratio L/B.  The curves are normalized to unity at their maxima.  Reprinted by courtesy of EG&G ORTEC.

and the percent standard deviation in $I_{p-b}$ can be computed from Eq. (11.23) or (11.24) as

$$(\sigma\%)_{net} = \frac{\sigma_p}{P} \times 100\% \tag{11.27}$$

At least two different detection limit formulas can be derived for the energy dispersive spectrometer, corresponding to the background estimation techniques inherent in Eqs. (11.23) and (11.24). In both cases a spectrum is accumulated on the specimen to be tested for a preset livetime $t_\ell$.  Note that it is important to measure the elapsed real time t corresponding to the livetime $t_\ell$, since any comparison of detection limits for different excitation conditions, different instruments, or different techniques must be based on the same elapsed *real* time.  Since both background and peak counts are recorded simultaneously, the total counting time is t.  Next, an integration of the counts within a region of optimum width centered on the anticipated peak position is performed, thus yielding $N_t$. This is followed by a sampling of the background in the spectrum to calculate the estimated background B under the peak.  To detect the presence of the peak, the difference

$$\delta = N_t - B \tag{11.28}$$

is examined. If $\delta$ is larger than the detection threshold $\delta_{MDL}$, the element is claimed to be present. The detection threshold is derived by considering the case where no peak is present. If a 95% confidence threshold is desired, then

$$\delta_{MDL} = 1.645 \; \sigma_\delta \tag{11.29}$$

For a background estimate corresponding to Eq. (11.23) where a large number of channels are utilized, the error due to the background estimation procedure is negligible and Eq. (11.29) becomes

$$\delta_{MDL} = 1.645 \; \sqrt{B} \tag{11.30}$$

In terms of counts, the detection limit becomes

$$
\begin{aligned}
P_{MDL} &= (N_t - B)_{MDL} \\
&= \delta_{MDL} \\
&= 1.645 \; \sqrt{B}
\end{aligned}
\tag{11.31}
$$

This can be converted to true intensities corrected for deadtime losses by writing

$$I_{p-b} = \frac{P}{t_\ell} \tag{11.32}$$

$$I_b = \frac{B}{t_\ell} \tag{11.33}$$

So that

$$(I_{p-b})_{MDL} = 1.645 \left(\frac{I_b}{t_\ell}\right)^{1/2} \tag{11.34}$$

If concentration is related to intensity by

$$mC = I_{p-b} = I_p - I_b \tag{11.35}$$

then Eq. (11.34) becomes

$$C_{MDL} = \frac{1.645}{m} \left(\frac{I_b}{t_\ell}\right)^{1/2} \tag{11.36}$$

For the single-standard method discussed in Sec. 11.2.1, Eq. (11.36) can be written

$$
\begin{aligned}
C_{MDL} &= \frac{1.645\ C_{std}}{[(I_{p-b}/I_b)I_{p-b}t_\ell]^{1/2}} \\
&= \frac{1.645\ C_{std}}{[(P/B)(P/t_\ell)t_\ell]^{1/2}}
\end{aligned}
\tag{11.37}
$$

where the intensities or counts are measured on the standard of concentration $C_{std}$. Equations (11.36) and (11.37), along with Eq. (11.23), apply to the case where a very large number of channels is used to estimate the background.

Where the number of channels integrated to estimate the background is equal to the number of channels integrated in the peak, Eq. (11.30) becomes

$$
\delta_{MDL} = 1.645\ \sqrt{2B}
\tag{11.38}
$$

leading to the result

$$
\begin{aligned}
C_{MDL} &= \frac{1.645\ \sqrt{2}}{m}\left(\frac{I_b}{t_\ell}\right)^{1/2} \\
&= \frac{2.33\ C_{std}}{[(I_{p-b}/I_b)I_{p-b}t_\ell]^{1/2}} \\
&= \frac{2.33\ C_{std}}{[(P/B)(P/t_\ell)t_\ell]^{1/2}}
\end{aligned}
\tag{11.39}
$$

This result corresponds to the use of Eq. (11.24).

With an energy spectrometer, the detection test is frequently based on a visual inspection of the spectrum. In this mode the analyst visually scans the spectrum and quickly establishes a background estimate based on a mental averaging of a large number of background channels. A small perturbation above this level which also has the shape of a peak is easily recognizable. Visual inspection yields a detection limit closely approaching the definition in Eqs. (11.36) and (11.37).

For both Eqs. (11.37) and (11.39), it can be shown that the
percent standard deviation in the concentration at the detection
threshold $C_{MDL}$ is greater than 60.8% and approaches 60.8% as the
background approaches infinity.  To achieve a 10% standard deviation
in the analyzed concentration, a concentration at least six times
the $C_{MDL}$ value is required.  This result is identical to that given
in Sec. 11.2.1 for the wavelength spectrometer.  The general comments in Sec. 11.2.1 regarding the calculation and interpretation
of minimum detection limits should be reviewed and applied to energy
dispersive systems at this point.

### 11.2.3  The Minimum Analyzable Limit

The *minimum detectable limit* defines the concentration level above
which it is possible to say with confidence that the element is
present.  In practical analytical problems one finds that detection
is necessary but seldom sufficient.  Reporting detection usually
stimulates the question "About how much?"  Hence, it becomes necessary to define the concentration level at which it is possible to
say roughly how much is present.  This concentration is called the
*minimum analyzable limit*.  Here again, the precise choice of numbers
to define the limit is somewhat arbitrary.  Frequently the analyzable limit is defined as the concentration which can be measured
with a percent standard deviation of 10%.  That is,

$$\frac{\sigma_{C_{MAL}}}{C_{MAL}} \times 100\% = 10\% \tag{11.40}$$

where $C_{MAL}$ is the concentration corresponding to the minimum analyzable limit.  Specific values for $C_{MAL}$ can be derived from Eqs.
(11.2), (11.23), or (11.24) where appropriate.

For the wavelength-dispersive spectrometer, it can be shown
that

$$\frac{C_{MAL}}{C_{MDL}} \to 6.1 \tag{11.41a}$$

as  $I_b t_b \to \infty$ $\tag{11.41b}$

and $\dfrac{C_{MAL}}{C_{MDL}} \to 43$                                        (11.42a)

as   $I_b t_b \to 1$                                                (11.42b)

provided equal counting times $t_p = t_p = t/2$ are employed.

For the energy-dispersive spectrometer with equal background and peak integration intervals [corresponding to Eqs. (11.24) and (11.39)], similar results are obtained

$\dfrac{C_{MAL}}{C_{MDL}} \to 6.1$                                        (11.43a)

as     $B \to \infty$                                            (11.43b)

and $\dfrac{C_{MAL}}{C_{MDL}} \to 43$                                        (11.44a)

as     $B \to 1$                                               (11.44b)

However, where the background can be estimated with superior precision [Eqs. (11.23) and (11.37)], the result becomes

$\dfrac{C_{MAL}}{C_{MDL}} \to 6.1$                                        (11.45a)

as     $B \to \infty$                                            (11.45b)

and $\dfrac{C_{MAL}}{C_{MDL}} \to 61$                                        (11.46a)

as     $B \to 1$                                               (11.46b)

From the above discussion it should be clear that quantitative analysis requires concentrations well above the minimum detectable limit. The factor to be applied to a $C_{MDL}$ value to estimate the minimum analyzable limit is a function of the background level and can readily vary over a range of 6 to 60. Considerable caution should be employed in using minimum detection limits to infer minimum analyzable limits.

Table 11.1  False Detection Probabilities for Various Confidence Factors

| | Probability of Falsely Claiming Element Present | | |
| --- | --- | --- | --- |
| | Sequential Wavelength-Dispersive Spectrometer | Energy-Dispersive Spectrometer | |
| $F_{MDL}$ | | $\sigma_\delta = \sqrt{B}$ | $\sigma_\delta = \sqrt{2B}$ |
| 1.645 | 61% | 5.0% | 12% |
| 2 | 34% | 2.2% | 7.9% |
| 1.645 $\sqrt{2}$ = 2.33 | 12% | 1.0% | 5.0% |
| 2 $\sqrt{2}$ = 2.83 | 7.9% | 0.23% | 2.2% |
| 3 | 6.7% | 0.14% | 1.7% |
| 1.645 × 2 = 3.29 | 5.0% | | |

## 11.2.4  Other $C_{MDL}$ Confidence Levels

Several choices of confidence limits are encountered in the literature resulting in slightly different formulas for $C_{mdl}$. Some of the common choices are summarized in Table 11.1, together with the probability of falsely reporting the element present. The detection limit formula is assumed to be of the form

$$C_{MDL} = \frac{F_{MDL}C_{std}}{\sqrt{(I_{p-b}/I_b)}I_{p-b}t} \tag{11.47}$$

where t is either real time or livetime depending on the spectrometer. For the energy-dispersive spectrometer, the left-hand column corresponds to Eq. (11.23), while the right-hand column corresponds to Eq. (11.24). Table 11.1 allows conversion of published $C_{MDL}$ values to the same basis outlined in Secs. 11.2.1 and 11.2.2., where the 5.0% false presence claim probability was chosen.

## 11.2.5  Effects of Instrument Drift on Minimum Detection Limits

In the derivation of the formulas describing the minimum detectable concentration, it has been assumed that the only source of variance in peak and background counts is that due to counting statistics.

Consequently, it was found that the detectable limit is inversely
proportional to the square root of the counting time.  Hence, when
a total analysis time of 10 min is inadequate to detect a trace
concentration, it is tempting to improve detection limits by a factor of 8 by running overnight or by a factor of 19 by running over
the weekend when the instrument normally would be shut down.  On
fluorescence spectrometers where the peak and background intensities are measured simultaneously with *identical* efficiencies, this
is a simple exercise.  However, on scanning wavelength-dispersive
spectrometers where peak and background intensities are measured
sequentially, the instrument drift becomes an important limitation
on minimum detection limits for long counting intervals.  Instrument
drift encompasses x-ray generator intensity drift and drift in the
spectrometer detection efficiency.  In multichannel wavelength-
dispersive spectrometers where one channel monitors the peak while
another counts background, only the relative drift between the detection efficiencies of the two spectrometer channels is important.
The following discussion treats the limitations on detection limits
caused by instrument drift in sequential wavelength-dispersive
spectrometers.

Medium-to long-term drift can be caused by changes in the
following parameters:

1.  Room temperature
2.  Temperature of the x-ray tube and generator cooling water
3.  Changes in the instrument operating temperatures from
    power off to power on (warm-up), or when the x-ray tube
    power setting is changed
4.  Changes in the proportional counter detection efficiency
5.  Changes in detector or amplifier gain
6.  Pulse-height-selector window drift
7.  Diffraction crystal temperature changes
8.  Contamination of the x-ray tube anode
9.  Changes in the composition or shape of the specimen being
    analyzed
10. Changes in ac line voltages
11. Drifts in clock frequency on the livetime clock, or
    elapsed real time clock due to temperature or ac line
    frequency changes

In addition, excess short-term fluctuations can be caused by residual sine-wave ripple in the electronics at some multiple of the ac
line frequency, as a result of incomplete electronic filtering.  In
contrast to counting statistics, the above sources of drift lead to
*systematic* errors in the data.  Although determining the parametric
relationships causing drift in a particular instrument can be a
difficult, time-consuming, and frustrating task, useful insight into
the long-term effects of drift can be gained by considering a few
simplified models.

The measured intensity at time t will be assumed to be related
to the drift by

$$I(t) = I(0)[1 + D(t)] \qquad\qquad (11.48)$$

where $I(0)$ is the intensity at time $t = 0$, $D(t)$ is the drift as a
function of time; and $I(t)$ is the intensity at time t.  Several
simple functional forms are useful for describing $D(t)$.  In each
case, the rate of drift is defined by $\omega$, while the amplitude and
direction is given by a.

A linear relationship

$$D(t) = a\omega t \qquad\qquad (11.49)$$

is appropriate whenever the characteristic drift period is large
compared to the total measurement time.  It may also describe the
effects of constant rate contamination of the x-ray tube anode, or
linear changes in the specimen composition.  Drifts in intensity
subsequent to changes in x-ray tube power or during the instrument
warm-up period, are usually of the exponential form

$$D(t) = a[1 - \exp(-\omega t)] \qquad\qquad (11.50)$$

Drifts due to changes in room temperature are frequently cyclic in
nature, reflecting the periodic changes in room heat load from day
to night.  Although an accurate description would require a Fourier
series expansion, the major effects can be understood using a sine
function,

$$D(t) = a[\sin \omega(t + t_o) - \sin \omega t_o] \tag{11.51}$$

The additional parameter $t_o$ accommodates a phase shift at $t = 0$. For small values of $\omega t$, both the sinusoidal and exponential forms can be approximated by the linear equation.

In Sec. 11.2.1 it was shown that detection of a peak involves the difference between a peak and background count

$$\delta = N_p - N_b \tag{11.52}$$

and the limit of detection is determined by the standard deviation in $\delta$ when both $N_p$ and $N_b$ are samples of the same background (no peak present). That is, the limit due to counting statistics is proportional to

$$\sigma_\delta = \sqrt{2I_b t_b} \tag{11.53}$$

The systematic effect of drift on the difference $\delta$ can be calculated by examining the case where $N_p$ and $N_b$ are sequential samples of the same background. The difference due to drift is given by

$$\delta_D = \int_{t=0}^{t_b} I_b(t)\ dt - \int_{t=t_b}^{2t_b} I_b(t)\ dt$$

$$= I_b(0) \int_{t=0}^{t_b} D(t)\ dt - I_b(0) \int_{t=t_b}^{2t_b} D(t)\ dt \tag{11.54}$$

where the subscript b has been added to the intensities in Eq. (11.48). If $\delta_D$ is small compared to $\sigma_\delta$, counting statistics control the detection limit. On the other hand, for $\delta_D > \sigma_\delta$, drift sets the detection limit. The results for the various drift models are

Linear:

$$\delta_D = -a\omega t_b^2 I_b(0) \tag{11.55}$$

Exponential:

$$\delta_D = -\frac{a}{\omega} [1 - \exp(-\omega t_b)]^2 I_b(0) \tag{11.56}$$

$$(\delta_D)_{max} = - \frac{aI_b(0)}{\omega} \qquad\qquad \text{for } t_b \to \infty \qquad (11.57)$$

Sinusoidal:

$$\delta_D = \frac{2a}{\omega} \cos \omega(t_b + t_o)(\cos \omega t_b - 1)I_b(0) \qquad (11.58)$$

The properties of the sinusoidal function become clearer if $t_b$ is expressed in terms of n, the number of complete cycles spanned, and a phase angle $\phi_1 < 2\pi$.

$$\omega t_b = 2\pi n + \phi_1 \qquad (11.59)$$

$$\omega t_o = \phi_o \qquad (11.60)$$

where $\phi_o < 2\pi$. With this change, Eq. (11.58) becomes

$$\delta_D = \frac{2a}{\omega} \cos(\phi_1 + \phi_o)(\cos \phi_1 - 1)I_b(0) \qquad (11.61)$$

Consequently, the bias introduced by the drift depends only on the phases of the start and the end of the counting interval and is independent of the number of complete cycles spanned. The maximum bias error from the sinusoidal function is obtained for $\phi_1 = \pi$ and $\phi_o = 0$.

$$(\delta_D)_{max} = \frac{4a}{\omega} I_b(0) \qquad (11.62)$$

The relative importance of drift can be determined from the ratio $\delta_D/\sigma_\delta$. If the ratio is small (preferably much less than one), the drift contribution to detection limits is small.

For exponential drift, the ratio is

$$\frac{\delta_D}{\sigma_\delta} = \frac{-\frac{a}{\omega}[1 - \exp(-\omega t_b)]^2 I_b(0)}{(2I_b t_b)^{1/2}} \qquad (11.63)$$

$$\frac{(\delta_D)_{max}}{\sigma_\delta} = \frac{-aI_b(0)/\omega}{(2I_b t_b)^{1/2}} \qquad (11.64)$$

For sinusoidal drift, the ratio is

$$\frac{\delta_D}{\sigma_\delta} = \frac{(2a/\omega)\cos\,\omega(t_b + t_o)\,(\cos\,\omega t_b - 1)\,I_b(0)}{(2I_b t_b)^{1/2}}$$

$$= \frac{(2a/\omega)\cos(\phi_1 + \phi_o)\,(\cos\,\phi_1 - 1)\,I_b(0)}{(2I_b t_b)^{1/2}} \qquad (11.65)$$

$$\frac{(\delta_D)_{max}}{\sigma_\delta} = \frac{4aI_b(0)/\omega}{(2I_b t_b)^{1/2}} \qquad (11.66)$$

In both cases, the effect of drift can be important for counting intervals short compared to $1/\omega$, but it is reduced by very large values of $\omega t_b$.

With linear drift, the ratio

$$\frac{\delta_D}{\sigma_\delta} = \frac{-a\omega t_b^2\,I_b(0)}{(2I_b t_b)^{1/2}} \qquad (11.67)$$

increases rapidly with increasing counting times. Consequently, linear drift or exponential and sinusoidal drift at low values of $\omega t_b$ can have a drastic effect on detection limits as the counting interval is increased. On wavelength-dispersive spectrometers, with a simple sequential counting scheme, drift usually prevents improvements in detection limits for counting times in excess of 30 min.

The effects of drift can be reduced by alternately counting the peak and background regions with a large number of very short intervals. If n intervals of length $t_b/n$ are used, then the ratio becomes

$$\frac{\delta_D}{\sigma_\delta} = \frac{n[-a\omega(t_b/n)^2\,I_b(0)]}{n^{1/2}[2I_b(t_b/n)]^{1/2}}$$

$$= \frac{-a\omega t_b^2\,I_b(0)}{n(2I_b t_b)^{1/2}} \qquad (11.68)$$

Clearly, the importance of drift has been reduced by the factor n

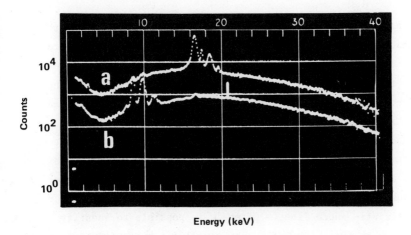

**Energy (keV)**

Figure 11.6  The scattered x-ray tube spectra obtained with a 6-mm-thick lucite specimen:  a depicts a molybdenum anode x-ray tube and b depicts a tungsten anode x-ray tube.  X-ray tube voltage was 50 kV and spectrum a has been shifted up by one decade for viewing. Reprinted by courtesy of EG&G ORTEC.

with respect to Eq. (11.67).  This counting strategy can be useful where long counting times are desirable.

Ultimately, the contribution due to drift is controlled by the nature, amplitude, and rate of drift.  These characteristics must be determined for the individual instrument before the maximum useful counting interval can be estimated.


11.3  SOURCES OF BACKGROUND

As determined in Sec. 11.2, the minimum detectable limit is controlled by the sensitivity for the element and the background contribution.  For trace element analysis it is important to understand the sources of background in order to minimize their contributions.

The most significant contribution to background is due to the x-ray tube spectrum scattered by the specimen.  This is particularly true when an unfiltered, or broadband, excitation spectrum is used on thick, low-atomic-number specimens.  Figure 11.6 illustrates the situation using a thick lucite specimen and two different anode

materials.  Scattering of the bremsstrahlung continuum leads to a
high background level at all energies.  Scattering of the characteristic anode lines from the specimen is also an important source
of interference.  Since both coherent and incoherent scattering are
involved, there is a doubling of the number of lines, or at least a
broadening of the characteristic lines, depending on the line energies and the spectrometer resolution.  Frequently, the x-ray tube
spectrum contains unwanted characteristic lines from materials used
in the anode and window construction.  These lines become interfering peaks as they scatter from the specimen just like the major
characteristic anode lines.  Often it is possible to identify the
x-ray tube as the source of interfering lines by confirming the
pattern of coherent and incoherent scattering.  For specimens with
a high average atomic number, or for very low energy lies, the
photoelectric interaction dominates, and the scattered x-ray tube
spectrum is significantly reduced in intensity.

In specimens having a high degree of atomic order or crystal
structure, interfering diffraction peaks become possible.  The angle
between the path from the x-ray tube to the specimen and the path
from the specimen to the spectrometer becomes the Bragg angle $2\theta$.
With broadband excitation, some wavelength can be found in the x-ray
tube spectrum to satisfy Bragg's law for the d spacings in the
specimen crystal structure.  For these wavelengths, diffraction
peaks will be recorded in the spectrum.  Due to the range of $2\theta$
values provided by the large solid angles in the instrument, the
diffraction peaks will be broad.  In liquids or glasses having short-
range structure ordering, broad diffraction humps can be observed
in the spectrum.  Interfering peaks due to diffraction effects can
be removed by using monochromatic excitation.

Background and contaminant lines can also be produced by the
specimen chamber and the materials holding the specimen in place.
This is particularly important where very thin specimens are analyzed, because the intensity contribution from the specimen itself

Figure 11.7  A transmitted radiation trap for reducing background scattered from the specimen chamber with thin specimens.

will be very low.  In addition, most of the excitation radiation passes through the thin specimen to strike the chamber.

Fluorescence of the chamber walls can provide contaminant lines which pass back through the thin specimen to the x-ray spectrometer.  Radiation scattered from the chamber walls and the specimen holder can also increase the level of background.  With thin specimens, or specimens of limited quantity, it is important to minimize the material in close proximity to the specimen.  In instruments where this is not possible, the addition of a transmitted radiation trap behind the thin specimen is helpful.  Figure 11.7 shows a transmitted radiation trap for insertion in a specimen holder.  The trap consists of a "honeycomb" built with small-diameter, thin-wall metal tubes.  The trap is placed directly behind the thin specimen to prevent transmitted primary x-rays from scattering back toward the x-ray spectrometer.  The material in the trap is chosen to avoid interference with lines of analytical interest in the specimen.

    Beyond the x-ray tube and specimen chamber, the x-ray spectro-
meter itself can contribute to the background. With the wavelength
spectrometer, the most important contribution is due to second- and
higher order diffraction [1,2]. For example, consider examining a
water specimen for trace amounts of iron. The spectrometer would
be set to the Fe Kα wavelength, $\lambda_{Fe} \approx 1.937$ Å. At this wavelength
one would expect to find the iron intensity superimposed on the
scattered bremsstrahlung continuum of the same wavelength. However,
without pulse-height selection, the spectrometer records an additional background component. The spectrometer crystal will also
diffract radiation from the specimen at wavelengths $\lambda = \lambda_{Fe}/n$, where
$n = 1, 2, 3, 4, \ldots$ If a tube voltage of 40 kV is used, higher order diffraction up to $n = 6$ can sample the x-ray tube continuum
scattered from the specimen. Since the scattered continuum is more
intense as $\lambda = 2\lambda_{min}$ is approached, the intensities recorded for
higher order diffraction can be substantial in spite of the lower
crystal reflectivity. Because the wavelengths for higher order
diffraction are different, use of a pulse-height selector is effective in removing this contribution to background. Note, however,
that the pulse-height selector does not remove the additional deadtime caused by the higher order lines. Clearly, it is important to
use the pulse-height selector for trace element analysis.

    Fluorescence of the Bragg crystal and scattering from the cry-
stal and its mounting can, in principle, contribute to the detected
background. Fluorescence of the crystal is particularly important
at long wavelengths due to the use of crystals containing elements
above atomic number 10 [3]. However, careful collimation, proper
choice of crystal, and use of the pulse height selector minimize
these sources in modern instruments.

    With energy dispersive spectrometers, the detector system pro-
vides the limiting background contribution when monochromatic excitation is used for trace analysis. Figure 11.8 illustrates the
effect on a pure-water specimen. For simplicity, it is assumed that
the monochromatic excitation source is the 17.4-keV molybdenum Kα

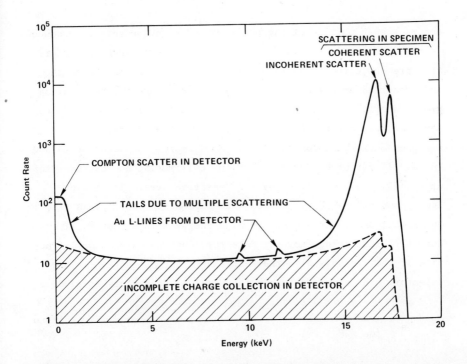

Figure 11.8  Background contributions in the energy-dispersive spectrometer with monochromatic excitation at 17.4 keV. The specimen is of low atomic number, such as distilled water. Reprinted by courtesy of EG&G ORTEC.

line from a graphite monochromator on a molybdenum anode x-ray tube [4,5]. The excitation x-rays are coherently and incoherently scattered from the water specimen to produce the intense peaks at 17.4 and 16.8 keV, respectively. The width of the coherent peak reflects the detector resolution at 17.4 keV. The incoherent peak is much broader due to the range of scattering angles included about the nominal 90° scattering angle. The low-energy tail on the incoherent peak extending down to about 10 keV is primarily due to multiple Compton scattering in the specimen. The remainder of the specimen constitutes background caused by interactions in the Si(Li) detector. The major background, represented by the cross-hatched area, is due to incomplete charge collection in the Si(Li) detector. This

occurs when a portion of the positive and negative charges produced
in the detector by the 16.8- and 17.4-keV x-rays recombine before
they are collected.  The result is a pulse of abnormally low amplitude recorded at a lower than normal energy.  The intensity of background due to incomplete charge collection is a function of detector quality and x-ray energy.  Generally, lower energy lines cause
a higher background contribution.  Some improvements can be gained
(primarily for high-energy x-rays) by collimating the detector and
using only the central 50% of its sensitive area.  At the low-
energy end of the spectrum lies the Compton shoulder.  This rise in
the background is caused by the high-energy x-rays Compton (incoherently) scattering from the detector, leaving only a small fraction of their energy with the recoiling Compton electron in the
detector.  The energy at which the Compton edge occurs is given by

$$E_e = \frac{E_o}{(mc^2/2E_o) + 1} \tag{11.69}$$

where $E_o$ is the energy of the x-ray incident on the detector and
$mc^2$ = 511 keV is the energy equivalent to the rest mass of an electron.  Both the detector resolution and multiple scattering tend to
smear out this sharp definition of the Compton edge energy in a
practical spectrum.  Contaminant gold L lines are also commonly
observed at the trace concentration level with Si(Li) detectors.
Incoming x-rays fluoresce the thin conductive layer of gold
usually employed as the front detector contact.  The resulting gold
x-rays can be detected in the Si(Li) diode.  The intensity of these
lines is a function of the thickness of the gold layer, and it may
vary significantly from detector to detector.  The gold L lines are
often broadened on the high-energy side due to the ejected photoelectrons recoiling from the gold layer into the detector's sensitive
volume.  Other materials surrounding the detector can be fluoresced
in a similar manner to yield contaminant lines.  For example,

aluminum is sometimes observed, as is lead, both of these elements
being common construction and collimator materials.  If major concentration peaks are present in the spectrum, one must also consider
the presence of sum peaks and the silicon escape peaks as discussed
in Chaps. 4 and 8.


## 11.4  METHODS FOR IMPROVING DETECTION LIMITS

### 11.4.1  The Primary Beam Filter

As discussed in Sec. 11.3, the major background source limiting
trace element analysis is usually the scattered x-ray tube continuum.  A simple and effective means of removing this limitation is
the use of a primary beam filter.  The principle is illustrated in
Fig. 11.9 for a chromium anode x-ray tube.  In Fig. 11.9(a) the
chromium K lines and the continuum hinder trace analysis in the
long-wavelength region.  Figure 11.9(b) illustrates the transmission
of a thin aluminum filter inserted between the x-ray tube and the
specimen.  In Fig. 11.9(c), the resulting filtered x-ray tube spectrum is depicted.  The aluminum filter transmits the short-wavelength radiation, but strongly attenuates the chromium lines and
all longer wavelengths.  Thus, a low-background region is created
for trace analysis of elements with long wavelengths.  These elements will be excited by the shorter wavelength continuum passing
through the filter.  Note that the sensitivity m for the trace
elements will be reduced unless the tube current can be increased
to compensate.  However, it is clear from Eq. (11.10) that the detection limits will be improved if the ratio $I_b^{1/2}/m$ is reduced.
Such an improvement is often possible.  Choice of the optimum filter thickness is important.  If the filter is too thick, the ratio
$I_b^{1/2}/m$ will increase [1].  The primary beam filter is commonly used
with chromium anode x-ray tubes to suppress the chromium K lines
from the tube and permit analysis of chromium in the specimen.

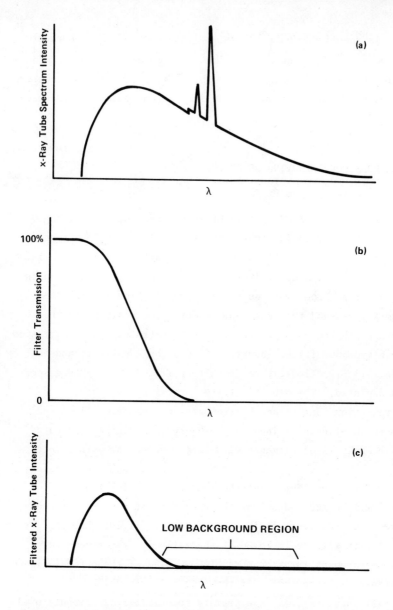

Figure 11.9  The use of a primary beam filter to reduce scattered
lines and background from the x-ray tube.  Reprinted by courtesy of
EG&G ORTEC.

## 11.4.2  Monochromatic Excitation

With energy dispersive fluorescence spectrometers, the background
suppression technique can be carried to the extreme represented by
monochromatic excitation sources.  These methods have been discussed in Chap. 3.  Improvements in the minimum detection limits by
factors ranging from 2 to 10 can be achieved over a limited range
of elements.  For energies above approximately 4 keV, the regenerative monochromator filter and secondary fluorescer methods provide
the best detection limits at somewhat higher tube power.  For lower
energies, unfiltered Mo, Rh, Ag, or W anodes operating at 15 kV and
high currents, provide the best detection limits.

Monochromatic excitation provides good trace element sensitivity only over a restricted range of elements close to the selected
excitation energy.  This occurs because the cross section for ionizing the appropriate shell in the atom decreases rapidly as the
excitation energy is increased above the absorption energy of the
analyte element.  Figure 11.10 illustrates the sensitive range for
simultaneous trace element analysis as a function of monochromatic
excitation energy.  Two bands are defined, one for analysis of the
$K\alpha$ line and the other for analysis of the $L\alpha$ line.  The high-atomic-
number boundary on each band is controlled by interference with the
incoherent scattered peak.  A good rule of thumb is to assume that
the highest energy line which can be analyzed is 3/4 of the excitation energy.  The absorption edge for this line must also lie below
the excitation energy.  For each energy in Fig. 11.10, the sensitivity at the upper limit of the band $m_u$ is calculated from the
Shiraiwa and Fujino equations [6] given in Table 2.2.  The lower
limit of the band is defined as the atomic number where the sensitivity has decreased to $m_u/10$.  Thus, for excitation with the 8-keV
copper line marked with arrows in Fig. 11.10, the sensitivity for
trace-element analysis varies by a factor of 10 from atomic number
25 to atomic number 20.  For each element, a horizontal line has

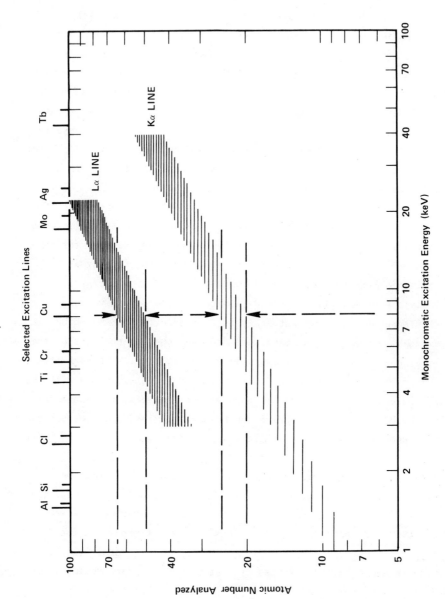

Figure 11.10  Sensitive range for trace analysis in water with monochromatic excitation.
Reprinted by courtesy of EG&G ORTEC.

been drawn within the limits of the band.  At a selected energy, the
number of lines included between the upper and lower boundaries of
the band indicate the number of trace elements which can be analyzed
simultaneously with high sensitivity.  At high atomic numbers, particularly using the Lα line, a large number of trace elements can be
analyzed simultaneously using monochromatic excitation.  At low
atomic numbers, very few trace elements can be analyzed simultaneously with a single excitation energy.  Since Fig. 11.10 does not
include detector window absorption effects, the sensitive range is
even more restricted for light elements than has been illustrated.

Although detection limits can be substantially improved using
monochromatic excitation, it should be clear from Fig. 11.10 that
analysis of a wide range of elements will require several analyses
with different excitation energies.  In analyzing a particular
specimen, it is the *total* analysis time required to quantify all
the elements which is important.  The question of total analysis
time should be addressed when choosing the optimum excitation
methods for a particular specimen type.  The severely restricted
sensitive range for light elements demonstrated in Fig. 11.10 is one
of the reasons why broadband excitation is usually more effective
for low-atomic-number elements.

## 11.4.3  Thin Specimens

With energy-dispersive spectrometers and a low-atomic-number matrix,
careful control of specimen thickness can sometimes improve detection limits [7,8].  This technique is applicable where the excitation spectrum contains intense components at energies much higher
than the energy of the analyte line.  With very thick specimens the
detected intensity of the scattered excitation radiation is high.
This causes two problems.  First, the maximum excitation intensity
which can be used is limited by the counting rate of the scattered
radiation rather than the analyte line.  Second, the intense high-
energy scattered radiation  produces a significant background under
the analyte line due to incomplete charge collection in the Si(Li)
detector.  The first effect limits the achievable sensitivity m for
the analyte line, while the second effect increases the background

intensity $I_b$. Both effects combine to degrade the minimum detection limits.

The situation is improved by reducing the specimen thickness. The principle can be most easily understood by treating the case of monochromatic excitation. As shown in Chap. 2, the intensity of the fluoresced analyte line for a specimen of finite thickness is

$$I_i(E_i) = \frac{\eta(E_i)}{4\pi \sin \psi_1} Q_{if}(E_o)$$

$$\times \left( \frac{1 - \exp\{-\rho T[\mu(E_o) \csc \psi_1 + \mu(E_i) \csc \psi_2]\}}{\mu(E_o) \csc \psi_1 + \mu(E_i) \csc \psi_2} \right)$$

$$\times I_o(E_o) \tag{2.29}$$

where

$$Q_{if}(E_o) = W_i \tau_{Ki}(E_o) \omega_{Ki} f$$

$$\approx W_i \tau_i(E_o) \frac{r_K - 1}{r_K} \omega_{Ki} f \tag{2.26}$$

The intensity of the excitation radiation at energy $E_o$ is $I_o(E_o)$, $\eta(E_i)$ is the detector efficiency at the energy $E_i$, $\psi_1$ is the incidence angle for the excitation radiation, $\psi_2$ is the takeoff angle for the fluoresced radiation, $\mu(E_o)$ is the mass absorption coefficient of the specimen for the excitation radiation, $\mu(E_i)$ is the mass absorption coefficient of the specimen for the analyte line energy $E_i$, $\tau_i(E_o)$ is the photoelectric mass absorption coefficient of the analyte for the excitation energy $E_o$, $W_i$ is the weight fraction of the analyte in the specimen, $r_k$ is the absorption edge jump ratio, $\omega_{ki}$ is the fluorescence yield, f is the fraction of the fluoresced intensity in the analyte line analyzed, $\rho$ is the specimen density, and T is the specimen thickness. This equation includes only primary fluorescence. Secondary and tertiary fluorescence can be ignored if the specimen is composed of trace elements in a low-atomic-number matrix.

The intensity of either the coherent or incoherent scattered radiation is given by

$$I_{sc}(E_{sc}) = \frac{\eta(E_{sc})}{\sin \psi_1}$$

$$\text{(11.70)}$$

$$\times \frac{\frac{d\sigma_{sc}}{d\Omega} I_0(E_0)(1 - \exp\{-\rho T[\mu(E_0) \csc \psi_1 + \mu(E_{sc}) \csc \psi_2]\})}{\mu(E_0) \csc \psi_1 + \mu(E_{sc}) \csc \psi_2}$$

where the appropriate values must be inserted for the parameters
with the sc subscript, depending on whether coherent or incoherent
scattering is being calculated. $d\sigma_{sc}/d\Omega$ is the probability in units
of square centimeters per gram per steradian for scattering the
excitation photon through the angle $2\theta = \psi_1 + \psi_2$ into the detector
[9]. The mass absorption coefficient of the specimen for the scattered radiation is $\mu(E_{sc})$. For coherent scattering, $\mu(E_{sc}) = \mu(E_0)$.
For incoherent scattering, $\mu(E_{sc})$ will be slightly larger than
$\mu(E_0)$ since the incoherently scattered x-rays have a slightly lower
energy.

What is of interest is the dependence of the ratio of fluoresced intensity to scattered intensity on the specimen thickness T.
From Eqs. (2.29) and (11.70), it is clear that the specimen thickness dependence is totally contained in the terms within the parentheses. To demonstrate the effect, it will be assumed that
$\psi_1 = \psi_2 = \psi$, and the approximation $\mu(E_0) \approx \mu(E_{sc})$ will be used.
Furthermore, the analyte energy is typically half the excitation
energy which yields the approximate result $\mu(E_i) \approx 7\mu(E_0)$ for a low-
atomic-number matrix containing only trace elements. Inserting
these approximations into Eqs. (2.29) and (11.70), it can be shown
that the ratio of fluoresced intensity to scattered intensity is
proportional to the ratio $_T I_i / _T I_{sc}$ given by

$$\frac{_T I_i}{_T I_{sc}} = \frac{1 - \exp[-8\mu(E_0)\rho T \csc \psi]}{1 - \exp[-2\mu(E_0)\rho T \csc \psi]}$$

$$\text{(11.71)}$$

The numerator

$$_T I_i = 1 - \exp[-8\mu(E_0)\rho T \csc \psi]$$

$$\text{(11.72)}$$

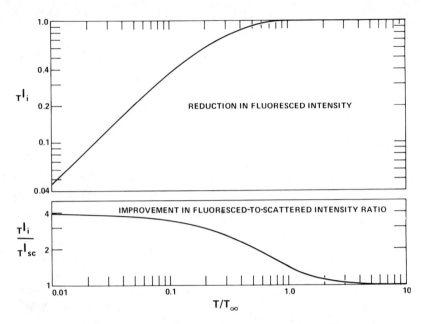

Figure 11.11   The effect of reducing specimen thickness on scattered
and fluoresced intensity.  See text for details.  Reprinted by
courtesy of EG&G ORTEC.

defines the thickness dependence of the fluoresced intensity, while
the denominator $_T I_{sc}$ describes the thickness dependence of the
scattered intensity.  The infinite thickness definition for the
analyte line requires that $_T I_i = 0.99$.  From Eq. (11.72) this yields
an infinite thickness

$$T_\infty = \frac{-\ln 0.01}{[\mu(E_o) \csc \psi_1 + \mu(E_i) \csc \psi_2]\rho}$$

$$\approx \frac{4.61 \sin \psi}{8\mu(E_o)\rho} \tag{11.73}$$

Figure 11.11 plots Eqs. (11.71) and (11.72) for various ratios
of specimen thickness to the infinite thickness given in Eq. (11.73).
By reducing the specimen thickness to 1/100 of the infinite thickness, the ratio of fluoresced intensity to scattered intensity can
be improved by a factor of 4.  However, the fluoresced intensity is
reduced by a factor of 22.  If the tube current can be increased by

a factor of 22, the original fluoresced intensity will be restored, but the peak-to-background ratio will be improved by a factor of 4. If another factor-of-4 increase in tube current is possible, the spectrometer will be back to the condition where the scattered intensity is at the counting rate limit for the spectrometer, with the analyte line intensity increased by a factor of 4. Hence, the detection limits will improve by an overall factor of 4. In most cases, it is not possible to increase the x-ray tube current by a factor of 88 (i.e., $4 \times 22$). It should be noted that a specimen thickness of $0.1T_\infty$ requires an increase in tube current by only a factor of 7.3 over the thick-specimen case and yields an improvement in detection limits by a factor of 3.4. This represents a more readily attainable objective. In general, trace elements are commonly analyzed over an energy range extending from 1/4 to 3/4 of the monochromatic excitation energy. Obtainable improvements in detection limits using the thin specimen method will be better at the low-energy end of the range and poorer at the high-energy limit. An improvement of detection limits by a factor between 1 and 4 can be expected with the thin-specimen method. With thin specimens, the fluoresced intensity is a function of specimen thickness. Monitoring the intensity of the scattered lines provides a convenient means of correcting for specimen thickness variations.

## 11.4.4 Concentration Methods

It seems almost trite to say that detection limits can be improved by concentrating the analyte before analysis. However, this approach is one of the most fruitful methods for improving the analytical precision on trace elements. Trace-element concentrations in water solutions can be increased substantially by chemical precipitation [10], or filtering through filter paper loaded with an ion-exchange resin [11]. Analysis of the precipitate or the filter paper presents a much larger quantity of the analyte to the fluorescence spectrometer without the scattering normally contributed by the water. Analyte concentration by water removal can also be

achieved by "freeze-drying." Lubozynski et al. obtained an improvement in detection limits by a factor of 3 for arsenic in whole
blood using this technique [12]. Concentration by water removal can
also be achieved through evaporation [10], a simpler but more time-
consuming method. For organic matrices, ashing or digestion followed by another method of concentration can be employed [10]. For
water solutions, electrolysis [10] and extraction [10,13] are sometimes useful. In general, one can conclude that judicious use of
wet chemistry can significantly improve trace-element analysis.
Detection limits often can be improved by one to two orders of
magnitude. On the other hand, one must monitor the process for
concentration distortion. For example, in speeding up evaporation
with heat, some of the desired elements may be unknowingly driven
off. Test standards and addition of internal standards can be used
to check for such distortions.

Mechanical methods of concentration can also be used. One of
the most interesting examples is the monitoring of air pollution
particulates by drawing large volumes of air through a filter paper.
The filter paper is analyzed directly in a fluorescence spectrometer. Detection limits ranging from 5 to 200 $ng/m^3$ of air are
feasible. References 8 and 14 to 18 deal with this application.


## 11.5  SAMPLES OF LIMITED QUANTITY

Sometimes the analyst is presented with a very small amount of material for analysis. The problem can range from having enough to
analyze but not enough to grind, to having a barely visible particle. If the latter is the case, the analyst may be well advised
to abandon the fluorescence spectrometer and analyze the particle
with an electron-beam microprobe or a scanning electron microscope
equipped with an x-ray spectrometer. These instruments are designed
for analyzing very small volumes of material with detection limits
as low as $10^{-14}$ to $10^{-15}$ g. The corresponding detection limits
for fluorescence spectrometers lie in the range of $3 \times 10^{-6}$ to

$6 \times 10^{-8}$ g.  For detecting extremely small masses of material, the electron-beam microprobe or the scanning electron microscope with an x-ray spectrometer are vastly superior to the x-ray fluorescence spectrometer.  The fluorescence spectrometer performs best in bulk analysis or analyzing a large area ($\sim 3$ cm$^2$).

Forensic studies often require the analysis of very small samples with the further restriction that the sample represents material evidence and must not be modified in any way.  This proviso usually rules out accurate quantitative analysis.  However, for forensic purposes, qualitative or crude semiquantitative analysis is often all that is necessary.  The small specimen can be supported on a thin plastic film in the fluorescence spectrometer.  To minimize scattered background from the large area of supporting film, a small-diameter collimator can be placed over the x-ray tube window.  This restricts the excitation radiation to regions containing the specimen and helps to improve peak-to-background ratios. Alternatively, the x-ray spectrometer can be collimated to view only the specimen.  However, the spacial resolution using the latter method is normally somewhat worse.  By adding a specimen stage with micrometer motion in the x and y directions, the capability of specimen scanning can be added to the collimated system.

Particle size effects are frequently a problem with limited quantity samples.  Normally there is not enough material available to permit grinding to reduce the particle size effects.  One method of handling this problem is to take the sample into solution.  The solution can be deposited on a thin (3.8 μm) plastic film and dried to form a thin-film deposit.  The thin plastic film is an ideal supporting medium for analysis in the fluorescence spectrometer. An alternative method is to deposit the solution on a filter paper, a drop at a time, allowing the solvent to evaporate.  The area for deposit can be defined by a wax impregnated ring on the filter paper.  Where a somewhat larger amount of material is provided, particle size effects can be eliminated by the glass fusion technique outlined in Chap. 7.

Finely powdered specimens of limited quantity can be sprinkled on Scotch tape.  The tape will retain a thin layer of powder for analysis in the spectrometer.  Another method is to support the powder between two layers of thin plastic film.  At least one commercially available specimen cup provides for this method.

With thin specimens or specimens of limited quantity, much of the background is caused by the excitation radiation striking materials surrounding the specimen.  It is important to keep all materials near the specimen to an absolute minimum, particularly the low-atomic-number materials which are efficient scatterers.  It is much more difficult to control the vertical position of the specimen in the fluorescence spectrometer on materials supported by mylar films or filter paper.  For this reason one can expect an error contribution in the range of 1 to 5% due to sample position variations.  Frequently, the error due to counting statistics overshadows the sample position error on limited quantity specimens.

## 11.6  INTERFERING PEAKS

Much of the discussion in this chapter assumes only trace elements are present in a low-atomic-number matrix, and considers the analyte line to be free of interference from other peaks.  These conditions are not always fulfilled.  Interference from other lines significantly degrades detection limits, particularly when the interfering peak is from an element of major concentration.  In an energy dispersive spectrometer, even though the intense peak does not overlap the trace-element peak, it can increase the background for the small peak if the major peak has a higher energy.  The increased background is due to incomplete charge collection, as discussed in Sec. 11.3.

When both overlapping peaks are weak and approximately equal in intensity, simple mathematical expressions for peak shapes and background can be used in a least-squares-fitting procedure to extract the individual intensities.  For example, a linear or quadratic background is often used with gaussian peak shapes in energy dispersive spectrometers.  When the interfering peak has a very high intensity, then its shape must be known very accurately in the region where the trace-element peak occurs.  With an incorrect peak shape, the least-squares-fitting method will produce a large error in the trace peak intensity.  One of the best solutions to this problem is to record reference spectra of the two interfering elements using single-element standards.  These experimentally determined peak shapes can be used to extract the correct intensities from the unknown.  Either the least-squares-fitting method can be used, or the proper amount of the more intense peak can be subtracted from the unknown spectrum to leave the trace peak as a residue.  This reference spectrum method can be used in some cases to subtract the background.  A blank standard composed of the matrix with no trace elements is analyzed to establish the background spectrum.  This spectrum is subtracted from the unknown to yield the trace element peaks without background.  Note that background subtraction or mathematical filtering techniques perpetrated on the acquired spectrum do not remove the statistical error caused by the background in the original spectrum.  Detection limits and analytical precision must *never* be computed from the spectrum after background has been removed.  A similar comment applies to estimation of analytical precision on overlapping peaks which have been unraveled.  In using the reference spectrum method, the statistical error will be increased by $\sqrt{2}$ if the reference spectrum contains the same number of counts as the unknown.  To avoid this contribution, the reference spectrum is normally counted for a much longer time and multiplied by the appropriate constant to normalize it to the unknown spectrum.

The problem of interfering peaks and changing background must be closely monitored for accurate trace-element analysis.

## 11.7  QUANTITATIVE MODELS FOR TRACE ANALYSIS

Trace analysis permits some simplification of the quantitative
models used to calculate concentrations from measured intensities.
In the worst case, where the specimens contain both trace and major
elements of varying concentrations, one can at least neglect the
effect of the trace elements on the major-element intensities.
Changes in enhancement and absorption of the trace-element intensities due to changes in the major-element concentrations will, of
course, need to be accounted for.

### 11.7.1  The Linear Calibration Curve

A more ideal situation is where the matrix is of constant composition and only the trace-element concentrations vary from specimen
to specimen.  Here, only the constant matrix absorption needs to be
considered for each element.  Interelement effects can be ignored,
and a simple linear equation such as Eq. (11.8) can be used to relate concentration to measured intensity separately for each element.  Probably the most desirable method for generating the calibration curves is by making up standards of known composition.
Several standards are required for each element, bracketing the expected range of concentration.  Since trace elements are involved,
the set of standards for one element can include the range of concentrations for all the elements.  Consequently, somewhere between
4 and 10 standards are adequate.

Sometimes only one sample is to be analyzed quantitatively.
If sufficient quantity is available, the sample can be homogenized
and split into three or more identical samples.  The first sample
is analyzed qualitatively and a rough estimate of the concentration
of each element is made.  The second sample is spiked with known
amounts of each element to bring the concentration to approximately
10 times the estimated concentration in the unknown.  Based on the
measured intensity on the second specimen, at least one more sample
is prepared with spiked concentrations providing another point on
the desired calibration curve.  The differences between the

intensities measured on the first, second, and third specimens permit a calculation of the concentration in the unknown. More spiked standards can be made up to better define the calibration curve where necessary.

## 11.7.2 Fundamental Parameters Calculations

Low-atomic-number specimens containing only trace elements considerably simplify theoretical calculations based on the fundamental parameters. Enhancement effects can be neglected, and the effective wavelength approximation becomes more accurate. Fundamental methods are particularly simple to apply if monochromatic excitation is used. A periodic calibration of the spectrometer efficiency with known standards is all that is required to calculate unknown concentrations directly from Eq. (2.31) [4,5].

## 11.7.3 Absorption Corrections: Method One

Fundamental parameter calculations require knowledge of the mass absorption coefficients for the specimen. If the mass absorption coefficients of the specimen for the excitation and fluoresced radiation are unknown, they can be measured by inserting a known thickness of the sample between the x-ray tube and specimen or the specimen and detector, respectively. The mass absorption coefficient $\mu$ is calculated from the Beer-Lambert law

$$I = I_o \exp(-\mu\rho T) \tag{11.74}$$

where $I$ is the counting rate measured with the absorption specimen in place and $I_o$ is the measured rate after removal. The mass thickness $\rho T$ is computed by dividing the weight of the absorption specimen by its measured area. The intensity measurements $I$ and $I_o$ are made on the analyte line with the regular specimen in the normal specimen position. To measure the mass absorption coefficient of the specimen for the excitation radiation, the intensities $I$ and $I_o$ are measured with and without the absorption specimen placed between the x-ray tube and the specimen. For the mass absorption coefficient of the specimen for the analyte line, the absorption

specimen is inserted between the specimen and the detector. This
method works well only for small values of $\mu\rho T$ where a reasonably
high counting rate I can be obtained. This means high-energy lines
(short wavelengths), thin absorption specimens, and low-atomic-
number matrices. The method is detailed in Sec. 10.6.2.

11.7.4  Absorption Corrections:  Method Two

Giauque et al. [19] have described a convenient method of making absorption corrections for thin specimens excited by monochromatic
radiation. The method is particularly suited to trace analysis in
thin films. With a slight modification, Eq. (2.29) for primary
fluorescence can be expressed as

$$I_i(E_i) = \frac{1}{4\pi \sin \psi_1} I_o(E_o) K_i M_i K_a \tag{11.75a}$$

where

$$K_i = \tau_i(E_o) \frac{r_k - 1}{r_k} \omega_{ki} f\eta(E_i) \tag{11.75b}$$

and

$$K_a = \frac{1 - \exp\{-[\mu(E_o) \csc \psi_1 + \mu(E_i) \csc \psi_2]\rho T\}}{[\mu(E_o) \csc \psi_1 + \mu(E_i) \csc \psi_2]\rho T} \tag{11.75c}$$

$I_i(E_i)$ is the counting rate measured for the analyte line. $\eta(E_i)$
is the detection efficiency for the analyte line at energy $E_i$, including the transmission of the x-ray path from the specimen to the
detector.

The mass per unit area of the analyte in the specimen is $M_i$,
while $\mu(E_o)$ and $\mu(E_i)$ are the specimen mass absorption coefficients
for the excitation radiation and analyte line, respectively. All
other parameters have the same meaning as in Eq. (2.29). The value
for $K_i$ either can be calculated theoretically, or measured experimentally on single-element, thin-film standards. Giauque et al.
report excellent agreement between both methods [19]. The thin-
film standards are made by depositing a thin layer of the element
on an aluminum substrate, and they are sufficiently thin to make

the absorption correction term $K_a$ become unity.  Similarly, $I_o(E_o)/(4\pi \sin \psi_1)$ can be determined by a single measurement on a thin-film standard and may need to be repeated from time to time to compensate for long-term drift in the x-ray generator.

For absorption corrections, targets are made up to match the analyte elements in the thin specimen.  The target must contain the analyte and have no elements in sufficient concentration to enhance the intensity of the analyte line.  Furthermore, the target material should be chosen to minimize scattering of the excitation radiation. In some cases, a single multielement target can be used if it is thin enough to satisfy the requirements outlined above.  Three measurements are made on the analyte line intensity.  With the specimen in place and the target directly behind it, the intensity $I_{TS}$ is measured.  Next, the specimen is removed and the intensity $I_T$ from the target alone is measured.  Last, the target is removed and the intensity $I_s$ is measured from the specimen alone.  These intensities yield the value

$$A = \frac{I_{TS} - I_s}{I_T} = \exp\{-[\mu(E_o) \csc \psi_1 + \mu(E_i) \csc \psi_2]\rho T\} \qquad (11.76)$$

and the absorption correction term, Eq. (11.75c), becomes

$$K_a = \frac{1 - A}{\ln(1/A)} \qquad (11.77)$$

With a well-calibrated spectrometer and a complete set of thin single-element standards, this method is capable of analyzing any new unknown without the need to develop a new set of standards similar in composition to the unknown.  Reasonably good accuracy has been demonstrated for this method on a variety of thin-specimen types [19].

REFERENCES

1.  R. Jenkins, *An Introduction to X-Ray Spectrometry*, Heyden, London, 1974.
2.  R. Jenkins and J. L. deVries, *Analyst, 94*:447 (June 1969).

3.  R. Jenkins and J. L. DeVries, *Practical X-Ray Spectrometry*, Springer-Verlag, New York, 1973.
4.  C. J. Sparks, Jr., and J. C. Ogle, in *Proceedings of the First Annual NSF Trace Contaminants Conference*, Oak Ridge National Laboratory, August 8-10, 1973, p. 421.
5.  C. J. Sparks, Jr., O. B. Cavin, L. A. Harris, and J. C. Ogle, in *Trace Substances in Environmental Health - VII. 1974. A Symposium* (D. D. Hemphill, ed.), Proceedings of the University of Missouri's 7th Annual Conference on Trace Substances in Environmental Health, University of Missouri, Columbia, p. 295.
6.  Toshio Shiraiwa and Nobukatsu Fujino, *Jap. J. Appl. Phys., 5(10)*:886 (October 1966).
7.  J. R. Rhodes, in *Energy Dispersion X-Ray Analysis: X-Ray and Electron Probe Analysis, ASTM Special Publication STP 485*, American Society for Testing and Materials, Philadelphia, 1971, p. 243.
8.  J. R. Rhodes, *Amer. Lab., 57* (July 1973).
9.  B. E. Warren, in *X-Ray Diffraction*, Addison-Wesley, Reading, Mass., 1969.
10. H. Huberman, G. Warner, and F. Widman, *Norelco Reporter, 20(3)*:10 (1973).
11. W. J. Campbell, T. E. Green, and S. L. Law, *Amer. Lab., 28* (June 1970).
12. M. F. Lubozynski, R. J. Baglan, G. R. Dyer, and A. B. Brill, *Intern. J. Appl. Radiation Isotopes, 23*:487 (1972).
13. Frank J. Marcie, *Environ. Sci. Technol., 1(2)*:165 (1967).
14. D. C. Camp, J. A. Cooper, and J. R. Rhodes, *X-ray Spectrom., 3*:47 (1974); D. C. Camp, A. L. VanLehn, J. R. Rhodes, and A. H. Pradzynski, *X-ray Spectrom., 4*:123 (1975).
15. F. S. Goulding and Joseph M. Jaklevic, *X-Ray Fluorescence Spectrometer for Airborne Particulate Monitoring*, Environmental Protection Technology Series, EPA-R2-73-182, (April 1973), U.S. Environmental Protection Agency, Washington, D.C.
16. J. Walinga and A. H. C. Hendriks, *Can. Res. Develop., 36* (March-April 1974).
17. Robert D. Giauque, Lilly Y. Goda, and Roberta B. Garrett, *X-Ray Induced X-Ray Fluorescence Analysis of Suspended Air Particulate Matter*, LBL-2951, UC-11, TID-4500-R61, National Technical Information Service, U.S. Department of Commerce, Springfield, Va., July 1974.
18. J. A. Cooper, *Review of a Workshop on X-Ray Fluorescence Analysis of Aerosols*, Battelle report BNWL-SA-4690, June 1, 1973, Battelle Pacific Northwest Laboratories, Richland, Wash.
19. Robert D. Giauque, Fred S. Goulding, Joseph M. Jaklevic, and Richard H. Pehl, *Anal. Chem., 45*:671 (1973).

# 12

## Radiation Health Hazards in X-ray Spectrometry

### 12.1  GENERAL

Although the potential health hazard in the use of x-radiation has
long been recognized, it is only since approximately 1968 or so that
strict legislation has been generally introduced in such a way as
to control radiation leakage problems in new and modified instrumentation and to protect the individual user.  In most parts of the
world, the degree to which the radiation laws are enforced is
directly dependent upon the number of government officials available to enforce them, and at the present time, this number is much
too small.  As a result, the safety control of x-ray installations
is poor, and much of the responsibility for adequate training and
protection of x-ray analysts falls under the responsibility of the
local safety officer and in many cases directly on the user himself.
The undesirability of this situation can be judged from the fact
that the number of reported radiation accidents is high, and the
number of accidents not reported is probably even higher.  In the
United States, figures for 1972 indicate accident rates of the order
of 1 per 100 or so installations per year [1,2].  Since there are
probably in excess of 10,000 x-ray spectrometers in the world,

roughly two accidents occur every week. It is, therefore, in the
best interests of all x-ray spectroscopists to acquaint themselves
with the potential health hazards involved, with their legal and
moral responsibilities to their co-workers and themselves, and with
the correct control, protection, and monitoring of x-radiation.


## 12.2  UNITS OF MEASUREMENT FOR IONIZING RADIATION

The standard unit for measuring the quantity of ionizing radiation
is the roentgen (R). The roentgen is defined as a radiation flux
that will produce $2.08 \times 10^9$ ion pairs per 0.001293 g (i.e., 1 cm$^3$)
of dry air at standard temperature and pressure. This number of
ion pairs is equivalent to an energy of 84 erg.

The absorption of the radiation depends on the nature of the
absorbing material; thus the actual energy deposited (i.e., ionization produced in the material) can differ considerably for different materials. For this reason, two other units of measurement were
devised which measure only the energy deposited in the absorbing
material. One of these is the roentgen-absorbed-dose (rad) which
is defined as that amount of ionizing radiation that deposits 100
erg/g of energy in the absorber. The second unit is the roentgen-
equivalent-man (rem) which is the absorbed dose in rads corrected
for the equivalent absorption of the radiation in living tissue.
Thus the dose in rem = (dose in rad × RBE), where the RBE is the
relative biological effectiveness of the radiation.

Hence, whereas the roentgen is defined and measured in terms
of ion pairs produced in air, the rad and rem are defined and
measured in terms of the energy deposited in the absorbing substance. For all practical purposes for radiation produced by x-ray
spectrometers operating up to 100 kV, the dose in rad = dose in rem.

The common measurement of x-ray intensity is the roentgen per
hour (R/h) or milliroentgen per hour (mR/h). By multiplying the
radiation intensity by the total exposure time, the total dose is
obtained. Normal background radiation levels are of the order of

0.01 to 0.1 mR/h.  At this rate, one would expect to receive
0.05 × 24 = 1.2 mR of exposure per day or 1.2 mR/day × 365, approximately 500 mR per year from terrestrial radiations.  Studies have
shown that an occupational exposure of 5 rem per year or 3 rem for
any 13-week interval is considered acceptable for regular safe
operation of x-ray equipment.

## 12.3  THE POTENTIAL HEALTH HAZARD

X-rays are energetic electromagnetic radiations that ionize matter
with which they interact, by ejecting electrons from their atoms.
The extent of the ionization, absorption, and molecular change on
a material depends on the quantity (radiation flux or photon rate)
and the quality (the spectral distribution of the photon energy) of
the radiation.  Living organisms which are exposed to various doses
of ionizing radiation can be injured by such exposures, and death
may result from severe exposures.  It is thus imperative that all
operators of x-ray instruments be knowledgeable in their use in
order to protect themselves from injury.

Biological effects of x-radiation generally follows a definite
pattern, commencing with an initial or latent period, which is essentially a time lag between the exposure and the first appearance
of biological symptoms.  This time may be a matter of hours or
days (short-term or acute exposure), or years (long-term or chronic
exposure).  There follows a period of demonstrable effects on cells
and tissues which generally arise from the cessation of cell division or cell death.  The last stage is the recovery period.  The
extent of recovery depends mainly upon the dose received, but for
doses in excess of several hundred rem, recovery is rarely complete.
The residual damage may then give rise to long-term effects.

Skin erythema (reddening of the skin) occurs with a local dose
of approximately 300 rem.  Skin destruction (burns) results from
doses above 500 rem.  This order of magnitude of dose required to
damage living tissue is at first sight large by normal background

conditions, but realizing that x-ray spectrographic tubes generate
dose rates around 100 R/s, only a short exposure to the primary
beam of analytical instruments is required to cause severe skin
burns.

Fortunately, most of the accidents involving x-ray analytical
instrumentation result in local, rather than whole-body effects.
The extremities of the body, such as fingers and hands, are able to
survive dose rates at levels which, if delivered to the whole body,
or to sensitive parts of the body, might be fatal. For example,
whereas a whole-body dose of 700 rem is generally lethal in 100% of
cases, the same dose received by the fingers, although probably
giving severe burns, would give minimal whole-body effects and would
certainly not be fatal. For this reason, when formulating maximum
permissible doses as applied to occupational radiation workers, a
distinction is generally made between whole-body and body-extremity
dose rates. *Whole body* is defined as including head, trunk, active
blood-forming organs, lens of eyes, and gonads. *Body extremities*
refer to hands, forearms, feet, and ankles.

Current regulations in the United States require that no member
of the public may receive more than 500 mrem/year whole-body exposure from nonmedical x-ray equipment. On the other hand, a
designated radiation worker may receive up to 5 rem/year whole-body
exposure and 75 rem/year to hands and forearms. In order to record
this exposure value, radiation workers wear film badges or similar
dosimeters since it is assumed that they will be working around
operating radiation sources and it is potentially possible to exceed
this permissible exposure.

Each person who operates, services, or tests "live" x-ray
equipment is designated an occupational radiation worker. This
designation of radiation worker requires that he should be assigned
a film badge to record his lifetime occupational radiation exposure,
that the radiation dose he receives may exceed that of the general
public by a given amount, and that he should be instructed in radiation safety and follow specified regulations.

Areas in which the radiation level may exceed 500 mrem/year are called *controlled radiation areas* from which the general public is excluded.  Only radiation workers can enter controlled radiation areas *on a regular basis* while the equipment is operating [3].  The owner of x-ray-generating equipment generally has the responsibility of posting the correct signs and enforcing the local radiation requirements.

## 12.4  ENFORCEMENT OF SAFETY REGULATIONS

The authority for the enforcement of radiation safety regulations varies from country to country, but is is generally represented by a governmental department on a national level.  In the United States the Radiation Control for Health and Safety Act of 1968 provides that the Secretary of Health, Education and Welfare shall establish an "electronic product radiation control program which shall include the development and administration of performance standards to control the emission of electronic product radiation."  X-ray diffraction and spectrographic equipment, which are discussed in this report, are included within the scope of the Act.  The primary responsibility for implementing and enforcing the provisions of the Act has been delegated to the Food and Drug Administration's Bureau of Radiological Health.  This responsibility entails the establishment of an electronic product radiation control program that involves both formal regulatory activities and a variety of efforts in basic research and development.  The law specifically directs the Bureau of Radiological Health to study the conditions of exposure to electronic product radiation and the resulting biological effects.  The Bureau also has established and is maintaining liaison with interested persons both inside and outside of government and is supporting research and training in methods of minimizing unnecessary exposure to electronic product radiation.  While these general activities are of considerable importance in their own right, they are intended primarily to support the central purpose of the law, which is to regulate and control the emission of electronic product radiation which represents a hazard to human health and safety.

## 12.5 MONITORING OF X-RADIATION

Radiation monitoring devices fall into two major categories, "active" devices and "passive" devices. Active devices give immediate warning of dangerous radiation levels but, in turn, are not generally useful in providing a continuous record of exposure. Passive devices, on the other hand, record exposure levels but do not give an immediate warning of exposure.

There are many active detectors of ionizing radiation including Geiger tubes, scintillation detectors, ionization chambers, solid-state detectors, and gas proportional detectors. When these detectors are incorporated into an electronic indicating unit, they form an active radiation-sensing instrument or survey unit which indicates that x-rays are or are not present at that instant. These active detectors are distinguished from photographic film, thermoluminescent dosimeters, and pocket dosimeters, which are passive detectors, i.e., they must be examined sometime after the exposure to obtain an exposure value. To date, only two of the active detector types are commonly used in radiation monitoring instruments. These are the Geiger detector and the ionization chamber.

Geiger detector survey meters indicate counts (photons) per minute which must be converted into milliroentgens per hour by an appropriate calibration procedure. The meter scale on Geiger survey instruments may show both counts per minute and milliroentgens per hour. The counts per minute scale indicates the actual detector response whereas the milliroentgens-per-hour scale has been determined by calibrating the counts-per-minute scale with a radioactive isotope source. Thus, unless the meter has been calibrated using the particular x-ray source you are measuring, one *cannot* use the milliroentgen-per-hour scale to indicate true exposure values.

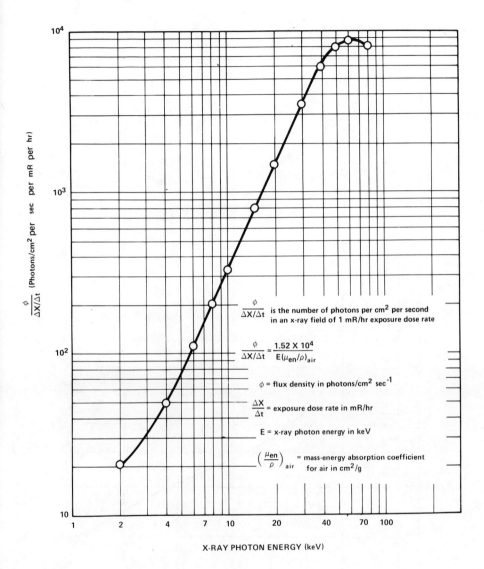

Figure 12.1  X-ray flux density per mR-per-hour exposure dose rate.
Reprinted by courtesy of EG&G ORTEC.

Geiger counters have relatively large deadtimes (~250 μs); hence
they will read too low at high (> 10,000 counts/min) counting
rates.

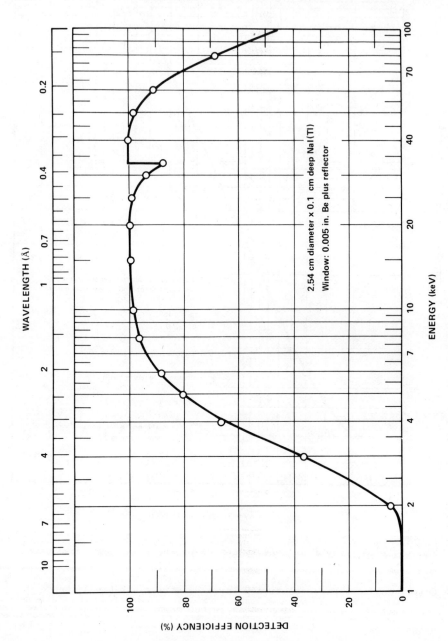

Figure 12.2   NaI(Tl) detector efficiency.   Reprinted by courtesy of EG&G ORTEC.

The scintillation counter can also be conveniently used as a radiation monitor, provided that it has been properly calibrated [4]. Figures 12.1 to 12.3 illustrate the procedure for such a calibration. First, it is necessary to calculate a curve of x-ray flux per mR/h exposure dose rate, as shown in Fig. 12.1. Second, a curve of the efficiency of the detector over the energy range must be prepared for the scintillation counter which is to be employed (Fig. 12.2). Finally, the two curves must be combined, as in Fig. 12.3, to give a curve of counting rate vs. photon energy, expressed in terms of milliroentgen per hour. This curve may now be directly employed to establish the milliroentgen rate *provided* that the energy source is essentially monochromatic. For example, a counting rate of 7000 counts/s from Mo Kα (20 keV) would correspond to 1 mR/h.

The currently acceptable level of radiation from an x-ray spectrometer is taken as 0.5 mR/h at a distance of 5 cm from the surface of the equipment.

Ionization-chamber survey meters indicate true radiation exposure (milliroentgens per hour) over the energy range for which they are designed. These meters usually have a slow response time before the meter reaches its final value. This makes them unsuitable for rapid surveys where the requirement is to detect small radiation leaks. In addition, ionization survey meters are quite fragile and require frequent calibration. Nevertheless, they are valuable because they indicate true exposure rates in milliroentgens per hour, and because they give accurate readings over large x-ray energy ranges.

It should be appreciated that although most commercially available spectrometers incorporate continuous x-ray output, constant potential generators, the use of pulsed x-ray tube generators may be encountered, particularly in energy dispersive spectrometers. Survey meters that only indicate rates are *not* suitable for measuring the x-ray exposure from pulsed sources. This is due to the fact that the deadtime of the meter is very long compared to the

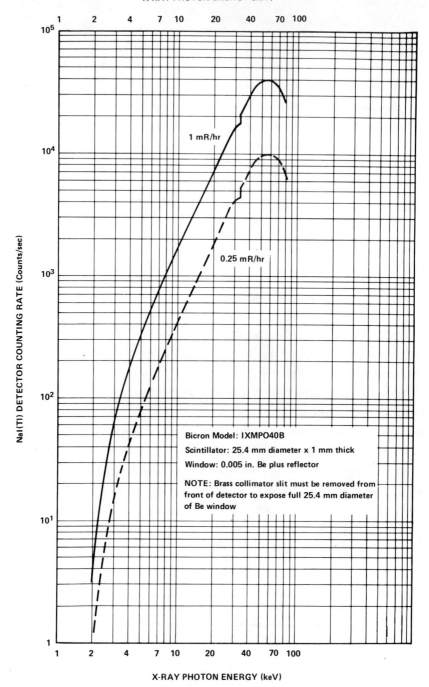

Figure 12.3  Calculated exposure rate equivalent for the NaI(Tl) detector.  Reprinted by courtesy of EG&G ORTEC.

x-ray pulse length (usually about 200 ns).   Hence, an integrating
survey meter is required which will accumulate the exposure from
a specific number of pulses.   This, in turn, enables one to determine the exposure (in roentgen units) per pulse.   A dose rate can
be calculated by multiplying the roentgen per pulse times the number of pulses per unit time.

## 12.6   USE OF FILM BADGES

The film badge is by far the most popular (and useful) passive
monitor used by the x-ray spectroscopist.   The film badge contains
(generally) two pieces of film and several absorbers through which
the radiation must pass before reaching the film.   Study of the
blackening of the film behind these absorbers thus gives a measure
of both the quantity and quality of radiation absorbed.   The major
function of the film badge is to protect the wearer from accumulated
whole-body overexposure (although it must be emphasized that is will
not protect the wearer from accidental exposures).   About every two
or four weeks the film badge is renewed and  the previous one sent
for processing by the film badge supply company.   An exposure report is sent to the companies' radiation safety officer about two
weeks after processing and any above-background exposure noted.
Film badge companies thus provide incremental (weekly or bi-weekly)
records and integral (from time of first registration to date) records for all registered employees.   The U.S. National Committee on
Radiation Protection *recommends* that a radiation workers' exposure
be limited to not more than 3 rem in 13 consecutive weeks and their
lifetime exposure should not exceed $5 \times (A - 18)$ rem, where A is
the age of the worker.

Each employee classified as a radiation worker should be assigned a film badge to record his lifetime occupational exposure.
He must wear this badge at all times in a controlled radiation area
or near an operating x-ray source while performing his work.

The construction of the film requires that it be worn correctly
with the absorbers between the film and the potential x-ray source.
For all x-ray analytical installations, film badges are best worn on
the left breast, facing the source and clipped to the extreme outside clothing (i.e., not behind a necktie or coat lapel). They
should always be worn above the table top of the equipment so that
minimum obstruction would occur if a port is left open.


12.7  LOCATION OF X-RAY SPECTROMETERS

Although the majority of modern x-ray spectrometers are adequately
designed to protect the user from accidental exposure, it is still
good practice to position the unit in such a way as to bar indis-
criminate access to the immediate vicinity of the installation.
Ideally, the spectrometer should be located in a separate room
which is in turn equipped with radiation warning signs and warning
lamps.  Where this is not possible, an area of the laboratory should
be screened off with an obvious barrier, again fitted with warning
signs and warning lamps.  Such an area should be designated the
"Controlled Radiation Area" and should be off limits to all non-
classified personnel.  It is also good practice to provide the area
with a permanently mounted radiation monitor fitted with a thin
window (low-energy-sensitive) detector connected by a long cable
to allow easy and rapid surveying [1].


REFERENCES

1.  R. Jenkins and D. J. Haas, X-ray Spectrom., 4:33 (1975).
2.  Serious Accidents, U.S. Atomic Energy Commission Issue 338,
    December 1974.
3.  Radiation Safety for X-Ray Diffraction and Fluorescence Analysis
    Equipment, N.B.S. Handbook III, American National Standard
    N43.2. 1971.
4.  K. Z. Morgan and J. E. Turner, in Principles of Radiation
    Protection, Wiley, New York, 1967, pp. 130-132.
5.  Radiological Health Handbook, U.S. Dept. of Health, Education,
    and Welfare, Washington, D.C. (January 1970).

# Appendix 1

## Nomenclature

| Symbol | *Quantity Represented* (where the given symbol represents more than one quantity, the alternatives are listed in parenthesis) |
|--------|------------------------------------------------------------|
| $\lambda$ | wavelength |
| E | energy |
| h | Plank's constant |
| $\nu$ | frequency |
| c | velocity of light |
| Z | atomic number |
| A | atomic weight (gas gain), (activity) |
| N | Avogadro's number (number of counts) |
| V | voltage |
| g | gram |
| i | current |
| t | time |
| $\psi$ | scattering angle |
| $\lambda_{incoh}$ | Compton scattered wavelength |
| $f_e$ | coherent scattering amplitude |
| r | absorption jump ratio (distance from center of electron cloud) |
| $\rho(r)$ | electron density distribution |
| $i_{e(inc)}$ | incoherent scattering per electron |
| $f(\chi)$ | absorption correction function |
| $i_{inc}$ | sum of $i_{e(inc)}$ for all atomic electrons |

561

| Symbol | Quantity Represented |
|---|---|
| $\phi$ | binding energy |
| R | reflection coefficient (resistance) |
| RC | time constant |
| $\theta$ | angle between crystal planes and diffracted beam |
| $2\theta$ | Bragg diffraction angle |
| FWHM | peak width at half maximum height above background |
| $\Gamma$ | detector resolution |
| F | Fano factor |
| q | charge |
| $\varepsilon$ | average energy to produce one ion or one electron-hole pair |
| I | intensity |
| $\rho$ | density |
| x | thickness |
| i | analyte |
| j | interfering element |
| W | weight fraction |
| Q | excitation probability (charge) |
| $\omega$ | fluorescence yield |
| $\mu$ | mass absorption (attenuation) coefficient |
| $\alpha$ | total absorption (i.e., primary and secondary) |
| $\tau$ | photoelectric mass absorption coefficient |
| A | sine incident angle/sine takeoff angle |
| $\sigma$ | standard deviation |
| n | order of diffraction |
| g | probability of electron transfer in a series |
| G | geometric constant including effect of solid angle |
| C | concentration |

# Appendix 2

# Mass Absorption Coefficients

ELEMENTS

The use of mass absorption coefficients has been discussed throughout this text, especially with regard to the calculation of primary
and secondary absorption, interelement correction coefficients,
and the fundamental parameters method of quantitative analysis.
Unfortunately, at this time, there is no single, convenient, authori-
tative table of mass absorption coefficients available. Variations
between the data given in published mass absorption coefficient
tables may be large and will often depend upon the source and manner
of obtaining and refining the data (either experimental or mathematical). Table A.1 illustrates some variations of mass absorption
coefficients as a function of the data source, and this is fairly
typical of most of the published data.

It has been shown in Chap. 2 that the total mass absorption coefficient consists of a photoelectric and a scattering component,
which can be further divided into a coherent and incoherent fraction.
Some investigators have tabulated the total mass absorption coefficient, others photoelectric only, and others still who have measured
their data obtain, for experimental reasons, numbers which lie

563

Table A.1  Variation in Mass Absorption Coefficients (in $cm^2/g$)

| Radiation | Ref. 6 (photoelectric) | Ref. 4 | Ref. 9 | Ref. 8 | Ref. 1 | Ref. 7 | Ref. 10 | Ref. 6 (Total) |
|---|---|---|---|---|---|---|---|---|
| | | Absorber Si | | | | | | |
| 1 KeV (12.396 Å) | 1606 | -- | -- | -- | -- | -- | -- | 1608 |
| 5 KeV ( 2.479 Å) | 252 | -- | -- | -- | -- | -- | -- | 253 |
| 10 keV (1.240 A) | 33.7 | -- | -- | -- | -- | -- | -- | 34.4 |
| Si Kα | 365[a] | 295 | -- | -- | 321 | 315 | -- | -- |
| Cr Kα | 205[a] | 190 | 192 | 192 | 169 | 193 | 192 | 203 |
| Mo Kα | 6.0[a] | 6.8 | 6.7 | 6.7 | 6.7 | 6.4 | 6.1 | 6.5 |
| | | Absorber Cu | | | | | | |
| 1 KeV | 12540 | -- | -- | -- | -- | -- | -- | 12550 |
| 5 KeV | 190 | -- | -- | -- | -- | -- | -- | 193 |
| 10 KeV | 218 | -- | -- | -- | -- | -- | -- | 219 |
| Si Kα | 3300[a] | 2500 | -- | -- | 3808 | 2415 | -- | -- |
| Cr Kα | 155[a] | 167 | 154 | 154 | 158 | 159 | 146 | 155 |
| Mo Kα | 49[a] | 51 | 49.7 | 49.7 | 47 | 51 | 48 | 49.3 |

[a]Graphical interpolation.

Table A.2  Mass Absorption Coefficients of Some Common Compounds (in cm$^2$/g)

| Compound | Ag Kα | Mo Kα | Zn Kα | V Kα | Radiation 10 KeV | 20 KeV | 30 keV | Ref. |
|---|---|---|---|---|---|---|---|---|
| Water | 0.7 | 1.2 | 9.2 | 42.1 | -- | -- | -- | 8 |
| Water | -- | -- | -- | -- | 5.18 | 0.775 | -- | 5 |
| Air | 0.6 | 1.0 | 8.9 | 40 | 4.99 | 0.752 | 0.349 | 8 |
| Polyethylene | 0.38 | 0.6 | 3.4 | 16.2 | -- | -- | -- | 8 |
| Pyrex | -- | -- | -- | -- | 17.1 | 2.25 | 0.786 | 5 |
| Muscle | -- | -- | -- | -- | 5.27 | 0.793 | 0.373 | 5 |
| Bone | -- | -- | -- | -- | 20.3 | 2.79 | 0.962 | 5 |
| Concrete | -- | -- | -- | -- | 26.9 | 3.59 | 1.19 | 5 |
| SiO$_2$ | -- | -- | -- | -- | 19.0 | 2.49 | 0.859 | 5 |

| | Al Kα | Ca Kα | Cu Kα | Mo Kα |
|---|---|---|---|---|
| Coal | 583 | 46 | 5.2 | 0.59 |
| Mylar | 800 | 63 | 7.1 | 0.80 |
| Granite | 1022 | 332 | 38 | 4.4 |

between these values.  Another difficulty arises from the inconveni-
ence of the early  tables which were made exclusively for wave-
length instruments, i.e., in terms of $\lambda$.  Tables of $\mu(E)$ and $\mu(\lambda)$
are both needed today.  The most comprehensive and theoretically
reliable set of mass absorption data available today is that due to
McMasters et al. [6].  Unfortunately, these data are not in a convenient form for practical x-ray spectroscopy.  In order to obtain
$\mu(E_o)$ or $\mu(\lambda_o)$, the user must mathematically or graphically inter-
polate the tabular data.  Table A.2 lists sources of currently
available mass absorption coefficient data.

## REFERENCES

1.  R. D. Dewey, R. S. Mapes, and T. W. Reynolds, *A Study of X-Ray Mass Absorption Coefficients with Tables of Coefficients,* Reynolds Metals Co., Metallurgical Research Div., Richmond, Va., 1967.
2.  K. F. J. Heinrich, X-ray absorption uncertainity (including tables of mass absorption coefficients *The Electron Microprobe* (E. D. McKinley, K. F. J. Heinrich, and D. B. Wittry, eds.), John Wiley, New York, 1966, pp. 296-377.
3.  B. L. Henke and E. S. Ebisu, *Low-Energy X-ray Anal.,* 17:150-213 (1974).
4.  L. S. Birks, *Electron Probe Microanalysis,* Interscience, New York, 1963.
5.  J. H. Hubbell, *Photon Cross Sections, Attenuation Coefficients and Energy Absorption Coefficients from 10 KeV to 100 GeV,* Nat. Bur. Stand. NSRDS-NBS-29.
6.  W. H. McMaster, N. K. Del Grande, J. H. Mallett, and J. H. Hubbell, Compilation of X-Ray Cross Sections, Sec. 2, Rev. 1, University of Calif., Livermore, U.S. Atomic Energy Comm. Rep. UCRL-50174 (1969).
7.  N. V. Philips Gloeilampenfabrieken, Application Lab., Table of x-ray mass absorption coefficients, *Norelco Reporter,* 9(3) (1962).
8.  K. Sagel, Tabellen Zur Roentgenstrukturanalyse, Band VIII, Springer-Verlag, Berlin, Gottinger, Heidelberg, 1958.
9.  H. S. Peiser, H. P. Rooksby, and A. J. C. Wilson, *X-Ray Diffrac-Diffraction by Polycrystalline Materials,* The Institute of Physics, London, 1955.
10.  D. T. Cromer and D. Liberman, Relativistic Calculation of Anomalous Scattering Factors for X-rays, Los Alamos Scientific Laboratory, LA 4403 TID 4500, July 1970.
11.  J. A. Victoreen, The calculation of x-ray mass absorption coefficients, *J. Appl. Phys.,* 20:1141 (1949).

# Appendix 3

## Table of Conversions and Physical Constants

Avogadros number = $6.02257 \times 10^{23}$ (g mole)$^{-1}$

Energy(KeV) = $12.39804/\lambda$ (Å)

1 rad = $57.29578°$

1 in = 2.54001 cm

1 Å = $10^{-8}$ cm

1 micrometer ($\mu$m) = $10^{-3}$ mm = $10^4$ Å

1 cm$^2$ = 0.15500 in$^2$ = $10^{24}$ Barns

1 in$^2$ = 6.4516 cm$^2$

1 cm$^3$ = 0.061024 in$^3$ = 0.03381 fluid oz.

1 gal = 231 in$^3$ = 0.13368 ft$^3$ = 3.78542 liter

1 liter = 1000 cm$^3$ = 33.8142 fluid oz.

1 steradian = 0.079577 Total Solid Angle

1 g = $2.20462 \times 10^{-3}$ lbs = 0.03527 oz.

1 W = 1 joule/sec = 0.0569 Btu/min

1 g/cm$^3$ = 0.02613 lb/in$^3$ = 62.43 lb/ft$^3$

1 atm = 1.0133 bars = 14.696 lb/in$^2$ = 29.921 in Hg ($0°$C)

1 wt % = 10,000 ppm

1 wt % = $10\rho$ g/liter ($\rho$ = solution density)

1 wt % = $10,000\rho$ $\mu$g/ml ($\rho$ = solution density)

$1 \ \mu g/ml = 10^{-4} \rho^{-1}$ wt % ($\rho$ = solution density)

$1 \ \mu g/ml = \rho^{-1}$ ppm ($\rho$ = solution density)

1 Kx unit = 1.00202 $\overset{o}{A}$

# Appendix 4

## Atomic Weights and Densities

EXAMPLE

ATOMIC SYMBOL → H, 1.008 (ATOMIC WEIGHT), 1 (ATOMIC NUMBER), .000089 (DENSITY $gm/cm^3$)

**Group IA**

| Symbol | Atomic No. | Atomic Weight | Density |
|---|---|---|---|
| H | 1 | 1.008 | .000089 |
| Li | 3 | 6.941 | .53 |
| Na | 11 | 22.989 | .97 |
| K | 19 | 39.09 | .86 |
| Rb | 37 | 85.467 | 1.53 |
| Cs | 55 | 132.906 | 1.9 |
| Fr | 87 | (223) | |

**Group IIA**

| Symbol | Atomic No. | Atomic Weight | Density |
|---|---|---|---|
| Be | 4 | 9.0122 | 1.82 |
| Mg | 12 | 24.305 | 1.74 |
| Ca | 20 | 40.08 | 1.55 |
| Sr | 38 | 87.62 | 2.6 |
| Ba | 56 | 137.34 | 3.5 |
| Ra | 88 | 226.025 | 5.0 |

**Group IIIB**

| Symbol | Atomic No. | Atomic Weight | Density |
|---|---|---|---|
| Sc | 21 | 44.956 | 2.989 |
| Y | 39 | 88.906 | 4.469 |
| La | 57 | 138.906 | 6.145 |
| Ac | 89 | (227) | 10.1 |

**Group IVB**

| Symbol | Atomic No. | Atomic Weight | Density |
|---|---|---|---|
| Ti | 22 | 47.9 | 4.54 |
| Zr | 40 | 91.22 | 6.5 |
| Hf | 72 | 178.49 | 13.3 |

**Group VB**

| Symbol | Atomic No. | Atomic Weight | Density |
|---|---|---|---|
| V | 23 | 50.941 | 6.0 |
| Nb | 41 | 92.906 | 8.57 |
| Ta | 73 | 180.947 | 16.6 |

**Group VIB**

| Symbol | Atomic No. | Atomic Weight | Density |
|---|---|---|---|
| Cr | 24 | 51.996 | 7.19 |
| Mo | 42 | 95.94 | 10.2 |
| W | 74 | 183.85 | 19.3 |

**Group VIIB**

| Symbol | Atomic No. | Atomic Weight | Density |
|---|---|---|---|
| Mn | 25 | 54.938 | 7.43 |
| Tc | 43 | 98.906 | 11.5 |
| Re | 75 | 186.2 | 21.0 |

**Group VIIIB**

| Symbol | Atomic No. | Atomic Weight | Density |
|---|---|---|---|
| Fe | 26 | 55.847 | 7.87 |
| Co | 27 | 58.933 | 8.9 |
| Ni | 28 | 58.71 | 8.9 |
| Ru | 44 | 101.07 | 12.2 |
| Rh | 45 | 102.906 | 12.44 |
| Pd | 46 | 106.4 | 12.0 |
| Os | 76 | 190.2 | 22.5 |
| Ir | 77 | 192.22 | 22.5 |
| Pt | 78 | 195.09 | 21.4 |

**Group IB**

| Symbol | Atomic No. | Atomic Weight | Density |
|---|---|---|---|
| Cu | 29 | 63.546 | 8.96 |
| Ag | 47 | 107.868 | 10.49 |
| Au | 79 | 196.967 | 19.32 |

**Group IIB**

| Symbol | Atomic No. | Atomic Weight | Density |
|---|---|---|---|
| Zn | 30 | 65.38 | 7.13 |
| Cd | 48 | 112.4 | 8.65 |
| Hg | 80 | 200.59 | 13.55 |

**Group IIIA**

| Symbol | Atomic No. | Atomic Weight | Density |
|---|---|---|---|
| B | 5 | 10.81 | 2.3 |
| Al | 13 | 26.982 | 2.7 |
| Ga | 31 | 69.72 | 5.91 |
| In | 49 | 114.82 | 7.31 |
| Tl | 81 | 204.37 | 11.85 |

**Group IVA**

| Symbol | Atomic No. | Atomic Weight | Density |
|---|---|---|---|
| C | 6 | 12.011 | 2.22 (GRAPHITE) |
| Si | 14 | 28.086 | 2.33 |
| Ge | 32 | 72.59 | 5.36 |
| Sn | 50 | 118.69 | 7.3 |
| Pb | 82 | 207.2 | 11.34 |

**Group VA**

| Symbol | Atomic No. | Atomic Weight | Density |
|---|---|---|---|
| N | 7 | 14.007 | .001165 |
| P | 15 | 30.974 | 1.82 (YELLOW) |
| As | 33 | 74.922 | 5.73 |
| Sb | 51 | 121.75 | 6.62 |
| Bi | 83 | 208.981 | 9.8 |

**Group VIA**

| Symbol | Atomic No. | Atomic Weight | Density |
|---|---|---|---|
| O | 8 | 15.9994 | .001332 |
| S | 16 | 32.06 | 2.07 (YELLOW) |
| Se | 34 | 78.96 | 4.81 |
| Te | 52 | 127.6 | 6.24 |
| Po | 84 | (209) | 9.27 |

**Group VIIA**

| Symbol | Atomic No. | Atomic Weight | Density |
|---|---|---|---|
| F | 9 | 18.998 | .001696 |
| Cl | 17 | 35.453 | .00321 |
| Br | 35 | 79.904 | 3.12 (LIQUID) |
| I | 53 | 126.905 | 4.93 |
| At | 85 | (210) | |

**Group 0**

| Symbol | Atomic No. | Atomic Weight | Density |
|---|---|---|---|
| He | 2 | 4.0026 | .000166 |
| Ne | 10 | 20.17 | .000839 |
| Ar | 18 | 39.948 | .001663 |
| Kr | 36 | 83.8 | .00349 |
| Xe | 54 | 131.3 | .005495 |
| Rn | 86 | (222) | 4.4 |

**Lanthanides**

| Symbol | Atomic No. | Atomic Weight | Density |
|---|---|---|---|
| Ce | 58 | 140.12 | 6.767 |
| Pr | 59 | 140.907 | 6.773 |
| Nd | 60 | 144.24 | 7.007 |
| Pm | 61 | (145) | |
| Sm | 62 | 150.4 | 7.52 |
| Eu | 63 | 151.96 | 5.234 |
| Gd | 64 | 157.25 | 7.9 |
| Tb | 65 | 158.925 | 8.229 |
| Dy | 66 | 162.5 | 8.55 |
| Ho | 67 | 164.93 | 8.795 |
| Er | 68 | 167.26 | 9.066 |
| Tm | 69 | 168.934 | 9.321 |
| Yb | 70 | 173.04 | 6.965 |
| Lu | 71 | 174.97 | 9.84 |

**Actinides**

| Symbol | Atomic No. | Atomic Weight | Density |
|---|---|---|---|
| Th | 90 | 232.038 | 11.5 |
| Pa | 91 | 231.036 | 15.4 |
| U | 92 | 238.029 | 19.0 |
| Np | 93 | 237.048 | 20.4 |
| Pu | 94 | (244) | 19.8 |
| Am | 95 | (243) | 11.9 |
| Cm | 96 | (247) | |
| Bk | 97 | (247) | |
| Cf | 98 | (251) | |
| Es | 99 | (254) | |
| Fm | 100 | (257) | |
| Md | 101 | (256) | |
| No | 102 | (254) | |
| Lr | 103 | (257) | |

# Appendix 5

## Wavelengths and Energies

**EXAMPLE**

SYMBOL → Li 3 ← ATOMIC NUMBER
λ*Kα (Å)* → 240 / .054 ← E*(keV) Kα

λ*Kα : ELEMENTS 3–54
λ L*: ELEMENTS 55–93

Each cell lists: Symbol, atomic number, λ value, E* value.

| I A | II A | III B | IV B | V B | VI B | VII B | VIII B | VIII B | VIII B | I B | II B | III A | IV A | V A | VI A | VII A | O |
|---|---|---|---|---|---|---|---|---|---|---|---|---|---|---|---|---|---|
| H 1 | | | | | | | | | | | | | | | | | He 2 |
| Li 3; 240; .054 | Be 4; 113; .109 | | | | | | | | | | | B 5; 67; .184 | C 6; 44; .279 | N 7; 31.603; .393 | O 8; 23.707; .524 | F 9; 18.307; .675 | Ne 10; 14.62; .849 |
| Na 11; 11.909; 1.041 | Mg 12; 9.889; 1.255 | | | | | | | | | | | Al 13; 8.339; 1.487 | Si 14; 7.126; 1.739 | P 15; 6.155; 2.014 | S 16; 5.373; 2.307 | Cl 17; 4.729; 2.622 | Ar 18; 4.192; 2.957 |
| K 19; 3.744; 3.312 | Ca 20; 3.36; 3.69 | Sc 21; 3.032; 4.088 | Ti 22; 2.75; 4.508 | V 23; 2.505; 4.949 | Cr 24; 2.291; 5.411 | Mn 25; 2.103; 5.895 | Fe 26; 1.937; 6.4 | Co 27; 1.788; 6.925 | Ni 28; 1.659; 7.472 | Cu 29; 1.542; 8.041 | Zn 30; 1.437; 8.631 | Ga 31; 1.431; 9.243 | Ge 32; 1.256; 9.876 | As 33; 1.177; 10.532 | Se 34; 1.106; 11.210 | Br 35; 1.041; 11.907 | Kr 36; .981; 1.263 |
| Rb 37; .927; 13.375 | Sr 38; .877; 14.142 | Y 39; .831; 14.933 | Zr 40; .788; 15.746 | Nb 41; .788; 16.584 | Mo 42; .748; 17.443 | Tc 43; .674; 18.327 | Ru 44; .644; 19.235 | Rh 45; .614; 20.167 | Pd 46; .587; 21.123 | Ag 47; .561; 22.104 | Cd 48; .536; 23.109 | In 49; .514; 24.139 | Sn 50; .492; 25.193 | Sb 51; .472; 26.274 | Te 52; .453; 27.380 | I 53; .435; 28.512 | Xe 54; .418; 29.669 |
| Cs 55; 2.892; 4.286 | Ba 56; 2.776; 4.465 | La 57; 2.665; 4.651 | Hf 72; 1.569; 7.9 | Ta 73; 1.522; 8.144 | W 74; 1.476; 8.398 | Re 75; 1.433; 8.65 | Os 76; 1.391; 8.911 | Ir 77; 1.352; 9.168 | Pt 78; 1.313; 9.441 | Au 79; 1.277; 9.707 | Hg 80; 1.242; 9.981 | Tl 81; 1.207; 10.27 | Pb 82; 1.175; 10.55 | Bi 83; 1.144; 10.836 | Po 84; 1.114; 11.27 | At 85; 1.085; 11.425 | Rn 86; 1.057; 11.727 |
| Fr 87; 1.03; 12.035 | Ra 88; 1.005; 12.334 | Ac 89; .9799; 12.65 | | | | | | | | | | | | | | | |

**Lanthanides**

| Ce 58 | Pr 59 | Nd 60 | Pm 61 | Sm 62 | Eu 63 | Gd 64 | Tb 65 | Dy 66 | Ho 67 | Er 68 | Tm 69 | Yb 70 | Lu 71 |
|---|---|---|---|---|---|---|---|---|---|---|---|---|---|
| 2.561; 4.84 | 2.463; 5.033 | 2.37; 5.23 | 2.283; 5.429 | 2.199; 5.637 | 2.12; 5.847 | 2.046; 6.058 | 1.976; 6.273 | 1.909; 6.493 | 1.845; 6.718 | 1.785; 6.944 | 1.726; 7.182 | 1.672; 7.414 | 1.619; 7.656 |

**Actinides**

| Th 90 | Pa 91 | U 92 | Np 93 | Pu 94 | Am 95 | Cm 96 | Bk 97 | Cf 98 | Es 99 | Fm 100 | Md 101 | No 102 | Lr 103 |
|---|---|---|---|---|---|---|---|---|---|---|---|---|---|
| .956; 12.967 | .933; 13.286 | .911; 13.607 | .889; 13.944 | .868; 14.278 | .848; 14.616 | .8287; 14.958 | .8098; 15.308 | .7917; 15.658 | .774; 16.016 | .757; 16.375 | | | |

# Appendix 6

# Reference Books in X-ray Spectrometry

Adler, *X-ray Emission Spectrography in Geology,* Elsevier, Amsterdam, 1966.

Anderson (ed.), *Microprobe Analysis,* Wiley-Interscience, New York, 1973.

Azaroff, *X-ray Spectroscopy,* McGraw-Hill, New York, 1974.

Bearden, *X-ray Wavelengths,* U.S. Atomic Energy Commission, Division of Technical Information, Oak Ridge, Tenn., 1964.

Bertin, *Principles and Practice of X-ray Spectrometric Analysis,* 2nd ed., Plenum, New York, 1975.

Birks, *X-ray Spectrochemical Analysis,* Interscience, New York, 1959.

*Electron Microprobe Analysis,* Interscience, New York, 1963.

Blokhin, *Methods of X-ray Spectroscopic Research,* Pergamon, New York, 1965.

Clark, *Encyclopedia of X-rays and Gamma Rays,* Reinhold, New York, 1963.

Compton and Allison, *X-rays in Theory and Experiment,* Van Nostrand, New York, 1935.

Heilbron, H. G. J., *Moseley -- The Life and Letters of an English Physicist,* University of California Press, Berkeley, Calif., 1974.

Heinrich, *The Electron Probe,* Wiley, New York, 1966.

Jenkins, *An Introduction to X-ray Spectrometry,* Heyden, London, 1974.

Jenkins and de Vries, *Practical X-ray Spectrometry*, MacMillan, London, 1967.

*Worked Examples in X-ray Spectrometry*, MacMillan, London, 1970.

Kaelble, *Handbook of X-rays*, McGraw-Hill, New York, 1967.

Liebhavsky, Winslow, and Zemany, X-rays, electrons, and analytical chemistry, in *Spectrochemical Analysis with X-rays*, Wiley-Interscience, New York, 1972.

Muller, *Spektrochemische analysen mit Rontgenfluoreszenz*, Oldenbourq, Munchen, 1967.

Parrish, *X-ray Analysis Papers*, Centrex, Eindhoven, 1965.

Reynolds, Mass absorption tables, in *Handbook of X-ray and Microprobe Data*, Dewey, Mapes, and Reynolds, Polycrystal Book Service, Pittsburgh, 1967, pp. 323-339.

Russ, *Elemental X-ray Analysis of Materials*, EDAX Laboratories Division, EDAX International, Prairie View, Ill., 1972.

Sagel, *Tabellen zur Rontgen Emissions und Absorption Analyse*, Springer-Verlag, Heidelberg, 1959.

Theissen, *Quantitative Electron Probe Microanalysis*, Springer-Verlag, New York, 1965.

Van Olphen and Parrish (eds.), X-ray and electron methods of analysis, in *Progress in Analytical Chemistry*, Plenum, New York, 1968.

Woldseth, *All You Ever Wanted to Know about X-ray Energy Spectrometry*, Kevex Corporation, Burlingame, Calif., 1973.

*X-ray Emission and Absorption Wavelengths and Two Theta Tables*, (2nd ed.), ASTM Report DS37A, Philadelphia.

*X-ray Emission Wavelengths and KEV Tables for No. Diffractive Analysis*, ASTM Report No. DS46, Philadelphia.

**Appendix 7**

**Wavelength Tables for K, L, and M Series**

## Principal Emission Lines of X-Ray Spectra (Emission Wavelengths in Å Units)
### K Series

Transition (shell) and approximate intensity (relative to $K\alpha_1$) for each column:

| Line | α° | α₁ (L$_{III}$) | α₂ (L$_{II}$) | β₁ (M$_{III}$) | β₂ (M$_{II}$) | β₃ (N$_{II}$,N$_{III}$) | β₄ (N$_{IV}$,N$_{V}$) | β₅ (M$_{IV}$,M$_{V}$) | — (O$_{II,III}$) | K Absorption Edge |
|---|---|---|---|---|---|---|---|---|---|---|
| Approximate Intensity (Rel. $K\alpha_1$) | 150 | 100 | 50 | 15 | 15 | 5 | <1 | <1 | <1 | |

**Elements Z = 3–44**

| Line | Z | α° | α₁ | α₂ | β₁ | β₂ | β₃ | β₄ | β₅ | — | K Abs. Edge |
|---|---|---|---|---|---|---|---|---|---|---|---|
| Li | 3 | 240. | | | | | | | | | 226.953 |
| Be | 4 | 113. | | | | | | | | | |
| B | 5 | 67. | | | | | | | | | |
| C | 6 | 44. | | | | | | | | | 43.767 |
| N | 7 | 31.603 | | | | | | | | | 31.052 |
| O | 8 | 23.707 | | | | | | | | | 23.367 |
| F | 9 | 18.307 | | | | | | | | | |
| Na | 11 | 11.909 | | | 11.617 | | | | | | |
| Mg | 12 | 9.889 | | | 9.558 | | | | | | 9.512 |
| Al | 13 | 8.339 | 8.338 | 8.341 | 7.981 | | | | | | 7.951 |
| Si | 14 | 7.126 | 7.125 | 7.127 | 6.769 | | | | | | 6.744 |
| P | 15 | 6.155 | | | 5.804 | | | | | | 5.787 |
| S | 16 | 5.373 | 5.372 | 5.375 | 5.032 | | | | | | 5.018 |
| Cl | 17 | 4.729 | 4.728 | 4.731 | 4.403 | | | | | | 4.397 |
| A | 18 | 4.192 | 4.191 | 4.194 | 3.886 | | | | | | 3.871 |
| K | 19 | 3.744 | 3.742 | 3.745 | 3.454 | 3.442 | | | | | 3.437 |
| Ca | 20 | 3.360 | 3.359 | 3.362 | 3.089 | 3.074 | | | | | 3.070 |
| Sc | 21 | 3.032 | 3.031 | 3.034 | 2.780 | 2.764 | | | | | 2.758 |
| Ti | 22 | 2.750 | 2.749 | 2.753 | 2.514 | 2.498 | | | | | 2.497 |
| V | 23 | 2.505 | 2.503 | 2.507 | 2.285 | 2.270 | | | | | 2.269 |
| Cr | 24 | 2.291 | 2.290 | 2.294 | 2.085 | 2.071 | | | | | 2.070 |
| Mn | 25 | 2.103 | 2.102 | 2.105 | 1.910 | 1.897 | | | | | 1.897 |
| Fe | 26 | 1.937 | 1.936 | 1.940 | 1.757 | 1.745 | | | | | 1.744 |
| Co | 27 | 1.791 | 1.789 | 1.793 | 1.621 | 1.609 | | | | | 1.608 |
| Ni | 28 | 1.659 | 1.658 | 1.661 | 1.500 | 1.489 | | | | | 1.488 |
| Cu | 29 | 1.542 | 1.540 | 1.544 | 1.392 | 1.382 | | | | | 1.381 |
| Zn | 30 | 1.437 | 1.435 | 1.439 | 1.296 | 1.285 | | | | | 1.284 |
| Ga | 31 | 1.341 | 1.340 | 1.344 | 1.207 | 1.196 | | | | | 1.195 |
| Ge | 32 | 1.256 | 1.255 | 1.258 | 1.129 | 1.117 | | | | | 1.116 |
| As | 33 | 1.177 | 1.175 | 1.179 | 1.057 | 1.049 | | | | | 1.045 |
| Se | 34 | 1.106 | 1.105 | 1.109 | 0.992 | 0.984 | | | | | 0.980 |
| Br | 35 | 1.041 | 1.040 | 1.044 | 0.933 | 0.926 | | 0.866 | | | 0.920 |
| Kr | 36 | 0.981 | 0.980 | 0.984 | 0.879 | 0.871 | | | | | 0.866 |
| Rb | 37 | 0.927 | 0.926 | 0.930 | 0.829 | 0.817 | | | | | 0.816 |
| Sr | 38 | 0.877 | 0.875 | 0.880 | 0.783 | 0.771 | | | | | 0.770 |
| Y | 39 | 0.831 | 0.829 | 0.833 | 0.740 | 0.728 | | 0.727 | | | 0.727 |
| Zr | 40 | 0.788 | 0.786 | 0.791 | 0.701 | 0.690 | | | | | 0.688 |
| Nb | 41 | 0.748 | 0.747 | 0.751 | 0.665 | 0.654 | | 0.620 | | | 0.653 |
| Mo | 42 | 0.710 | 0.709 | 0.713 | 0.632 | 0.621 | | | | | 0.620 |
| Tc | 43 | 0.674 | 0.673 | 0.676 | 0.602 | | | | | | |
| Ru | 44 | 0.644 | 0.643 | 0.647 | 0.572 | 0.562 | | | | | 0.560 |

**Elements Z = 45–92**

| Line | Z | α° | α₁ | α₂ | β₁ | β₂ | β₃ | β₄ | β₅ | — | K Abs. Edge |
|---|---|---|---|---|---|---|---|---|---|---|---|
| Rh | 45 | 0.614 | 0.613 | 0.617 | 0.546 | 0.546 | 0.535 | | | | 0.534 |
| Pd | 46 | 0.587 | 0.585 | 0.590 | 0.521 | 0.521 | 0.510 | | | | 0.509 |
| Ag | 47 | 0.561 | 0.559 | 0.564 | 0.497 | 0.498 | 0.487 | | | | 0.486 |
| Cd | 48 | 0.536 | 0.535 | 0.539 | 0.475 | 0.476 | 0.465 | | | | 0.464 |
| In | 49 | 0.514 | 0.512 | 0.517 | 0.455 | 0.455 | 0.445 | | | | 0.444 |
| Sn | 50 | 0.492 | 0.491 | 0.495 | 0.435 | 0.436 | 0.426 | | | | 0.425 |
| Sb | 51 | 0.472 | 0.470 | 0.475 | 0.417 | 0.418 | 0.408 | | | | 0.407 |
| Te | 52 | 0.453 | 0.451 | 0.456 | 0.400 | 0.401 | 0.391 | | | | 0.390 |
| I | 53 | 0.435 | 0.433 | 0.438 | 0.384 | 0.385 | 0.376 | | | | 0.374 |
| Xe | 54 | 0.418 | 0.416 | 0.421 | 0.369 | | 0.360 | | | | 0.359 |
| Cs | 55 | 0.402 | 0.401 | 0.405 | 0.355 | 0.355 | 0.346 | | | | 0.345 |
| Ba | 56 | 0.387 | 0.385 | 0.390 | 0.341 | 0.342 | 0.333 | | | | 0.332 |
| La | 57 | 0.373 | 0.371 | 0.376 | 0.328 | 0.329 | 0.320 | | | | 0.319 |
| Ce | 58 | 0.359 | 0.357 | 0.362 | 0.316 | 0.317 | 0.309 | | | | 0.307 |
| Pr | 59 | 0.346 | 0.344 | 0.349 | 0.305 | 0.305 | 0.297 | | | | 0.296 |
| Nd | 60 | 0.334 | 0.332 | 0.337 | 0.294 | 0.294 | 0.287 | | | | 0.285 |
| Pm | 61 | 0.322 | 0.321 | 0.325 | | 0.283 | | | | | |
| Sm | 62 | 0.311 | 0.309 | 0.314 | 0.274 | 0.274 | 0.267 | | | | 0.265 |
| Eu | 63 | 0.301 | 0.299 | 0.304 | 0.264 | 0.265 | 0.258 | | | | 0.256 |
| Gd | 64 | 0.291 | 0.289 | 0.294 | 0.255 | 0.256 | 0.249 | | | | 0.246 |
| Tb | 65 | 0.281 | 0.279 | 0.284 | 0.246 | 0.246 | 0.239 | | | | 0.238 |
| Dy | 66 | 0.272 | 0.270 | 0.275 | 0.237 | 0.238 | 0.231 | | | | 0.230 |
| Ho | 67 | 0.263 | 0.261 | 0.266 | | 0.231 | | | | | 0.223 |
| Er | 68 | 0.255 | 0.253 | 0.258 | 0.222 | 0.223 | 0.217 | | | | 0.215 |
| Tu | 69 | 0.246 | 0.244 | 0.250 | 0.215 | 0.216 | | | | | 0.209 |
| Yb | 70 | 0.238 | 0.236 | 0.241 | 0.208 | 0.209 | 0.203 | | | | 0.202 |
| Lu | 71 | 0.231 | 0.229 | 0.234 | 0.202 | 0.203 | 0.197 | | | | 0.195 |
| Hf | 72 | 0.224 | 0.222 | 0.227 | 0.195 | 0.196 | 0.190 | | | | 0.189 |
| Ta | 73 | 0.217 | 0.215 | 0.220 | 0.190 | 0.191 | 0.185 | | | | 0.184 |
| W | 74 | 0.211 | 0.209 | 0.213 | 0.184 | 0.185 | 0.179 | | | | 0.178 |
| Re | 75 | 0.204 | 0.202 | 0.207 | 0.179 | 0.179 | 0.174 | | | | 0.173 |
| Os | 76 | 0.198 | 0.196 | 0.201 | 0.173 | 0.174 | 0.169 | | | | 0.168 |
| Ir | 77 | 0.193 | 0.191 | 0.196 | 0.168 | 0.169 | 0.164 | | | 0.163 | 0.163 |
| Pt | 78 | 0.187 | 0.185 | 0.190 | 0.163 | 0.164 | 0.159 | | | 0.158 | 0.158 |
| Au | 79 | 0.182 | 0.180 | 0.185 | 0.159 | 0.160 | 0.155 | | | 0.153 | 0.153 |
| Hg | 80 | 0.176 | 0.174 | 0.180 | | 0.151 | | | | | 0.149 |
| Tl | 81 | 0.172 | 0.170 | 0.175 | 0.150 | 0.151 | 0.147 | | | | 0.144 |
| Pb | 82 | 0.167 | 0.165 | 0.170 | 0.146 | 0.143 | 0.138 | | | | 0.141 |
| Bi | 83 | 0.162 | 0.161 | 0.165 | 0.142 | | | | | | 0.137 |
| Th | 90 | 0.135 | 0.133 | 0.138 | 0.117 | 0.118 | 0.114 | | | | 0.113 |
| U | 92 | 0.128 | 0.126 | 0.131 | 0.111 | 0.112 | 0.108 | | 0.116 | 0.113 | 0.107 |

Compiled by P. WILLIAM ZINGARO, Application Laboratory, PHILIPS ELECTRONIC INSTRUMENTS, Mount Vernon, New York.

## Principal Emission Lines of X-Ray Spectra (Emission Wavelengths in Å Units)

### L Series

| Line | α1 | α2 | β1 | β2 | β6 | β3 | β4 | γ1 | γ5 | γ2 | γ3 | l | η | LI | LII | LIII |
|------|----|----|----|----|----|----|----|----|----|----|----|----|----|----|-----|------|
| Transition | Mv | Miv | Miv | Nv | Ni | Miii | Mii | Niv | Ni | Nii | Niii | Mi | Mi | | (L Absorption Edges) | |
| Approx. Intensity (Rel. Lα1) | 100 | 10 | 80 | 20 | 30 | 30 | 20 | 40 | <1 | 10 | 10 | 30 | 10 | | | |
| Cl 17 | | | | | | | | | | | | 67.84 | 67.25 | | | |
| A 18 | | | | | | | | | | | | 56.212 | 56.813 | | | |
| K 19 | | | | | | | | | | | | 47.835 | 47.325 | | | |
| Ca 20 | 36.393 | | 36.022 | | | | | | | | | 41.042 | 40.542 | | 42.184 | |
| Sc 21 | 31.393 | | 31.072 | | | | | | | | | 35.671 | 35.200 | | 35.200 | 35.561 |
| Ti 22 | 27.445 | | 27.074 | | | | | | | | | 31.423 | 30.942 | | | |
| V 23 | 24.309 | | 23.898 | | | | | | | | | 27.826 | 27.375 | | 27.29 | |
| Cr 24 | 21.713 | | 21.323 | | | | | | | | | 24.840 | 24.339 | | | |
| Mn 25 | 19.489 | | 19.158 | | | | | | | | | 22.315 | 21.864 | 16.7 | 17.9 | 20.7 |
| Fe 26 | 17.602 | | 17.290 | | | | | | | | | 20.201 | 19.73 | | | |
| Co 27 | 16.000 | | 15.698 | | | | | | | | | 18.358 | 17.86 | | | |
| Ni 28 | 14.595 | | 14.308 | | | | | | | | | 16.693 | 16.301 | | | |
| Cu 29 | 13.357 | | 13.079 | | | | | | | | | 15.297 | 14.940 | | 13.010 | 13.289 |
| Zn 30 | 12.282 | | 12.009 | | | | | | | | | 14.081 | 13.719 | | 11.861 | 12.130 |
| Ga 31 | 11.313 | | 11.045 | | | | | | | | | 12.976 | 12.620 | | | |
| Ge 32 | 10.456 | | 10.194 | | | | | | | | | 11.944 | 11.608 | | | |
| As 33 | 9.671 | | 9.414 | | | | | | | | | 11.069 | 10.732 | 8.108 | 9.124 | 9.367 |
| Se 34 | 8.990 | | 8.735 | | | | | | | | | 10.293 | 9.959 | 7.505 | 8.417 | 8.645 |
| Br 35 | 8.375 | | 8.126 | | | | | | | | | 9.583 | 9.253 | | | |
| Kr 36 | | | | | | | | | | | | | | | | |
| Rb 37 | 7.318 | 7.325 | 7.075 | 6.821 | 6.788 | 6.984 | 6.754 | 6.045 | | | | 8.363 | 8.042 | 5.997 | 6.643 | 6.864 |
| Sr 38 | 6.863 | 6.870 | 6.623 | 6.403 | 6.367 | 6.519 | 6.297 | 5.644 | | | | 7.836 | 7.517 | 5.582 | 6.172 | 6.387 |
| Y 39 | 6.449 | 6.456 | 6.211 | 6.018 | 5.983 | 6.094 | 5.875 | 5.283 | | | | 7.356 | 7.040 | 5.233 | 5.756 | 5.962 |
| Zr 40 | 6.070 | 6.077 | 5.836 | 5.668 | 5.632 | 5.710 | 5.497 | 4.953 | 5.384 | | | 6.918 | 6.606 | 4.867 | 5.378 | 5.583 |
| Nb 41 | 5.725 | 5.732 | 5.492 | 5.346 | 5.310 | 5.361 | 5.151 | 4.654 | 5.036 | | | 6.517 | 6.210 | 4.581 | | 5.223 |
| Mo 42 | 5.406 | 5.414 | 5.176 | 5.048 | 5.013 | 5.048 | 4.837 | 4.380 | 4.726 | | | 6.150 | 5.847 | 4.299 | 4.719 | 4.913 |
| Tc 43 | | | | | | | | | | | | | | | | |
| Ru 44 | 4.846 | 4.854 | 4.620 | 4.523 | 4.487 | 4.487 | 4.288 | 3.897 | 4.182 | | | 5.503 | 5.204 | | 4.179 | 4.369 |
| Rh 45 | 4.597 | 4.605 | 4.374 | 4.289 | 4.253 | 4.242 | 4.045 | 3.685 | 3.944 | | | 5.217 | 4.922 | 3.626 | 3.942 | 4.129 |
| Pd 46 | 4.368 | 4.376 | 4.146 | 4.071 | 4.034 | 4.016 | 3.822 | 3.489 | 3.725 | 3.792 | 3.799 | 4.952 | 4.660 | 3.428 | 3.724 | 3.908 |
| Ag 47 | 4.154 | 4.162 | 3.935 | 3.870 | 3.834 | 3.806 | 3.616 | 3.307 | 3.523 | 3.605 | 3.611 | 4.707 | 4.418 | 3.254 | 3.514 | 3.698 |
| Cd 48 | 3.956 | 3.965 | 3.739 | 3.681 | 3.644 | 3.614 | 3.426 | 3.137 | 3.336 | 3.430 | 3.437 | 4.480 | 4.193 | 3.084 | 3.326 | 3.504 |
| In 49 | 3.752 | 3.781 | 3.555 | 3.507 | 3.470 | 3.436 | 3.249 | 2.980 | 3.162 | 3.268 | 3.274 | 4.269 | 3.983 | 2.926 | 3.147 | 3.325 |

| Element | | | | | | | | | | | | | | | | | | | | | | | |
|---|---|---|---|---|---|---|---|---|---|---|---|---|---|---|---|---|---|---|---|---|---|---|---|
| Sn 50 | 3.156 | 2.982 | 2.778 | | 4.071 | 3.789 | | 2.778 | 3.085 | 2.835 | 3.001 | | 3.270 | 3.155 | 3.115 | 3.121 | | 3.600 | 3.609 | 3.385 | 3.175 | 3.306 | 3.344 |
| Sb 51 | 3.000 | 2.830 | 2.639 | | 3.888 | 3.607 | | 2.639 | 2.932 | 2.695 | 2.852 | | 3.115 | 3.005 | 2.973 | 2.979 | | 3.448 | 3.448 | 3.226 | 3.023 | 3.152 | 3.190 |
| Te 52 | 2.856 | 2.687 | 2.510 | | 3.716 | 3.438 | | 2.510 | 2.790 | 2.567 | 2.712 | | 2.971 | 2.863 | 2.839 | 2.847 | | 3.290 | 3.299 | 3.077 | 2.882 | 3.009 | 3.046 |
| I 53 | 2.719 | 2.553 | 2.389 | | 3.557 | 3.280 | | 2.389 | 2.657 | 2.447 | 2.582 | | 2.837 | 2.730 | 2.713 | 2.720 | | 3.148 | 3.157 | 2.937 | 2.751 | 2.874 | 2.912 |
| X 54 | 2.592 | 2.429 | 2.274 | | | | | 2.274 | | | | | | | | | | | | | | | |
| Cs 55 | 2.474 | 2.314 | 2.167 | | 3.267 | 2.994 | | 2.167 | 2.417 | 2.233 | 2.348 | | 2.593 | 2.485 | 2.478 | 2.492 | | 2.892 | 2.902 | 2.683 | 2.511 | 2.628 | 2.666 |
| Ba 56 | 2.363 | 2.204 | 2.068 | | 3.135 | 2.862 | | 2.068 | 2.309 | 2.134 | 2.242 | | 2.482 | 2.382 | 2.376 | 2.387 | | 2.776 | 2.785 | 2.567 | 2.404 | 2.516 | 2.555 |
| La 57 | 2.259 | 2.103 | 1.973 | | 3.006 | 2.740 | | 1.973 | 2.205 | 2.041 | 2.141 | | 2.379 | 2.275 | 2.282 | 2.290 | | 2.665 | 2.674 | 2.458 | 2.303 | 2.410 | 2.449 |
| Ce 58 | 2.164 | | 1.890 | | 2.892 | 2.620 | | 1.890 | 2.110 | 1.955 | 2.048 | | 2.282 | 2.188 | 2.188 | 2.195 | | 2.561 | 2.570 | 2.356 | 2.208 | 2.311 | 2.349 |
| Pr 59 | 2.077 | 1.924 | 1.811 | | 2.784 | 2.512 | | 1.811 | 2.020 | 1.874 | 1.961 | | 2.190 | 2.100 | 2.100 | 2.107 | | 2.463 | 2.473 | 2.259 | 2.119 | 2.216 | 2.255 |
| Nd 60 | 1.995 | 1.843 | 1.735 | | 2.675 | 2.409 | | 1.735 | 1.935 | 1.797 | 1.878 | | 2.103 | 2.009 | 2.016 | 2.023 | | 2.370 | 2.382 | 2.166 | 2.035 | 2.126 | 2.166 |
| Pa 61 | | | | | | | | 1.855 | | | | | | | | | | 2.283 | | 2.081 | | | |
| Sm 62 | 1.845 | 1.702 | 1.598 | | 2.482 | 2.218 | 1.779 | 1.606 | 1.708 | 1.655 | 1.726 | | 1.946 | 1.856 | 1.862 | 1.870 | | 2.199 | 2.210 | 1.998 | 1.882 | 1.962 | 2.000 |
| Eu 63 | 1.776 | 1.626 | 1.536 | | 2.395 | | | 1.544 | | 1.591 | 1.657 | | 1.875 | 1.788 | 1.792 | 1.800 | | 2.120 | 2.131 | 1.920 | 1.812 | 1.887 | 1.926 |
| Gd 64 | 1.709 | 1.561 | 1.477 | | 2.312 | 2.049 | | 1.485 | | 1.529 | 1.592 | | 1.807 | 1.723 | 1.723 | 1.731 | | 2.046 | 2.057 | 1.847 | 1.746 | 1.815 | 1.853 |
| Tb 65 | 1.648 | 1.501 | 1.421 | | 2.234 | | 1.577 | 1.427 | 1.518 | 1.471 | 1.530 | | 1.742 | 1.659 | 1.659 | 1.667 | | 1.976 | 1.986 | 1.777 | 1.682 | 1.747 | 1.785 |
| Dy 66 | 1.579 | 1.438 | 1.365 | | 2.158 | 1.895 | | 1.374 | 1.462 | 1.423 | 1.473 | | 1.681 | 1.599 | | | | 1.909 | 1.920 | 1.710 | 1.623 | 1.681 | 1.720 |
| Ho 67 | 1.535 | 1.390 | 1.318 | | 2.086 | 1.826 | | 1.323 | 1.406 | 1.364 | 1.417 | | 1.622 | | | | | 1.845 | 1.856 | 1.647 | 1.567 | 1.619 | 1.658 |
| Er 68 | 1.482 | 1.339 | 1.269 | | 2.019 | 1.757 | | 1.276 | | 1.315 | 1.364 | 1.494 | 1.567 | 1.494 | 1.494 | 1.494 | | 1.785 | 1.796 | 1.587 | 1.514 | 1.561 | 1.601 |
| Tu 69 | 1.433 | 1.288 | 1.2?2 | | 1.955 | 1.695 | | | 1.355 | 1.274 | 1.316 | | 1.515 | | | | | 1.726 | 1.738 | 1.530 | 1.463 | 1.505 | 1.544 |
| Yb 70 | 1.386 | 1.243 | 1.181 | 1.831 | 1.094 | 1.635 | | 1.185 | 1.307 | 1.222 | 1.268 | | 1.466 | 1.395 | 1.384 | 1.392 | 1.372 | 1.672 | 1.682 | 1.476 | 1.416 | 1.452 | 1.491 |
| Lu 71 | 1.342 | 1.198 | 1.140 | 1.776 | 1.836 | 1.478 | | 1.143 | 1.260 | 1.179 | 1.222 | | 1.419 | 1.350 | 1.336 | 1.343 | 1.392 | 1.619 | 1.630 | 1.424 | 1.370 | 1.402 | 1.441 |
| Hf 72 | 1.298 | 1.154 | 1.099 | 1.723 | 1.782 | 1.523 | 1.663 | 1.103 | 1.215 | 1.144 | 1.185 | 1.437 | 1.374 | 1.306 | 1.291 | 1.299 | 1.328 | 1.569 | 1.580 | 1.374 | 1.327 | 1.353 | 1.392 |
| Ta 73 | 1.256 | 1.113 | 1.061 | 1.672 | 1.728 | 1.471 | 1.612 | 1.065 | 1.173 | 1.105 | 1.144 | | 1.331 | 1.264 | 1.247 | 1.254 | 1.287 | 1.522 | 1.533 | 1.327 | 1.285 | 1.307 | 1.346 |
| W 74 | 1.215 | 1.074 | 1.024 | | 1.678 | 1.421 | | 1.028 | 1.132 | 1.068 | 1.098 | 1.339 | 1.290 | 1.224 | 1.204 | 1.212 | 1.247 | 1.476 | 1.487 | 1.282 | 1.245 | 1.263 | 1.302 |
| Re 75 | 1.177 | 1.037 | .990 | | 1.630 | 1.374 | | .993 | 1.094 | 1.032 | 1.061 | 1.293 | 1.252 | 1.186 | 1.165 | 1.172 | 1.208 | 1.433 | 1.444 | 1.238 | 1.206 | 1.220 | 1.260 |
| Os 76 | 1.140 | 1.001 | .956 | | 1.585 | 1.328 | | .959 | 1.057 | .998 | 1.025 | | 1.213 | 1.149 | 1.126 | 1.133 | 1.171 | 1.391 | 1.402 | 1.197 | 1.169 | 1.179 | 1.218 |
| Ir 77 | 1.105 | .967 | .923 | | 1.541 | 1.285 | | .928 | 1.022 | .966 | .991 | 1.166 | 1.179 | 1.115 | 1.090 | 1.097 | 1.137 | 1.352 | 1.363 | 1.158 | 1.135 | 1.141 | 1.179 |
| Pt 78 | 1.072 | .934 | .893 | | 1.499 | 1.243 | | .897 | .988 | .934 | .958 | 1.128 | 1.143 | 1.082 | 1.054 | 1.062 | 1.072 | 1.313 | 1.325 | 1.120 | 1.102 | 1.104 | 1.141 |
| Au 79 | 1.040 | .903 | .864 | 1.414 | 1.460 | 1.202 | 1.352 | .867 | .956 | .905 | .927 | 1.090 | 1.111 | 1.050 | 1.021 | 1.028 | 1.072 | 1.277 | 1.288 | 1.083 | 1.070 | 1.068 | 1.106 |
| Hg 80 | 1.009 | .872 | .836 | | 1.422 | 1.164 | | .839 | .925 | .876 | .897 | 1.041 | 1.080 | 1.019 | .986 | .996 | 1.041 | 1.242 | 1.253 | 1.049 | 1.040 | 1.034 | 1.072 |
| Tl 81 | .979 | .844 | .808 | 1.279 | 1.385 | 1.127 | 1.342 | .812 | .895 | .848 | .863 | 1.056 | 1.050 | .981 | .957 | .964 | 1.012 | 1.207 | 1.218 | 1.015 | 1.010 | 1.001 | 1.039 |
| Pb 82 | .950 | .815 | .782 | 1.244 | 1.350 | 1.092 | 1.308 | | .867 | .822 | .840 | 1.022 | 1.021 | .953 | .927 | .934 | .983 | 1.175 | 1.186 | .982 | .983 | .969 | 1.007 |
| Bi 83 | .924 | .789 | .757 | 1.210 | 1.317 | 1.058 | | .840 | | .815 | .814 | .989 | .993 | .926 | .900 | .905 | .957 | 1.114 | 1.155 | .952 | .955 | .939 | .977 |
| Po 84 | | | | | | | | .799 | | .790 | .796 | | .967 | | | | | 1.114 | 1.125 | .922 | .929 | .909 | .948 |
| Fr 87 | | | | | 1.167 | .908 | | .680 | .717 | .675 | .788 | | .871 | | | | | 1.030 | 1.017 | .840 | .858 | | |
| Ra 88 | .803 | .670 | .644 | 1.080 | 1.115 | .855 | 1.011 | .640 | .675 | .635 | .682 | .844 | .828 | .817 | .807 | .776 | .838 | 1.005 | 1.017 | .814 | .836 | .803 | .841 |
| Th 90 | .761 | .630 | .606 | | 1.091 | .830 | | .611 | .655 | .617 | .642 | | .808 | .775 | .765 | .730 | | .956 | .968 | .766 | | .755 | .793 |
| Pa 91 | | | | | 1.067 | .806 | | .594 | .635 | .598 | .624 | | .789 | .755 | .746 | .708 | | .9?3 | .945 | .742 | .774 | .732 | .770 |
| U 92 | .722 | .592 | .569 | .964 | | | 1.035 | .601 | .635 | .605 | | .989 | .726 | .736 | .726 | .687 | | .911 | .923 | .720 | .755 | .710 | .748 |
| Np 93 | | | | | | | | | | .597 | | | .789 | | | | .931 | .889 | | .698 | .735 | | |

Compiled by P. WILLIAM ZINGARO, Application Laboratory, PHILIPS ELECTRONIC INSTRUMENTS, Mount Vernon, New York.

## Principal Emission Lines of X-Ray Spectra (Emission Wavelengths in Å Units)

### M Series

Transition / Relative Intensity labels (left margin): $M_I$, $M_{II}$, $M_{III}$, $M_{IV}$, $M_V$

| Line | $\alpha_1$ $N_{VII}$ (100) | $\alpha_2$ $N_{VI}$ (10) | $O_{III}$ | $\delta$ $N_{III}$ (15) | $N_{III}$ | $\delta_2$ $N_{II}$ | $O_{III}$ (2) | $O_{II}$ (5) | $\beta$ $N_{VI}$ (50) | $O_{IV,V}$ | $O_I$ (2) | $\gamma$ $N_V$ (5) | $N_{IV}$ | $N_I$ (2) | $O_{IV}$ | $N_{IV}$ (4) | $N_I$ | $P_{III}$ | $O_{III}$ (10) | $N_{III}$ | $N_{II}$ |
|---|---|---|---|---|---|---|---|---|---|---|---|---|---|---|---|---|---|---|---|---|---|
| 50 Sn |  |  |  |  |  |  |  |  |  |  |  | 17.9 |  |  |  | 16.9 | 20.1 |  |  |  |  |
| 51 Sb |  |  |  |  |  |  |  |  |  |  |  | 16.9 |  |  |  | 15.9 | 18.8 |  |  |  |  |
| 52 Te |  |  |  |  |  |  |  |  |  |  |  | 15.9 |  |  |  |  | 17.6 |  |  |  |  |
| 53 I |  |  |  |  |  |  |  |  |  |  |  | 15.0* |  |  |  |  |  |  |  |  |  |
| 54 Xe |  |  |  |  |  |  |  |  |  |  |  | 14.2* |  |  |  |  |  |  |  |  |  |
| -55 Cs |  |  |  |  |  |  |  |  |  |  |  | 13.5* |  |  |  |  |  |  |  |  |  |
| 56 Ba |  |  |  | 20.6 |  |  | 15.7 | 15.88 |  |  |  | 12.72 |  |  |  |  |  |  |  |  |  |
| 57 La | 14.85 |  |  | 19.4 |  |  |  |  | 14.48 |  |  | 12.05 |  |  |  |  |  |  |  |  |  |
| 58 Ce | 14.01 |  |  | 18.3 |  |  |  |  | 13.72 |  |  | 11.51 |  |  |  |  |  |  |  |  |  |
| 59 Pr | 13.32 |  |  | 17.3 |  |  |  |  | 13.03 |  |  | 10.98 |  |  |  |  |  |  |  |  |  |
| 60 Nd | 12.65 |  |  | 16.4 |  |  |  |  | 12.41 |  |  | 10.48 |  |  |  |  |  |  |  |  |  |
| 61 Pm | 12.11* |  |  | 15.6* |  |  |  |  | 11.80* |  |  | 10.01* |  |  |  |  |  |  |  |  |  |
| 62 Sm | 11.45 |  |  | 14.88 |  |  |  |  | 11.25 |  |  | 9.58 |  |  |  |  |  |  |  |  |  |
| 63 Eu | 10.94 |  |  | 14.19 |  |  |  |  | 10.73 |  |  | 9.19 |  |  |  |  |  |  |  |  |  |
| 64 Gd | 10.44 |  |  | 13.54 |  |  |  |  | 10.23 |  |  | 8.83 |  |  |  |  |  |  |  |  |  |
| 65 Tb | 9.98 |  |  | 12.95 |  |  |  |  | 9.77 |  |  | 8.47 |  |  |  |  |  |  |  |  |  |
| 66 Dy | 9.57 |  |  | 12.40 |  |  |  |  | 9.34 |  |  | 8.13 |  |  |  |  |  |  |  |  |  |
| 67 Ho | 9.18 |  |  | 11.84 |  |  |  |  | 8.95 |  |  | 7.85 |  |  |  |  |  |  |  |  |  |
| 68 Er | 8.80 |  |  | 11.35 |  |  |  |  | 8.57 |  |  | 7.53 | 7.58 |  |  |  |  |  |  |  |  |
| 69 Tm | 8.46 |  |  | 10.90* | 9.67 |  |  |  | 8.23 |  |  |  |  | 8.45 |  |  |  |  |  |  |  |
| 70 Yb | 8.13 |  |  | 10.46 |  |  |  |  | 7.89 |  |  | 7.01 | 7.87 | 7.87 |  |  |  |  |  |  |  |
| 71 Lu | 7.82 |  |  | 10.06* |  |  |  |  | 7.59 |  |  | 6.754 |  |  |  |  |  |  |  |  |  |
| 72 Hf | 7.52 | 7.28 |  | 9.67* |  |  |  |  | 7.29 |  |  | 6.530 |  |  |  |  |  |  |  |  |  |
| 73 Ta | 7.24 |  | 7.28 | 9.30 | 9.67 | 9.67 |  |  | 7.01 | 5.662 |  | 6.299 | 6.340 | 7.60 |  | 5.558 |  |  |  | 5.387 |  |
| 74 W | 6.969 | 6.978 | 6.990 | 8.94 | 8.88 | 9.31 |  |  | 6.743 |  | 5.822 | 6.079 | 6.121 | 7.35 |  | 5.346 | 6.626 |  | 4.430 | 5.161 |  |
| 75 Re | 6.715 |  |  | 8.61 | 8.56 | 8.97 |  |  | 6.491 |  | 5.616 | 5.873 | 5.919 |  |  |  |  |  |  |  |  |
| 76 Os | 6.48 |  |  | 8.29 | 8.22 | 8.65 |  |  | 6.254 |  |  | 5.670 | 5.712 | 6.882 |  |  |  |  |  |  |  |
| 77 Ir | 6.249 | 6.262 |  | 8.00 | 7.63 | 8.34 |  |  | 6.025 | 4.859 | 4.866 | 5.489 | 5.529 | 6.655 |  | 4.945 | 5.802 |  |  | 4.779 |  |
| 78 Pt | 6.034 | 6.045 | 5.975 | 7.72 | 7.36 | 8.05 |  |  | 5.816 | 4.684 | 4.866 | 5.308 | 5.346 | 6.442 |  | 4.770 |  |  |  | 4.621 |  |
| 79 Au | 5.828 | 5.842 | 5.755 | 7.45 | 7.09 | 7.77 |  |  | 5.612 | 4.513 | 4.693 | 5.134 | 5.175 | 6.246 |  | 4.591 |  |  |  | 4.451 |  |
| 80 Hg | 5.633 |  |  | 7.198* |  | 7.51 |  |  | 5.420 |  |  | 4.968 |  |  |  | 4.423 |  |  |  | 4.291 |  |
| 81 Tl | 5.449 | 5.461 |  | 6.960 | 7.017 |  | 5.185 |  | 5.238 | 4.207 | 4.235 | 4.813 | 4.855 | 5.872 |  | 4.107 | 4.645 |  |  | 4.005 |  |
| 82 Pb | 5.275 | 5.288 | 5.157 | 6.726 | 6.788 |  | 4.994 |  | 5.065 | 4.061 |  | 4.664 | 4.705 | 5.692 |  | 3.960 |  |  |  | 3.864 |  |
| 83 Bi | 5.107 | 5.119 |  | 6.507 | 6.571 |  | 4.813 |  | 4.899 | 3.924 | 4.096 | 4.523 | 4.562 | 5.526 |  | 3.826 |  |  |  | 3.732 | 3.884 |
| 84 Po | 4.96* |  |  | 6.300* |  |  |  |  | 4.76* |  |  | 4.390* |  |  |  |  |  |  |  |  |  |
| 85 At | 4.80* |  |  | 6.101* |  |  |  |  | 4.60* |  |  | 4.258* |  |  |  |  |  |  |  |  |  |
| 86 Rn | 4.66* |  |  | 5.912* |  |  |  |  | 4.46* |  |  | 4.132* |  |  |  |  |  |  |  |  |  |
| 87 Fr | 4.52* |  |  | 5.780* |  |  |  |  | 4.31* |  |  | 4.010* |  |  |  |  |  |  |  |  |  |
| 88 Ra | 4.39* |  |  | 5.557* |  |  |  |  | 4.18* |  |  | 3.893* |  |  |  |  |  |  |  |  |  |
| 89 Ac | 4.26* |  |  | 5.392* |  |  |  |  | 4.05* |  |  | 3.781* |  |  |  |  |  |  |  |  |  |
| 90 Th | 4.129 | 4.142 | 4.901 | 5.234 |  | 5.329 |  | 3.800 | 3.933 | 3.125 | 3.276 | 3.671 | 3.170 | 4.559 | 2.613 | 3.005 | 3.530 |  | 2.437 | 2.928 |  |
| 91 Pa | 4.014 | 4.027 | 4.615 | 5.081 |  | 5.182 |  | 3.683 | 3.819 | 3.032 | 3.238 | 3.570 | 3.607 | 4.441 | 2.522 | 2.904 | 3.434 |  | 2.299 | 2.747 |  |
| 92 U | 3.902 | 3.916 | 4.615 | 4.936 |  | 5.040 |  | 3.569 | 3.708 | 2.942 | 3.109 | 3.472 | 3.514 | 4.321 | 2.438 | 2.811 | 3.322 | 2.248 | 2.299 | 2.747 | 2.909 |

* Calculated value.

Compiled by R. Jenkins, *Application Laboratory*, Philips Electronic Instruments, Mount Vernon, New York.

# Index

## About the Authors

RON JENKINS is Principal Scientist for X-ray analysis at Philips
Electronic Instruments, Inc., Mahwah, New Jersey. An author of six
books and over one hundred papers on X-ray spectrometry and diffrac-
tometry, Dr. Jenkins has lectured throughout the world on all aspects of X-ray analysis. He is actively involved in developing
instrumentation and methodology for X-ray diffraction and spectroscopic procedures and is currently serving as Secretary to Commission V4 of the I.U.P.A.C., defining spectroscopic nomenclature.
Dr. Jenkins is a Fellow of the Institute of Physics, the American
Institute of Chemists, and the Institute of Petroleum, and he is a
member of the Society for Analytical Chemistry and the Society of
Applied Spectrometry. He received his tertiary education at Oxford
Polytechnic, England, where he studied for the Licentiateship of the
Royal Institute of Chemistry, and he received his Ph.D. degree
(1980) in chemical physics at the Polytechnic Institute of New York.

R. W. GOULD is Professor of Materials Science and Engineering at
the University of Florida, Gainesville, where he teaches courses in
X-ray diffraction and spectrometry. He received his Ph.D. degree
(1964) in materials science and engineering from the University of
Florida. An author of forty-six articles and three books,

Dr. Gould's research interests are in the fields of failure analysis, product liability and safety, product liability law, structure property correlations in materials, X-ray metallography, and underground corrosion. He is a member of numerous societies, including the American Society of Testing and Materials (ASTM), the American Institute of Mining and Metallurgical Engineers, and the American Society of Metals.

DALE GEDCKE is Senior Scientist at EG&G ORTEC, Oak Ridge, Tennessee, where he specializes in X-ray applications. He received his M.Sc. degree (1964) from the University of Ottawa, Ontario, and his Ph.D. degree (1967) in physics from the University of Alberta, Edmonton. Since 1969 he has been a pioneer in the design and application of energy-dispersive X-ray spectrometers in the areas of X-ray microanalysis and X-ray fluorescence analysis. An internationally recognized expert in his field, Dr. Gedcke has presented more than fifty-five papers and claims a number of unique inventions in X-ray and nuclear spectrometry.